KV-119-867
UNIVERSITY OF STRATHCLY
30125 00517686 1
ANDERSONIAN LIBRARY
WITHDRAWN
FROM
LIBRARY
STOCK
UNIVERSITY OF STRATHCLYDE

EFFECTIVE MEMBRANE PROCESSES – NEW PERSPECTIVES

This volume consists of papers presented at the 3rd International Conference on Effective Membrane Processes – New Perspectives, 12–14 May 1993 in Bath, UK, organised by BHR Group Limited.

TECHNICAL ADVISORY COMMITTEE

INTERNATIONAL CORRESPONDING MEMBERS

Organized and sponsored by
The Conference Division
BHR Group Limited
Cranfield
Bedford MK43 0AJ, UK
Tel: 0234 750422   Fax: 0234 750074

Co-sponsored by the European Society of Membrane Science and Technology

3RD INTERNATIONAL CONFERENCE ON

Effective Membrane Processes – New Perspectives

Edited by Dr R Paterson

BHR Group Conference Series
Publication No. 3

Papers presented at the *3rd International Conference on Effective Membrane Processes – New Perspectives*, organized and sponsored by BHR Group Limited, and held in Bath, UK, on 12–14 May 1993.

Mechanical Engineering Publications Limited
LONDON

ISBN 0 85298 871 0

A CIP catalogue record for this book is available from the British Library.

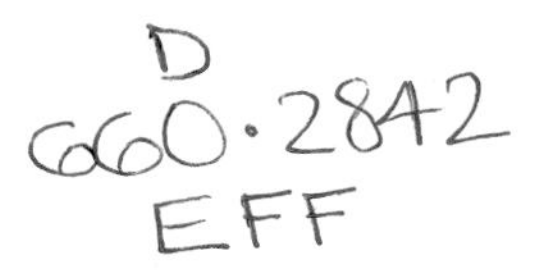

Produced by Technical Communications (Publishing) Ltd.

Printed by Information Press Ltd, Oxford, England.

CONTENTS

WATER AND EFFLUENT

PREFACE

PREFACE

The field of membrane technology is now a multi-billion US$ business world wide. This was not always so. Although membrane processes have a history of several hundred years and the basic principles of such as osmotic theory and diffusion were based upon membrane practice, the first serious applications, apart from simple filtration, were in electro-dialysis (by ion exchange membranes) in the early 1950s and reverse osmosis (using cellulose acetate membranes) in the 1960s. Industrial implementation is a matter of creating reliable and efficient technology.

In the past few years, the range of applications fully or partly implemented as industrial processes has greatly increased and now includes electro-dialysis, reverse osmosis, haemo-dialysis, microfiltration, ultrafiltration, nanofiltration, pervaporation, membrane distillation, facilitated transport by liquid membranes, gas separations and most recently membrane reactors, which combine catalysis and separation of products from reactants in a single step.

Application of membranes to gas separations is a major research and development project at present. At this time, 30% of all industrial nitrogen is produced by membrane processes and in the near future it seems certain that oxygen enriched air will supersede oxygen produced by fractional distillation of air for many purposes including medical ones. The range of applications increases rapidly day by day. Air may be dried, or petroleum removed from tank air in the forecourts of petrol stations or the bilges of oil tankers. Ultimately it is intended that separation of petroleum fractions will be a membrane and there are major long term research and development programmes to this end in several countries including Japan. A range of gas separations are now being developed successfully world wide and the Japanese (once more) have ambitious long-term research aims to remove carbon dioxide, the green house gas, directly from flue gases in the stack. These projects indicate the confidence which business and governments in the developed world now have in effective industrial membrane processes for good environmental and good economic reasons.

In all aspects of preparative technology and waste treatment, membrane processes are now replacing conventional ones. Their advantages are low energy consumption and low unit costs, which make point-of-use applications equally attractive to large scale operations.

Membranes are being used as ionic conductors in advanced battery technology and polymer gel membranes are now used routinely in controlled drug release devices.

A few years ago, there was a reticence by some parts of the chemical industry to implement these new technologies.

There were many causes. Primarily, there was a natural reluctance to abandon old "tried and true" technologies by senior plant engineers, partly because they had no training in or "feel" for this new technology and partly because there was a lack of information state of the art in membrane technology. Membrane processes need to be tailor made for particular applications and to be optimised, sometimes trading off enhanced short term performance for less fouling or longer membrane life or other operational factors which go to make a successful integrated process. The rewards are savings in energy and reduction of pollution for an improved environment. With such incentives, these difficulties are now being rapidly overcome. A new generation of membrane engineers is now emerging who can create effective new membrane applications in industry and where necessary integrate the old with the new technologies.

It is certain however that a combination of ever improving membrane technology and ever increasing demand for energy saving and environmental protection will make the adoption of integrated membrane separation technologies ever more essential to successful chemical, biochemical, medical and pharmaceutical industries.

It is in the tradition of this Conference to concentrate on technological innovation and effective membrane processes and the programme of this Conference is a true reflection of these aims. In this respect, we must thank the Technical Advisory Committee, our panel of referees and the professionalism of the Conference Organisers of BHR Group.

Russell Paterson
Conference Chairman
Yokohama National University, 12 March 1993

OPENING ADDRESS

Effective Membrane Processes : New Perspectives.

Enrico Drioli
Univ. of Calabria, Dept. of Chemistry-Sec. Chem. Engineering,
Arcavacata di Rende, Italy

Chemistry and chemical engineering have made important contributions to the welfare of people and have signed in positive the quality of our life more than the majority of all other disciplines. The most significant progresses in many such areas such as new materials, health case, nutrition, communication, transportation, and others, have been achieved as a result or with the contribution of chemical research and activities of chemical industry and the related ones.

The rapid social, economic, industrial transformations characterizing these years, requires an increasing constant contribution of chemical research for offering new positive answers to the new problems generated.

The role of membrane science and technology for what is indicated today as a Sustainable Development is becoming more and more evident and accepted at political and industrial level. If you read the joint program announcement of the National Science Foundation and the Council for Chemical Research, Inc. supporting pre-competitive research projects in chemical engineering and chemistry aimed at reducing pollution at its source, you will find correctly indicated the interest for improved membranes and technologies that integrate transport and reaction to enhance specificity, as an area of priority.

A similar attention is in the Research Institute of Innovative Technology for the Earth created in Japan in 1991, where specific research projects are in progress focalized for example on high-performance bioreactors, or on new membranes for CO_2 separation from air.

Unesco decided in 1992 to create in Sydney the Centre for Membrane Science and Technology to promote its application to improve the quality of life particularly in developping countries.

During the last ten years the ESMST, founded in 1982, has been active in Europe for improving the cooperation amongst European scientists and engineers and to stimulate the large scale industrial use of membranes. The Society is today well established and recognized as an effective organization promoting membrane technology not only in Europe but all over the world. The positive answer and the attention of the Commission of the European Communities to these efforts has to be acknowledged.

Various multinational research projects on membranes have been sponsored and supported in the frameworks of research programs of the EC (p.e. BRITE and BRITE/EURAM). During the last few months four different new joint European projects have been approuved mainly devoted to education in the field, demonstrating the intensified efforts of the transnational collaboration :

a) a TEMPUS program in which 9 different Universities agreed on the set-up of regional courses in membrane processes.
b) a COMETT program on the organization of an advanced joint training Programme in Membrane Science and Technology in a broad European context.
c) a HUMAN CAPITAL and MOBILITY NET WORK Programme on Functional Membranes. Thirtheen different laboratories in eight EC countries have combined in this network, focalized on intensive studies on membrane reactors and selective separations. A major objective will be to train postdoctoral fellows drawn from other than the country of the host laboratory.
d) a joint research project in the framework of the EC-Central-Eastern European countries cooperation, on Membrane Operation in the Agrofood Industry, involving seven european countries.

The scientific and technological results reached during the last twenty years in membrane operations, justify the attention and the expectation described. Membranology is today a set of different but omogeneous and similar separation processes and advanced reactors offering the widest spectrum of potential and realized applications that any other existing technology. They have an enormeous potentialities in particular to contribute to the solution of some of the most crucial problems of the world today, such as the energy problems, the rapid increase in the shortage of chemical species, the treatment of industrial waters, and the development of a clean industry, the realization of new artificial organs, etc.
In the following pages some of the results reached are summarized.

Reverse Osmosis has been the first process becoming objective of organized research efforts in the USA in the 60ths, for its potentiality in water desalination.
From a 1992 report of the Water Re-use Promotion Center in Japan,it appears that in 1990 the total capacity of RO desalination plants was of more than 4 million m^3/day, with a 4 times increase if compared with the one in the 70thies. The majority of the plants (67%) are treating brackish water or sea water (18%), for producing potable water (36%), industrial water (39%) or boiler feed water (10%). In Japan more than 79% of the water produced via RO is ultrapure water for the electronic industry. It is also interesting to realize the rapidly increasing number of waste water re-use installations which took place in Japan during the last ten years. More than 36% of the world desalination plants are located in this country, 24% in the USA, about 29% in the countries of the EC.

The use of membrane systems in the treatment of gas streams contributed to the re-organization of the industrial groups controlling the field. Air Products, Air Liquid and Ube, respectively in USA, Europe and Japan, are directly promoting membrane processes in the gas separation processes of their traditional interest.
Pure Nitrogen production from air with N_2 purity levels ranging from 90% to 99.9% and flow rate ranging from 200 to 60.000 scfh are commercially available as on-site units. Tenths of industrial plants are in operation for H_2 separation, purification and recovery from various hydrocarbons mixtures in the petrochemical industry or from N_2 in the NH_3 synthesis processes. Improuvements in combustion processes, control of emissions, CO_2 separation and recovery from air, are other examples of new important potential applications under investigation, which will benefit from advances in membranes modules technology, and in plants design.
Oxygen enrichment of air has been commercialized for biomedical applications; its potentialities in combustions, in improving biological processes etc. has been also studied. Pilot plants are in operation in a wide variety of other case in the petrochemical and chemical industry.

New glassy polymers are becoming available for membrane formation with interesting and in some case unexpected selectivities and permeabilities. Organo-inorganic polymers as the polyvinyltrimetylsilane and polyphosphazenes, polyacetylenic polymers as the poly-1-trimetylsilylpropyne are some important cases. This last one is characterized by a permeability of one order of magnitude higher than the silicon rubber, without any significant decrease in the O_2/N_2 selectivity. The structures of these polymers are hetereogeneous and contains a continuous network of submicroscopic voids which provide the pathways for smaller molecule trasport, resembling the zeolite structure.

The membrane separation of CO_2 and H_2S from methane is of interest in the recovery of CO_2 from gas produced at oil fields that are subject to CO_2 flooding for enhanced oil recovery and in the purification of natural gas. Both processes have been studied assuming an integration of the membrane operation with a traditional acid gas removal process DEA (diethanolamine) for reaching products gas of sufficiently high purity.

The use of membrane operations in the natural gas purification for removing H_2S or CO_2 but also He,N_2 and water vapor, has been suggested and experimented. The DEA process is again a traditional alternative. Also in this case the membrane unit reduces the acid gas content and total flow rate to the DEA process. An optimum fractional acid gas recovery can be identified for a 25% acid gas feed at 68% membrane acid gas recovery. The purity requirements of the products gas are important in the economic analysis. The generally low acid gas content (below 3%) indicates the necessity of membranes with much higher selectivity of the existing ones.
Gas separation membranes can be used in dehumidification and dehydration processes. Polymide membrane, p.e., in hollow fiber configuration, produced by the condensation polymerization of biphenyltetracarboxylic dianhydride and aromatic diamines, are used in the dehydration of organic

solvents by vapor separation (Ube Ind.,Japan). The feed vapor might be supplied for example to the membrane unit at a temperature of 120°C and at a pressure of 28 psig. A certain degree of vacuum is produced on the permeate side. Fooling problems are reduced, if compared to pervaporation, having a clean vapor phase as feed.

Pervaporation is a membrane process which combines the evaporation of volatile components with their permeation through a selective membrane. In pervaporation the feed is a liquid mixture and the permeate is a gas. The chemical potential gradient across the membrane, acting as driving force in pervaporation, is obtained by lowering the partial pressure of the different components in the permeate by applying a vacuum to the permeate side of the membrane. Mass transport in this membrane operation, as in gas separation, is determined by a solution-diffusion mechanism.

The process, initially studied and industrially developped in Europe, is today considered of particular interest in a variety of applications. Alcohol dewatering by pervaporation particularly at concentrations near the azeotropic values is already in operation.

The continuous removal of ethanol from fermentation broths, moreover, with recycling of microorganisms, can drastically reduce the ethanol inhibitory effects on the microorganisms.

In this application the pervaporation membranes will have to pass preferentially alcohols and not water; the contrary is requested to the membranes used in the concentration step; membranes for both applications are available today.

MTBE (methyltertiobutylether), used as a high octane blending component, contributing to the reduction of lead content in traditional gasoline, is generally produced by selective methoxylation of isobutene contained in C4 fractions.

CA pervaporation membranes have been suggested for separating the excess alcohol from ether. The TRIM (Air Products) pervaporation process, for example, is applicable specifically to the production of ethers, such as tert-amyl ether (TAME) and ethyl-butylether (ETBE).

Acceleration of esterification reactions, dehydration of vinegar,dehydration of acids, removal of aceton from waste water, are some other cases in which pervaporation has been tested commercially. In general pervaporation is an attractive alternative to distillation for separating azeotropic mixtures, close-boiling compounds, and also finds applications in removing small amounts of a component from a bulk liquid stream.

The combination of transport across a selective barrier with a chemical reaction, typical of various biological processes, is one of the most promising future applications of membrane processes.This combination can be realized by integrating traditional reactors with the most appropriate membrane operation or by using the membrane itself as a reactor. Synthetic polymeric or ceramic membranes provide an ideal support for catalysts and biocatalysts immobilization for the wide surface per unit volume available and for disclosing new immobilization procedures. Enzymes are retained in the reaction site, do not pollute the products and can be continuously re-used. Immobilization has also been proven to enhance enzyme stability. Moreover, provided that membranes of suitable molecular weight cut-off are used, chemical reaction and physical separation of catalysts (and/or substrates) from the products may take place in the same device. Selective removal of products through membrane yields effective conversion with product inhibited or thermodynamically unfavorable reactions. Substrate partition at the membrane/fluid interface can be used to improve the selectivity of the catalytic reaction towards the desired products with minimal side reactions. Membranes are also attractive to retain in the reaction volume expensive cofactors often required to carry on some enzymatic reactions. Controlled addition of one of the reactants supplied to the reaction zone, permits controlled sequential reactions and discloses other interesting possibilities.

Membrane bioreactors have been the first systems of this kind studied. Productions of fatty acids and glycerine from renewable resources (e.g., vegetable oils) and synthesis of glycerides from fatty acids and glycerine have been performed by use of lipase enzymes immobilized within the membrane pores or on the surface of asymmetric membranes as an enzymatic gel.

Bacterial and mammalian cells have been effectively immobilized by means of membranes. Cells of S.solfataricus, a thermophilic and acidophilic microorganism of about 1 μm diameter, have been successfully entrapped in the wall of polysulphone U.F. capillary membrane in the fiber formation process. Protein/lactose solutions (such as whey) can be processed with these immobilized cell membranes, to yield a protein rich, lactose poor retentate (useful as a sweetener).

Islets of Langerhans (the cellular aggregates responsible for pancreatic insulin secretion) have been immobilized in the shell of capillary membrane reactors in "shell-and-tube" configuration to simulate the pancreatic functions and protect the islets against the attack of immunocompetent proteins in the blood of the host.

The interest for new membrane reactors to be used in chemical and petrochemical processes, in liquid as well as in gas phase, is growing significantly. Ceramic membranes appear as ideal candidates for the development of these studies.

Catalyst design itself will be influenced in the future, as the membrane reactor concept becomes well established. Inorganic membrane reactors integrated in productive cycles, will offer new solutions.

Enzyme membrane reactors (EMRs) are an ideal mean for the transformation of wastes from the food industry into high added value products. Selectivity and efficiency of the catalytic action of enzymes make them ideal tools for the production of specific stereoisomers. Besides, catalysis may be carried out under mild reaction conditions thus preventing products denaturation. As enzymes are used in EMRs, bioseparation may occur at the same time (and at the same site) where the reaction takes place. A careful selection of the membranes permits the separation of pure products from enzymes and unconverted high molecular weight substrates. Products removal from the reaction mixture as they form may be extremely advantageous in case product inhibited enzymes were to be used, as it is the case for most hydrolytic enzymes.

An interesting application of this concept has been recently proposed for the transformation of cheese whey proteins into compounds for medical use. Cheese whey protein have been successfully transformed by means of proteolytic enzymes in oligopeptidic mixtures to replace glucose as the osmotic agent in solutions for peritoneal dialysis.

The integration of the described membrane separations and membrane reactors, is contributing to the design and realization of new productive cycles in various industrial sectors. They are characterized by a lower direct and indirect energy consumption, high productivity combined with high qualities, lower waste production and environmental impact, approaching at the best the ideal hybrid pollution-controlled productive systems.

The successes and rapid progresses also in bioartificial organs mainly based on membranes operations (artificial kidney, artificial pancreas, artificial retina, p.e.) open a future possibility of extending the strategy of hybrid membrane systems also in biomedical applications.

PHARMACEUTICAL

THE EXTRACTION OF DILUTE FERMENTATION PRODUCTS USING HIGH INTERNAL PHASE EMULSIONS (APHRONS)

D.C.Stuckey, K. Matsushita[a], A.H. Mollah, and A.I. Bailey
Department of Chemical Engineering & Chemical Technology, Imperial College of Science, Technology and Medicine, London SW7 2BY
[a]Nippon Mining Co., Tokyo, Japan.

SUMMARY

Currently, very little information is available in the literature on the properties and use of colloidal liquid and gas aphrons (CLA,CGA) in the pre–dispersed solvent extraction (PDSE) of dilute fermentation products. Liquid aphrons are sometimes referred to as high internal phase emulsions (HIPEs) since the volume ratio of the internal phase (solvent) to the continuous phase (aqueous) can be as high as 20:1. In this preliminary study it was found that the CLAs prepared were very stable over a long period (100 days), and did not coalesce. When they were pumped through a filter they did not rupture but seemed to break into smaller sizes. Electron micrographs supported the hypothesis that CLAs are stabilised by a soapy shell which makes them different from monolayer microemulsions Experiments with the PDSE of trimethyl acetic acid (TMAA) showed 85% removals in under a minute with over 80% solvent recovery. Finally, using a modified Lewis cell it was shown that the surfactants did not seem to either enhance or decrease both the partition coefficient and extraction flux. It appears that the advantage of using CLAs in PDSE is simply one of large surface area created with a very low power input.

INTRODUCTION

The downstream separation of valuable fermentation products is often extremely expensive and time consuming due to their dilute nature and lability. One conventional downstream process which is commonly used is solvent extraction. However, its use in biotechnology has a number of drawbacks, namely: the capital cost of mixer settlers, the power requirement for solvent dispersion, large solvent inventories, and potential toxicity problems if the cells come in contact with the solvent. One novel technique which can circumvent these problems is the use of colloidal liquid and gas aphrons (CLAs, CGAs) in pre–dispersed solvent extraction (PDSE). This technique involves forming a stable oil in water (o/w) microemulsion (CLA) using certain non–ionic and ionic surfactants, and since the volume of solvent used is small and surfactants are present, the energy input required is small. These solvent droplets are then dispersed into the pregnant solution, and due to their large surface area partition equilibrium is achieved very quickly. These droplets then need to be separated, and although the solvent is usually lighter than water, and would be expected to rise to the surface under gravity, this is usually very slow due to their small size. Hence one technique is to use oppositely charged gas aphrons (CGA) to float the CLAs to the surface. Due to charge neutralisation the CLAs rupture releasing a laden solvent on the surface of the aqueous phase which can be extracted to recover the desired solute. However, another technique would be to formulate mechanically robust CLAs which could then be separated by filtration, stripped and then recycled.

Colloidal gas aphrons (CGA) were first made by Sebba[1] under the name of micro foams, but further experiments proved that the CGA name was more appropriate[2] The small size of the gas bubbles (25–100μm), which gives them colloidal properties, enables them to be pumped from one vessel to another, and produces a system of considerable potential in a remarkable diversity of applications[3-10]. CGAs can be made from solutions with a great variety of ionic surfactants, and can contain up to 65% gas. A colloidal liquid aphron (CLA) has been defined as a liquid (oil) core globule with colloidal dimensions (1–20μm) encapsulated by a soapy (aqueous) shell and dispersed in a continuous aqueous phase[10]. The requirements for CLA preparation are: (a) the internal phase (oil) has to be immiscible with water, or its solubility has to be very small (the upper solubility limit at this time is not known), (b) the internal phase has to be divided into small globules, (c) the globules must be encapsulated in a soapy film (Figure 1).

To stabilize this soapy film, the water soluble surfactant solution is made of sodium dodecyl sulphate (SDS) or sodium dodecylbenzene sulphonate (SDBS). In addition, an oil soluble surfactant, usually nonionic, is necessary in the internal phase in order to maintain the spreading pressure of oil greater than the surface pressure produced by the surfactant dissolved in water[2]. When CLAs are added to water they easily disperse homogeneously, and hence with CLAs the oil (solvent) is the internal (discontinuous) phase and water is the continuous phase. Pre–dispersed solvent extraction (PDSE) using both liquid and gas aphrons is a novel method for the separation of solutes from aqueous solution on the basis of favourable partition coefficients, and one which has shown promise for the rapid extraction from extremely dilute solutions[7,9,11-13]. The large interfacial area provided by the aphrons makes them very useful in processes which are controlled by this parameter, such as mass transfer. A litre of colloidal liquid aphrons (CLAs) of 90% oil phase, 2μm in diameter, will contain 2 x 10^{14} aphrons and will provide an oil–water interface [14] of 2,700m^2. In addition, and in contrast with conventional solvent extraction, there is no need for a mixer–settling stage, which can be expensive. Therefore, and in particular for very dilute pregnant solutions, the power cost for product extraction can be significantly reduced. Furthermore, the ratio of extracting solvent to pregnant solution can be very low, of the order of one to a thousand or even lower. This results in high concentration factors, a small solvent inventory and a reduction in toxicity and fire risks. Although PDSE using aphrons has a number of advantages over conventional liquid–liquid extraction for dilute product extraction, it has only been applied in the extraction of orthodichlorobenzene and ethanol from water solutions in laboratory scale experiments[9,13].

The successful implementation of product extraction using pre–dispersed solvent extraction depends strongly on solvent selection. Treybal[15] and King[16] have documented several criteria that should be considered in the selection of solvents for extraction in general. Often, a compromise needs to be made between the desired physiochemical properties of the system, solvent toxicity, and economic considerations[17]. Furthermore, in the case of solvent extraction using high internal phase emulsions or aphrons, solvents must form stable aphrons. The effect of solvent type and polarity, surfactant type and concentration, and mixing speed on aphron stability and size has been examined in a previous piece of work[15].

The objective of this work was to study the feasibility of using PDSE for the recovery of trimethyl acetic (TMAA), and to investigate some of the parameters controlling the process. Hence some of the physical properties of aphrons were examined together with the ability of CGAs to separate and break CGAs. Finally some preliminary results were obtained for stripping TMAA.

MATERIAL AND METHODS

Reagents: The following reagents were used in the present work:

Decanol (decyl alcohol, 90%) (Aldrich); Polyoxyethylene alcohol (Softanol–30, Alcohol ethoxylates No. 3; Softanol–120, Alcohol ethoxylates No. 12) (BP);DTMAB (dodecyltrimethyl ammonium bromide, 99% Aldrich), SDS dodecyl sulphate, sodium salt, 98%) (Aldrich); NaOH, AnalaR (BDH). De–ionized water was used for the preparation of the mother liquor, and pH was controlled using a pH controller (BIOENGINEERING).

Colloidal Gas Aphron Preparation

The colloidal gas aphrons (CGA) were prepared using a high speed emulsifier (Silverson R, Model SRT–1). The surfactant solution was stirred at high speed (>4000 rpm) until a constant volume of creamy CGAs were generated. These CGAs can be kept dispersed under low stirring conditions, ie 500 rpm, and can also be pumped by means of a peristaltic pump without breaking. In this work, CGAs were characterized by their stability over time, and % gas content. A half life was employed as a measure of CGA stability, where half life was the time required for the reduction of half the initial volume of the CGAs with no stirring. The gas content was determined by subtracting the volume of surfactant solution from the total volume of CGAs generated in a graduated beaker.

Colloidal Liquid Aphron Preparation

The oil phase (50–200 ml) containing a non–ionic surfactant was gradually dropped (2–10 ml/min) into 10 ml of a foaming aqueous solution of an anionic surfactant under adequate mixing conditions using a magnetic stirrer. Initially, the solvent disperses easily in the aqueous phase, but after adding about two thirds of the total solvent, the mixture starts to become highly viscous, and it takes considerably longer to disperse the final volume of the solvent. In the end, a white creamy dispersion of CLAs is obtained. The PVR was calculated from the volume ratio of solvent added to the aqueous volume.

Mass Transfer Studies Using a Lewis Cell

Substrate transfer rates from the organic to the aqueous phase were calculated from transient concentration measurements in a modified Lewis cell[19]. This apparatus is widely used in mass transfer studies since it provides an accurately measurable flat aqueous–organic interface. Both liquid phases were stirred independently such that they were well mixed, while not disturbing the interface. The Lewis cell consisted of a cylindrical glass vessel operated with two different liquid volumes. The two phases were mixed by four–bladed impellers (6 cm diameter) mounted in the middle of the individual phases on a common top driven agitator shaft (motor: Heidolph Type RTR 50), and samples were taken from both phases for analysis. The dimensions of the Lewis cell and mixing conditions used were; vessel diameter 9.5 cm, cross–sectional area 70.9 cm^2, phase volumes 450 ml aqueous and 150 ml organic, stirring speed 70 rpm for quiescent runs, 300 rpm for turbulent runs, and 2000 rpm for homogenous runs.

Pre–dispersed Solvent Extraction (PDSE)

In order to evaluate PDSE a variety of CLAs were prepared from decanol using varying concentrations of Softanol–30 and SDS in the aqueous phase. A 1.5% (w/v) solution of TMAA was also prepared, and 25 ml of the CLAs gently dispersed in 75 ml of the TMAA for a specific contact time. At the end of this time a preparation of CGAs made from the cationic surfactant DTMAB were pumped in at a certain flowrate for a specific time. The CLA+CGA mixture was then allowed to separate for a specified time before the raffinate was drawn off, and the volumes of the phases measured and analysed for TMAA.

ANALYTICAL METHODS

Concentrations of TMAA were determined by titration using 0.1 and 0.01M NaOH, with phenolphthalein as an indicator. CLA size was determined by laser light scattering using a Malvern Particle Size Analyser (Master Particle Sizer M.3.0, Malvern Instruments), and a standard deviation of 10.5% was associated with the particle size measurement. The Sauter mean diameter which is a measure of the ratio of the total volume of particles to the total surface area was used to characterise the average size of the CLAs, and the instrument also gave a particle size distribution. The stability of the CLAs was determined by visually observing the clear solvent layer on the surface of the CLA bottle.

The viscosity of the CLA suspensions with different solvents was evaluated using a Bohlin rheometer at 20°c. The electrophoretic mobility of the CLAs at various spHs was determined using a Malvern Instrument Zeta Sizer 3.

The electron micrographs were prepared by freezing a sample of CLAs (decanol, 1% (w/v) Softanol 120, 0.5% (w/v) SDS, PVR=10) in a small rivet in slushy liquid nitrogen. This was then transferred to a cyrochamber on an SEM S200 where it was fractured, and sputter coated with gold. The sample was examined at – 170° c and photographs were taken.

RESULTS AND DISCUSSION

a) Physical properties of colloidal liquid aphrons.

Previous work has shown that CLAs can be formulated with quite polar solvents (e.g. pentanol), and that they become more stable as the HLB number (hydrophilic/lipophilic balance) of the non–ionic surfactant used increases[18]. In addition, it was found that CLAs could be formulated with a phase volume ratio (PVR–ratio of solvent phase volume to continuous aqueous phase volume) as high as 20 without coalescence which is markedly higher than with microemulsions, and which seems to indicate that the liquid aphrons are stabilized by more than a surfactant monolayer.

In order to use CLAs in PDSE more needs to be known about their stability, size distribution, viscosity, surface charge and structure.

1) Long term stability

It has been reported that CLAs can be stored in a stoppered bottle for a long time without visible deterioration[10]. To determine whether any coalescence occurred which would increase CLA size, particle size was determined over long intervals of time (Figure 2).

CLAs (PVR=10,0.5% SDS in water) were made with two oil phase surfactants: Softanol–30 and Softanol–120. It was found that the variation of particle size over time was within the range of measurement error, hence it was concluded that CLAs can be stored for a long time without coalescence. This property of CLAs demonstrates that they are extremely stable, and hence their structure is different from monomolecular layer microemulsions. Sebba has postulated that they may contain a soapy film which confers considerable stability since three interfaces have to be broken for coalescence to occur[19].

2) Size distribution

CLAs were prepared using decanol with 1.0% (w/v) Softanol–120, and 0.5% (w/v) SDS in the aqueous phase at a PVR of 10. The sample size distribution was then measured in a Malvern Particle Sizer, and the results are shown in Figure 3.

From the figure it can be seen that the particle sizes distribute themselves around a mean, but the distribution is skewed towards the smaller sizes. After these CLAs were pumped through a ceramic membrane and a ball valve a number of times the size distribution shifted towards the smaller sizes (Figure 4), and yet no solvent was released. These results are encouraging for the practical application of CLAs as it demonstrates their stability under conditions of high shear stress.

3) Viscosity

CLAs were prepared using dekalin, n–octane and trimethyl pentane with 1.0% (w/v) Softanol 30 and 0.5% (w/v) SDBS at a PVR=20 Their viscosity was then measured in a rheometer at 20°c under a variety of shear rates. As can be seen from Figure 5 the CLA suspensions exhibit normal shear thinning properties, and hence are amenable to pumping at high concentrations. At low concentrations (PVR<5) they behave in a Newtonian manner.

4) Electrophoretic mobility

CLAs were prepared with decanol, 1.0% (w/v) Softanol 120 and 0.5% (w/v) SDS and tested at two different concentrations and pHs for their Zeta potential and electrophoretic mobility. These properties are important in CLAs as the more charged the solvent droplets are, the more they will repel each other, and hence remain dispersed, and the easier they are to attract with oppositely charged CGAs. As can be seen in Table 1, in dilute suspensions (0.1g/l) and pH 8.4 they are quite strongly negative.

However, as may be expected, with increasing CLA concentration, and decreasing pH the zeta potential decreases due to compression of the Stern layer, and charge neutralisation.

5) Structure

Questions on the structure of CGAs and CLAs are still unresolved. Sebba[10] claims that both types of aphrons are stabilised by a thin aqueous soapy shell formed by the accumulation of both the ionic and non–ionic surfactants at the three interfaces (Figure 1). He cites as proof for his hypothesis the fact that CLAs are considerably stronger than monolayer microemulsions, and do not coalesce at PVRs as high as 20, remaining as the dispersed phase in contrast to microemulsions. In addition, he claims that dye present in the aqueous phase during CLA formulation is found in an aqueous shell during CLA formulation.

In an attempt to resolve this question a number of experiments were carried out with fluorescein dye and a fluorescent microscope to see whether a "soapy shell" existed. Unfortunately, all these experiments were inconclusive since the spherical shape of the CLAs inhibited good resolution with light transmission microscopy. Finally, freeze fracture electron microscopy was used with a CLA suspension made of decanol, 1% (w/v) Softanol –120 and 0.5% (w/v) SDS, PVR=10.

The results (Figure 6) seem to indicate that there is indeed a "soapy shell" around the solvent core, and that in the picture a piece has been broken out of the shell. However, this observation will need to be confirmed by other means, e.g. cryogenic microtome TEM.

While this result is interesting, it does beg the question about the structure of CLAs when they are dispersed in an aqueous phase in a ratio of 1:100. In this situation the concentration of the ionic detergent at the external interface should be considerably reduced, and hence the existence of a soapy shell is considerable more problematic. Nevertheless, even in dilute suspensions CLAs appear to be very stable, and do not coalesce. Obviously, more work is needed to resolve this question.

b) Solute extraction and CLA separation

In order to utilize PDSE effectively the solvent phase of the CLAs has to have a high partition coefficient for the solute, and for equilibrium to be reached quickly. In addition, the CLAs need to be able to be floated by oppositely charged CGAs, and for the CLAs to break easily to release the solute rich solvent. A model solute was needed to evaluate these phenomena, and trimethyl acetic acid (TMAA) was chosen since it had a moderate solubility in water, an ideal partition coefficient in decanol (24.1 at 21°c), was easy to analyse, and was stable. Although TMAA is not a very interesting product, it is a good model for a number of organic acids with similar molecular structure and weight.

Table 2 lists the composition of the CLAs and CGAs used, and the volume of CGAs pumped in for each experiment, while Table 3 shows the results of the experiments. It is clear from Table 3 that in every case but one, solvent recovery in the extract exceeded 80% (of the 25ml decanol added), while in many cases it approached, or even exceeded 100%.

This apparent anomaly was due to the fact that in some instances the interface was not strongly delineated, and hence some water was included in the solvent phase. In many instances between the solvent and aqueous phases was an emulsion which was relatively stable and was comprised partly of unbroken CLAs. While there are many variables involved in each experiment, there is a general trend towards greater solvent recovery as the surfactant concentrations in both the CLAs and CGAs decrease. This is not surprising since the CLAs are less stable at lower surfactant concentrations. In addition, the turbidity in the raffinate also appeared to decrease with decreasing surfactant concentration.

Finally, solvent phase recovery did not seem to be influenced in any discernible way by either the rate of addition of the CGAs, or the volume added. It appears that small amounts of CGAs (expt 6–30mls CGA) can separate almost the same volume of CLAs.

The data for extraction and solute recovery in Table 3 show that the partitioning of TMAA into decanol CLAs results in the same partition coefficient as with pure decanol (within experimental error). This was not unexpected, and with similar solvent recovery efficiencies for the range of CLA/CGA parameters tried led to an overall solute recovery of between 80 and 86%. This recovery did not appear to be influenced by the CLA contact time, even when it was only five minutes. These data show that CLAs can extract a solute quickly, and that CGAs can separate and break the CLAs to give a solute rich solvent layer. The question posed by the data, however, is how fast is equilibrium reached, and do the surfactants present appreciably decrease the rate of uptake of the solute into the solvent phase.

c) **Extraction rate and the influence of surfactants.**

1) **CLA extraction**

Since CLA extraction proceeded quickly, an experiment was carried out to determine the time required to reach equilibrium. Decanol CLAs were made with Softanol 30 (0.5% w/v) and SDS (0.25% w/v) at a PVR=12. To 20 ml of a 1.5% w/v TMAA solution, 6.7 ml of the CLAs were added and gently dispersed. Six samples were set up in this way, and after a certain contact time the samples were filtered, and the aqueous and solvent phases analysed for TMAA. These data are shown in Table 4 and reveal that due to the large surface area per unit volume of the CLAs, equilibrium is reached in less than a minute.

2) **Lewis cell**

In order to slow the extraction down to a measurable rate, a modified Lewis cell with an interfacial area of 70.9 cm^2 was used to evaluate the effect of the CLA surfactants on interfacial mass transfer. The aqueous phase contained 1.5% (w/v) of TMAA (450 ml) while 150 ml of decanol was used to keep the phase ratio constant at 1:3. Both phases were mixed at 70 rpm with a turbine impeller, and in the run with surfactants 0.5% (w/v) Softanol 30 was added to the decanol, and 0.25% (w/v) SDS to the aqueous phase.

From Figure 7 it can be seen that the addition of these surfactants did not either enhance or reduce the interfacial flux of the TMAA. If the "soapy shell" hypothesis of Sebba is true then it could be argued that the interface of a CLA, and that present in a Lewis cell, are very different and hence no inferences can be drawn from these experiments. Nevertheless, these data indicate that surfactants do not influence the extraction rate directly.

Finally, in order to show conclusively that the rate of extraction was dependent on surface area alone, the Lewis cell was operated with a turbulent interface (300 rpm), and fully homogenous (2000 rpm). It can be seen from the data in Table 5 that extraction rate increases with surface area, and at 2000 rpm in a fully homogenous system the extraction is as fast as with CLAs.

In summary, these preliminary experiments have shown that PDSE has the potential for rapidly extracting quite high levels of a dilute solute and the rate or extraction does not seem to be influenced by the surfactants present. In terms of structure, electron micrographs of the CLAs seem to confirm Sebba's soapy shell hypothesis,. However, more conclusive pictures are required before this is finally confirmed. While the properties of the CLAs and CGAs enable them to be handled easily at a commercial scale overall, it may be more efficient to separate CLAs from suspension by filtration rather than by CGA flotation and rupture. This is because the solvent then has to be stripped of the solute by contacting it with an aqueous phase at high pH. Since this operation is also controlled by surface area, it makes sense to maintain the integrity of the CLAs by filtering them, stripping them with another aqueous phase and then recycling them. This is one avenue that will be explored in future work.

Table 1
Effect of pH and CLA Concentration on Zeta potential

concentration g/l CLA	pH	Zeta potential mV
0.1	8.4	−49.0 ± 2.1
1.1	8.4	−45.0 ± 14.5
1.1	4.0	−35.6 ± 23.5

Table 4
Extraction rate of Trimethyl acetic acid by decanol-CLA (at 22°c)

Contact Period min	TMMA Conc. wt % Solv.	A_q	Partition Co-efficient	Recovery* %
1	4.7	0.17	28	88
3	4.6	0.18	26	87
5	4.6	0.18	26	87
10	4.6	0.18	26	87
30	4.7	0.17	28	88
60	4.6	0.18	26	87

* Due to the loss of some of the solvent phase during handling, recovery value is derived from a balance on the aqueous phase.

Table 5
Extraction of TMAA with decanol in a Lewis Cell

Stirring Time (min)	300 RPM Conc. (wt%) Organic	300 RPM TMAA (wt%) A_q	2000RPM Conc (wt%) Organic	2000RPM TMAA (wt%) A_q	Partition Coefficient 300 RPM	Partition Coefficient 2000 RPM	Recovery % 300 RPM	Recovery % 2000 RPM
1	0.79	1.23	4.8	0.18	0.62	26.6	14.7	89
3	-	-	4.8	0.18	-	26.6	-	89
5	1.79	0.95	4.7	0.19	2.1	24.7	36.6	87.3
10	4.0	0.38	4.7	0.19	10.5	24.7	74.3	87.3
30	4.4	0.27	4.7	0.19	16.3	24.4	81.7	87.3
60	4.7	0.19	-	-	24.7	-	87.3	-

Table 2
PDSE Extraction of Trimethyl acetic acid by n-Decanol CLA:
Experimental Conditions (21°c)

	CLA		CGA	CGA Flow	CGA Flowrate	CGA Contact Time
EXP	Softanol-30%(w/v)	SDS%(w/v)	DTMAB%(w/v)	(min)	(ml/min)	(min)
1	1.0	0.5	0.5			30
2	1.0	0.5	0.2	9	11	30
3	1.0	0.5	0.3	5	12	30
4	1.0	0.5	0.25	30	4	30
5	0.5	0.5	0.33	20	3	30
6	1.0	0.25	0.25	15	2	30
7	1.0	0.25	0.25	20	4	5
8	0.5	0.25	0.25	20	4	5
9	0.5	0.25	0.10	25	4	5
10	0.5	0.25	0.05	10	6	5
11	0.5	0.25	0.20	15	3	5
12	0.25	0.25	0.20	40	3	5
13	0.25	0.25	0.10	20	5	5
14	0.25	0.25	0.10	20	4	5
15	0.25	0.25	0.10	1.5	50	5

$[C]_{TMAA}$ = 1.5 wt.%, 75 ml pregnant aqueous phase, 25 ml CLAs

Table 3
PDSE Phase Separation and Extraction

EXP	Recovery (ml) EXT	Recovery (ml) RAF	Observation EXT	Observation RAF(Turbidity)	TMAA Conc. %(wt/v) Solv.	TMAA Conc. %(wt/v) Aq	Part. Coeff	Recovery* of TMAA(%)
1	22.5	122.9	C/E	++	4.9	0.18	27	80.3
2	19.4	125.7	C	++	3.8	0.16	24	82.1
3	20.6	147.5	C/E	+	4.0	0.18	22	76.4
4	28.2	195.0	( )	( )	4.3	0.14	31	75.7
5	22.4	132.4	C/E	++	4.3	0.19	23	77.6
6	22.4	108.8	( )	++	4.7	0.19	25	81.6
7	22.0	144.6	C/E 20/5	( )	4.2	0.19	22	75.6
8	21.8	155.6	C/E 20/5	C	4.7	0.19	25	73.4
9	22.5	169.3	C/E 20/8	C	4.0	0.18	22	72.9
10	25.2	133.1	C+E	C	4.2	0.19	22	77.5
11	25.6	118.9	C/E 15/15	C	4.5	0.18	25	81.0
12	26.6	196.4	C+E	++	3.7	0.18	21	68.6
13	27.0	160.2	C+E	+	3.7	0.18	21	74.3
14	24.5	147.6	C+E	+	3.5	0.18	19	76.4
15	24.5	145.5	C/E 20/5	C	3.6	0.18	20	72.7

C Clear
C/E Clear phase and emulsion with apparent interface
C+E Clear phase and emulsion without apparent interface
E Emulsion
Degree of turbidity ++, +, C
Partition coefficient of TMAA in n-decanol = 24.1 at 20 - 22 C
( ) = no observation recorded

*Due to the loss of some of the solvent phase during handling, recovery value is derived from a balance on the aqueous phase.

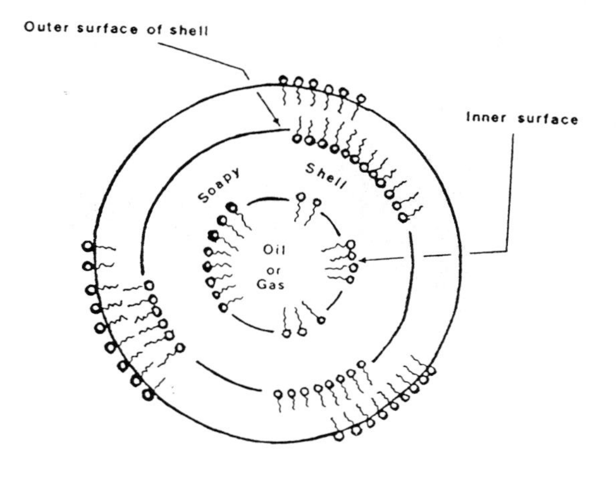

Figure 1. Structure of a CLA

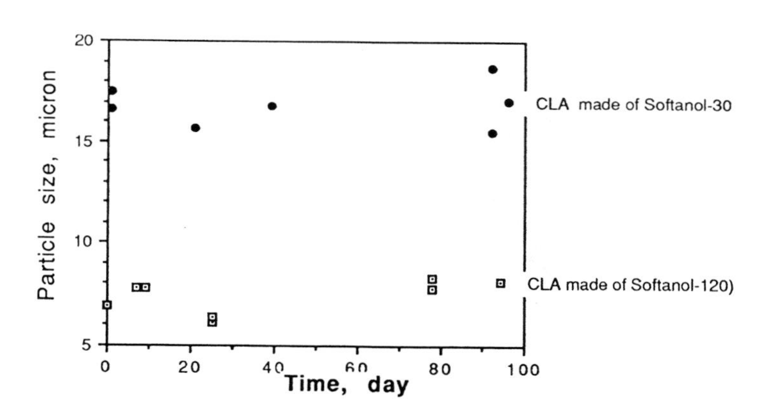

Figure 2. Variation of CLA size with time CLA: PVR = 10 SDS (0.5%) in water, 1% Softanol–30 & 120 in decanol.

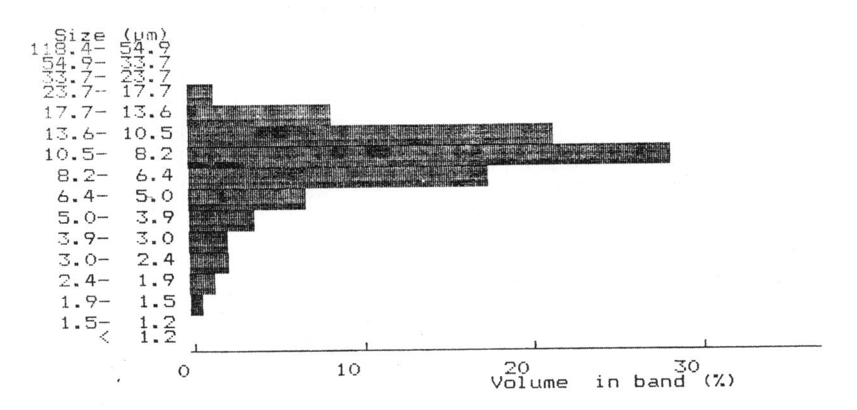

Figure 3. Size distribution of CLAs (D = 9.3 μ) before filtration

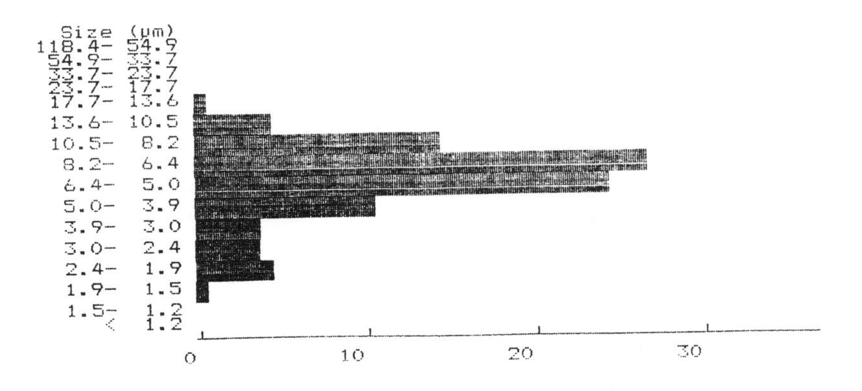

Figure 4. Size distribution of CLAs (D = 6.5 μ) after filtration

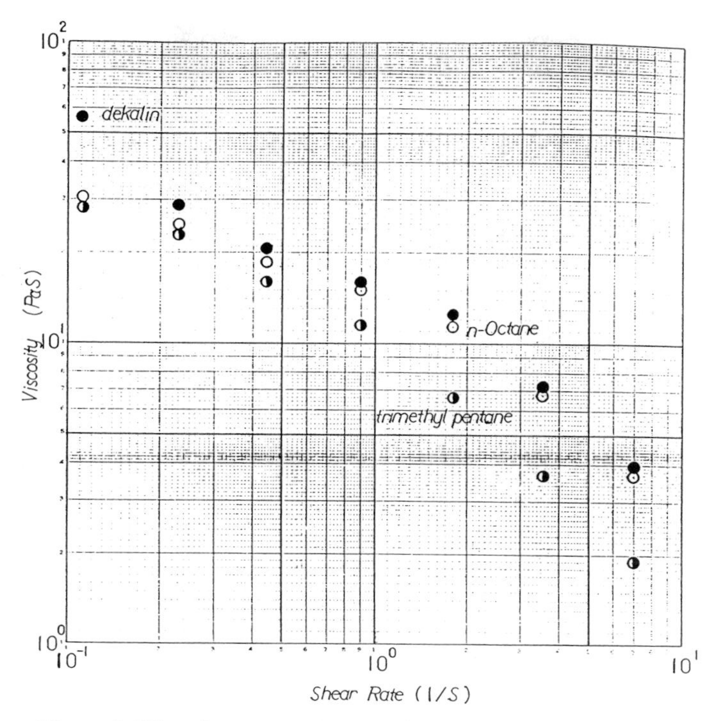

Figure 5. Viscosity versus shear rate for CLAs (PVR = 20)

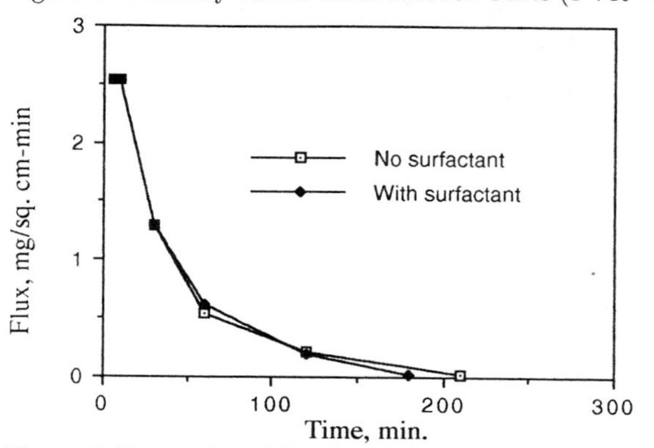

Figure 7. Extraction of TMAA using the Lewis cell (feed 300ml of 15g/L TMAA solution at pH 3, decanol:feed = 1:6).

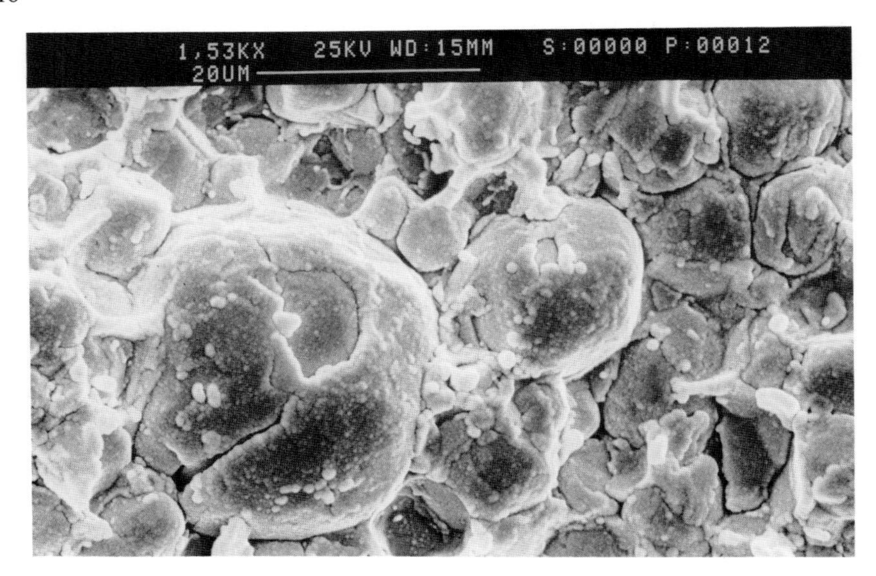

Figure 6. Freeze Fracture EM of CLAs

REFERENCES

[1] F. Sebba, Micro foams–an unexploited colloidal system. Colloid. Interface Sci. **35**, 643–646 (1971).

[2] F. Sebba, Colloidal dispersions and micellar behaviour (ed K.L. Mittal), Washington DC: A.C.S. Symposium Series No. **9**, 18–39 (1975).

[3] F. Sebba, Novel separations using aphrons, In: "Surfactant–based separation processes", J.F. Scamehorn and J.H. Harwell, ed. (1989).

[4] S. Ciriello, S.M. Barnett, and F.J. Deluise, Removal of heavy metals from aqueous solutions using microgas suspensions. Sep. Sci. Technol. **17**, 521–34 (1982).

[5] S.S. Honeycut, D.A. Wallis and F. Sebba, A technique for harvesting unicellular algae using colloidal gas aphrons. Biotech. Bioeng. Symp. No. **13**, 567–575 (1983).

[6] F. Sebba and R.H. Yoon, Interfacial phenomena in mineral processing (eds. B. Yarar and D.J. Spottiswood). New York: Engineering foundation, 161–72 (1982).

[7] D.L. Michelsen, D.A. Wallis, and F. Sebba, In–situ biological oxidation of hazardous organics. Environmental Progress, vol. **3** No. 2, 103–107 (1984).

[8] W.L. Auten and F. Sebba. The use of Colloidal Gas Aphrons (CGAs) for the removal of slimes from water by floc flotation. In: "Solid and liquid separations". J. Gregory (ed). Ellis Horwood, Chichester, England. 1141–52 (1984).

[9] D.A. Wallis, D.L. Michelsen, F. Sebba, J.K. Carpenter and D. Houle, Applications of aphron technology to biological separations. Biotechnol. Bioeng. Symp. No. **15**, 399–408 (1985).

[10] Sebba, Foams and Biliquid Foams–Aphrons, John Wiley & Sons (1987).

[11] F. Sebba, Predispersed solvent extraction. Separation Sci. Technol. **20**(5&6), 331–334 (1985).

[12] F. Sebba, Predispersed solvent extraction. Chem Ind. 16 March, 185–187 (1987).

[13] D.L. Michelsen, K.W. Ruttimann, K.R. Hunter and F. Sebba. Feasibility study on use of predispersed solvent extraction/floatation techniques for removal of organics from wastewaters. Chem. Eng. Commun. vol **48**, 155–163 (1986).

[14] F. Sebba, Polyaphrons in process engineering. The Chem. Eng. March, 12–14 (1986).

[15] R.E. Treybal, Mass transfer operations, 3rd ed. McGraw–Hill, New York (1980).

[16] C.J. King, Separation processes. 2nd ed. McGraw–Hill, New York (1981).

[17] L.E.S. Brink and J. Tramper, Optimization of organic solvent in multiphase biocatalysis. Biotechnol. Bioeng. **27**, 1258–1269 (1985).

[18] K. Matsushita, A.H. Mollah, D.C. Stuckey, C. del Cerro and A.I. Bailey, Pre–dispersed solvent extraction using aphrons for the recovery of dilute products: Aphron preparation, stability and size. Accepted by Colloids and Surfaces (1992).

[19] Sebba, F., Preparation and properties of polyaphrons (biliquid foams). Chem. Ind. 21 May, 1984, pp 367–372.

TUBULAR IONIC MEMBRANE : I- FACILITATED TRANSPORT OF α-ALANINE.

M.METAYER, D.LANGEVIN, M.LABBE and N.LAIR
C.N.R.S.- U.R.A. 500, Polymères, Biopolymères, Membranes
U.F.R. des Sciences, Université de Rouen, B.P. 118, 76134 Mt.-St.-Aignan

ABSTRACT

A transport-reaction process occurs when the counterion H^+ of cation-exchange membrane reacts with the zwitterion $^+NH_3$-$CH(CH_3)$-COO^- to form the cationic species $^+NH_3$-$CH(CH_3)$-COOH. The feed solution contained the zwitterion and the receiving solution was pure water. The membrane permselectivity was reduced by concentration polarization.

In the present case, facilitated transport has been carried out under well-defined hydrodynamic conditions. A tubular ionic membrane was set inside a concentric shell. One of the solutions circulated inside, the second outside. Two very different flow rates were used.

A numerical method has been used to correlate the concentration profiles in the source and receiving solutions, the interfacial concentration and the fluxes of α-Alanine.

INTRODUCTION

Membrane processes are developed with extractions and separations in view. The coupling of diffusion to a reversible chemical reaction is a possible way to increase both permeability and selectivity of a permeate denoted by S. According to these principles, transport-reaction coupling can be carried out with the usual ion-exchange membranes by judiciously selecting a counter-ion T^z as a carrier:

$$S + T^z \rightleftarrows ST^z$$

These membranes, initialy in the T^z form, separate a feed solution of S from a receiving solution. The latter could be pure water (facilitated transport) or a solution of T^z (facilitated extraction). The high exchange capacity allows the membrane to support high concentration gradient of ST^z, whence high permeability, while the selectivity is given by the reaction specificity. But the membrane permselectivity is reduced by the concentration polarization effects. Previously [1] , measurements were made using flat sheets and the hydrodynamic conditions were not known on both sides of the membrane. In facilitated extraction and for a given dilute feed solution, the flux initially increases with the concentration of the receiving solution and later remains constant. This constant value is called "critical flux" and is proportional to the concentration of the feed solution [2]. This critical flux which is very similar to critical current density in electrodialysis, was used to calibrate the thickness of an equivalent polarization layer. In the present case, facilitated transport of α-Alanine were carried out under well-defined hydrodynamic conditions. A tubular ionic membrane was set inside a concentric shell. The source solution circulated outside and the receiving solution inside.

THEORY

Tubular membrane and its coaxial shell are characterized by an active height h directed towards the z-axis and successive radii r_i , r_e , r_s which delimite an internal I.C. and an external E.C. compartments (see figure 1). The solutions circulate at the constant flow rates L_i in I.C. and L_e in E.C. . Counter-current as well co-current flows are considered. It is assumed that the flow is laminar and so the liquid velocity V(r) is a function of the radial distance r [3] and not of the axial coordinate z. In I.C. ($r < r_i$), V(r) is related to r_i and L_i by :

$$V(r) = 2.L_i.(1-\rho_i^2)/(\pi.r_i^2) \qquad (1)$$

$$\text{with } \rho_i = r/r_i$$

In E.C. ($r_e < r < r_s$) :

$$\left.\begin{array}{l} V(r) = 2.L_e.(1-\rho_s^2-u_0.Ln\ \rho_s)/[\pi.r_s^2.(1-\rho_0^2).v_0] \\ \text{where } \rho_0 = r_e/r_s\ ,\ \rho_s = r/r_s,\ 0 < \rho_s < 1 \\ u_0 = (1-\rho_0^2)/Ln\ \rho_0 \text{ and } v_0 = 1+\rho_0^2+u_0 \end{array}\right\} \quad (2)$$

In this system, the permeate flux density J has been expressed in the axial and radial directions. The axial component has been related to the liquid velocity V(r) and to the local concentration (the axial diffusion has been neglected) as follows :

$$J_A(r,z) = C(r,z).V(r) \quad (3)$$

The radial component has been considered as a diffusion term, defined by the Fick's first law :

$$J_R(r,z) = -D.\partial C(r,z)/\partial r \quad (4)$$

In the steady state, the Fick's second law reduces to :

$$\partial C/\partial t = -div.\ J = 0$$

On account of the cylindrical symmetry, the two flux components are connected by :

$$\partial[r.J_A(r,z)]/\partial z + \partial[r.J_R(r,z)/\partial r] = 0 \quad (5)$$

and, from eqns (3-5) :

$$V(r).\partial C(r,z)/\partial z = D.[\partial^2 C(r,z)/\partial r^2 + (1/r).\partial C(r,z)/\partial r] \quad (6)$$

It is assumed that the permeate diffusivity D is constant in the feed and receiving solutions.

Inside the membrane ($r_i \le r \le r_e$), only the radial diffusion has been considered, because the axial component is nil :

$$J_A(r,z) = 0$$

and, from eqn (5), $r.J_R(r,z)$ is independent of r that is why we write

$$\mathcal{J}_R(z) = 2.\pi.r.J_R(r,z),\ \text{when } r_i \le r \le r_e \quad (7)$$

$\mathcal{J}_R(z)$ is the local diffusion rate through the tubular membrane per unit length of cylinder and constant for a given axial coordinate z. Eqns (7) and (4), after integration allow one to relate this parameter to the interfacial concentrations :

$$\mathcal{J}_R(z) = 2.\pi.\overline{D}\ .\ \frac{\overline{C}\ (i) - \overline{C}\ (e)}{Ln\ (r_e/r_i)} \quad (8)$$

where $\overline{D}$ is the overall permeate diffusivity in the membrane. It is supposed constant. $\overline{C}$ (I) is an overall concentration inside the membrane, at the internal (I=i) or external (I=e) interface taken through unit volume of swollen membrane. $\overline{C}$ (I) takes into account the free (S) and the complexed (ST) permeate :

$$\overline{C}\ (I) = y_{ST}(I).\overline{C_X} + \overline{C_S}(I) \quad (9)$$

$\overline{C_X}$ is the concentration of ionic site balanced by T^z and ST^z inside the membrane. $\overline{C_S}(I)$ is the interfacial concentration of free permeate in the membrane and related to the interfacial molar concentration $C_S(I)$ of free S in the solution and to the molar distribution coefficient λ :

$$\overline{C_S}(I) = \lambda . C_S(I) \qquad (10)$$

and the ionic fraction $y_{ST}(I)$ of ST^z is related to the stability constant K^* of the species formed by :

$$K^* = \frac{y_{ST}(I)}{[1-y_{ST}(I)].a_S(I)} \qquad (11)$$

where $a_S(I)$ is the interfacial activity of S. The activity a_S is practically equal to the molal concentration of free S in the solution, on the hypothesis that the formation/dissociation of ST^z is an instantaneous equilibrium [4] :

$$a_S(I) \simeq m_S(I) \text{ and } m_S(I) = \rho . C_S(I) \qquad (11a)$$

where ρ is the density of the solution. Here m_S and C_S are expressed in m.mol g^{-1} and m.mol cm^{-3} respectively and $\rho \simeq 1$ g.cm^{-3} .

Eqns (7-11) define the interfacial conditions. Two other boundary conditions have to be taken into account :

$$J(r,z) = 0 \,, \quad r = 0 \text{ and } r = r_S \qquad (12)$$

The total flux Φ can be obtained by integration of the local diffusion rate :

$$\Phi = \int_{z=0}^{h} \mathcal{J}_R(z).dz \qquad (13)$$

In practice, the product $\overline{D}.\overline{C}(I)$, used in eqn.(8), can be expressed as follows :

$$\left.\begin{array}{l} \overline{D}.\overline{C}(I) = y_{ST}(I).\overline{P_X} + P.C(I) \\ \text{where } \overline{P_X} = \overline{D}.\overline{C_X} \text{ and } P = \lambda.\overline{D} \end{array}\right\} \qquad (14)$$

$\overline{P_X}$ is a parameter typical of the counterion permeability, and P the usual permeability coefficient of neutral α-Alanine.

Eqn (6) can be solved by remplacing the derivatives by finite-difference approximations, see [5], with z in place of t. We obtain :

$$\left.\begin{array}{l} C(r,z_+) = C(r,z).[1-2.Y(r)] + Y(r).f[C(r_+,z),C(r_-,z)] \\ \text{with } f[C(r_+,z),C(r_-,z)] = \{C(r_+,z).[1+E(r)] + C(r_-,z).[1-E(r)]\} \\ \quad z_+ = z + \delta z \,,\ r_+ = r + \delta r \,,\ r_- = r - \delta r \\ \quad Y(r) = D.\delta z/[v(r).(\delta r)^2] \text{ and } E(r) = \delta r/2r \end{array}\right\} \qquad (14)$$

Numerical stability of this solution is obtained with $Y \leq 1/2$.

The conditions at $r = 0$, $r = r_S$ are given by eqn (12) and may be written :

$$C(0,z_+) = C(\delta r,z_+) \quad \text{and} \quad C(r_S,z_+) = C(r_S-\delta r,z_+) \qquad (15)$$

At the interfaces $r=r_i$, $r=r_e$, eqns (4,7) become

$$\left.\begin{aligned} C(r_i,z_+) &= C(r_{i-},z_+) - \frac{E(r_i).\mathcal{J}_R(z_+)}{\pi.D} \\ C(r_e,z_+) &= C(r_{j+},z_+) + \frac{E(r_e).\mathcal{J}_R(z_+)}{\pi.D} \end{aligned}\right\} \quad (16)$$

A dichotomic method allows one to select the local diffusion rate :

1) $C(r_i,z_+)$, $C(r_j,z_+)$ are obtained from the first value of $\mathcal{J}_R(z_+)$ via eqn (16)

2) $y_{ST}(i)$, $y_{ST}(e)$ are connected with $C_S(i) = C(r_i,z_+)$, $C_S(e) = C(r_e,z_+)$ via eqns (11–11a)

3) and a new value of $\mathcal{J}_R(z_+)$ is obtained via eqn (8).

This numerical calculation has been applied to co-current and counter-current flows. Obviously in this case an iterative procedure must be applied :

1) Only I.C. and membrane are considered. The concentration $C(r_e,z_+)$ is assumed to be equal to the inside E.C. concentration. First values of $C(r_i,z_+)$ have been calculated.

2) Only E.C. and the membrane are considered. The concentrations $C(r_i,z_+)$ are given by the previous step. First values of $C(r_e,z_+)$ have been calculated.

3) Steps (1) and (2) are repeated, using the last values of $C(r_e,z_+)$ or $C(r_i,z_+)$, until a satisfactory convergence is obtained.

EXPERIMENTAL

Transport measurements were made using a tubular ionic membrane Nafion 811 X, 1100 E.W., perfluorosulfonated hollow fiber with internal radius r_i = 0.03125 cm and external radius r_e = 0.04375 cm. A sample was gently stretched in a coaxial glass shell (radius r_s = 0.15 cm) thermostated at 25 °C with a tubular glass jacket (see Fig.2).

The feed solution was α-alanine (zwitterion) solution circulating outside the fiber at the flow rate F = 20 or 200 ml h^{-1} while the receiving solution (pure water) was circulated inside the fiber at the same flow rate.

In facilitated transport measurements, the ionic fiber was converted to the H^+ form while Na^+ was the selected counter-ion for passive diffusion experiments.

The equilibrium properties of the membrane were determined using an ion-exchange column filled with small slices of Nafion fiber in H^+ form. The amount of absorbed amino-acid at equilibrium was deduced from the change in concentration of an alpha-alanine solution that circulated in closed loop through the column.

RESULTS AND DISCUSSION

Fig. 3 shows the sorption isotherm of the α-alanine by the Nafion fiber initially in the acid form. The fraction y_{ST} of the ionic sites occupied by the protonated α-alanine is represented as a function of the activity ($pS = -\log a_S$) of the neutral α-alanine in the equilibrium solution. The experimental values are in good agreement with the theoretical curve calculated from eqn. 11 by adjusting to 1250 the α-Alanine protonation constant K^* in the ion-exchanger.

This value of K^* is about six times the real constant K . This has been already observed with the formation of silver-ammonia complexes in a Nafion membrane and is described by a selectivity coefficient K_d such that $K^* = K_d\, K$ [6].

TABLE 1
Typical experiments (counter-current flows, T = 25°C)

Ionic Form	Flow rates F	E.C., Inlet Concentration	I.C., Oulet Concentration	Total Flux Φ
Na^+	20	0.132	0.0025	$1.4\ 10^{-5}$
	200	0.102	0.00027	$1.5\ 10^{-5}$
	20	0.502	0.0068	$3.8\ 10^{-5}$
	200	0.506	0.0010_4	$5.8\ 10^{-5}$
H^+	20	0.0072	0.0013	$7.2\ 10^{-6}$
	200	0.0078	0.00032_5	$1.8\ 10^{-5}$
	20	0.0090	0.0016	$8.8\ 10^{-6}$
	200	0.0100	0.00036	$2.0\ 10^{-5}$
	20	0.030_0	0.0050	$2.8\ 10^{-5}$
	200	0.030_0	0.00097	$5.4\ 10^{-5}$
	20	0.050_6	0.0077	$4.3\ 10^{-5}$
	200	0.050_8	0.0014	$7.7\ 10^{-5}$
(1)	cm^3/hour	m.mol/cm^3 (2)		m.mol/s

(1) Initial ionic form of membrane

(2) Here the external compartment E.C. is supplied with neutral α-alanine solution and the internal compartment I.C. with pure water. The aqueous phases are ionic species free. Counter-current flows are used, as shown on figure 1 and 2.

In transport experiments (see Table 1), ionic forms of the membrane and flow rates of the solutions were also investigated. A reactive ionic form (H^+) and a high flow rate are favourable to the transport efficiency. Flux was doubled when the flow rate increased from 20 up to 200 ml/hour. The effluent concentration of the receiving solution could reach up to one fifth of C_S. Previous measurements were performed using counter-current flows. For co-current flows, the same values of flux have been observed.

A computer simulation has been employed to predict fluxes and concentration profiles under the same experimental conditions. The total flux Φ has been calculated as a function of the feed concentration C_S, the flow rates F and the diffusivity of the counterions in the membrane characterized by the parameter $\overline{P_X}$ when the ion-exchanger is initially in the H^+ form (see eqn.14). Cocurrent flow has been studied in fig.4. Two ranges of C_S could be considered. With dilute solutions ($C_S < 0.01$ m.mol cm^{-3}), ϕ is independent of $\overline{P_X}$. The membrane permeability is high enough and the mass transfer is controlled by the radial diffusion inside I.C. and E.C. . With more concentrated solutions Φ becomes a function of $\overline{P_X}$ because the transport is partially controlled by the membrane. Curves have been computed at three diffusivities. The best agreement between experimental and calculated data are obtained with $\overline{P_X} = 4.10^{-6}$ cm^2 s^{-1} m.mol cm^{-3}. The relatively high value of $\overline{P_X}$ has been attributed to the high mobility of H^+ which is involved in interdiffusion with protonated α-Alanine. The numeric simulation has been extended to a passive diffusion, i.e. when the membrane is in a non-reactive form (Na^+). In this case the experimental data agree with $P = 2.5\ 10^{-7} cm^2 s^{-1}$. The passive diffusion is chosen as reference in order to calculate the facilitation factor or ratio of Φ in the reactive form to Φ in the non-reactive form. The values of the parameters used in computation are collected in Table 2.

TABLE 2

Parameters used in computed simulation

Active height of the fiber :	$h = 22.4$ cm
Internal radius of the fiber :	$r_i = 0.03125$ cm
External radius of the fiber :	$r_e = 0.04375$ cm
Internal radius of the glass shell :	$r_S = 0.15$ cm
Diffusion coefficient of S in solution : (S is the neutral α-alanine)	$D = 0.91\ 10^{-5}\ cm^2.s^{-1}$
Protonation constant :	$K^* = 1250$
Permeability coefficient (Na^+ form) :	$P = 2.5\ 10^{-7}\ cm^2.s^{-1}$
Parameter of the counterions permea- : bility (unless otherwise mentioned)	$\overline{P_X} = 4\ 10^{-6}\ cm^2.s^{-1}$ m.mol cm^{-3}
Flow rates (inside F_i and outside F_e) :	$F = F_i = F_e = 20$ or 200 ml/h

Curves in Fig.4 concerned only the co-current flow. Computation have been extended to the countercurrent flow (see fig.5). Similar values of fluxes (dotted lines) have been obtained, which confirms the previous experimental results. On the other hand, differences in the concentration profiles have been predicted. Fig.6 visualizes concentration polarization, shows the extent of this phenomenon (due to the high permeability of the membrane) and confirms the effects of the feed concentration C_S and of the flow rates on the solution/membrane control of mass transfer. These concentration profiles are computed at the oulet of the two compartments.

In order to complete the investigation, interfacial concentration profiles have been drawn in fig.7 . From one end to the other, the internal $C(r_i,z)$ and the external $C(r_e,z)$ concentrations change conversely for the co-current flow or not for the counter-current flow. In the former case, the local diffusion rate $\mathcal{J}_R(z)$ decreases progressively from the inlet to the outlet while in the latter, the change in $\mathcal{J}_R(z)$ is lower. However, the total fluxes Φ remain comparable due to a compensation effect. In the two cases the differences between $C(r_e,z)$ and $C(r_i,z)$ is strongly dependent on the concentration C_S and the flow rates.

CONCLUSION

In passive diffusion, the mass transfer is controlled by the membrane in which α-alanine diffuses with a constant permeability coefficient (about $2.5\ 10^{-7}\ cm^2.s^{-1}$).

In facilitated transport, diffusion is controlled by the solutions in which α-Alanine diffuses with a constant diffusion coefficient (about 10^{-5} $cm^2.s^{-1}$). The facilitation factor increases from 5 to 25 when the flow rates of the feed and receiving solutions increase from 20 to 200 ml/h. These results confirm the extent of the concentration polarization which decreases the efficiency of the transport, particularly when the feed solution is dilute and the flow rate is low. The good agreement between calculated and experimental diffusion rates proves that the hydrodynamic conditions were well-controlled.

REFERENCES

[1] D. LANGEVIN, M. METAYER, M.LABBE, B. EL MAHI, in Effective Industrial Membrane Processes - Benefits and Opportunities, Elsevier, (1991), 307-319.

[2] D. LANGEVIN, M. METAYER, M.LABBE, M. HANKAOUI and B. POLLET, in Membranes and membrane processes, Plenum Pub. Corp., (1986),306-318.

[3] F. COEURET and A. STORCK, Eléments de génie électrochimique, Lavoisier Tec.Doc, (1984), 146-7

[4] M. METAYER, D. LANGEVIN, B. EL MAHI and M. PINOCHE, J.Membrane Sci, 61 (1991), 191-213.

[5] J.CRANK, The mathematics of diffusion, Oxford, (1967),186-9

[6] M. METAYER, D. LANGEVIN, M. HANKAOUI, B. POLLET, M.LABBE, Reactive Polymers, 7 (1988),111-122.

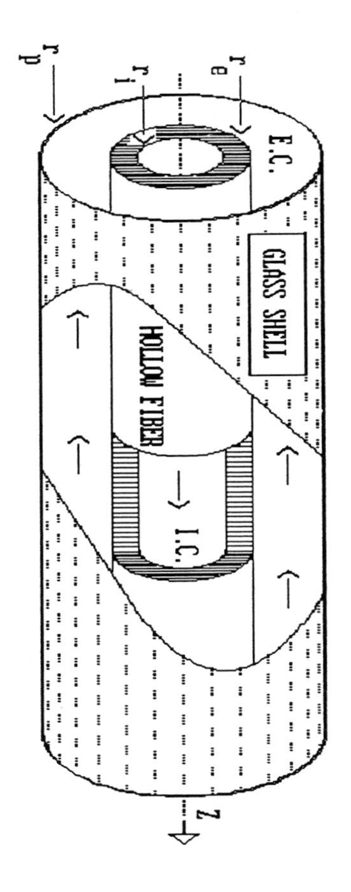

Figure 1:

Tubular membrane and its coaxial shell directed towards th z-axis. The internal and external radii of the fiber are denoted by r_i and r_e respectively. The internal radius of the glass shell is denoted by r_s.

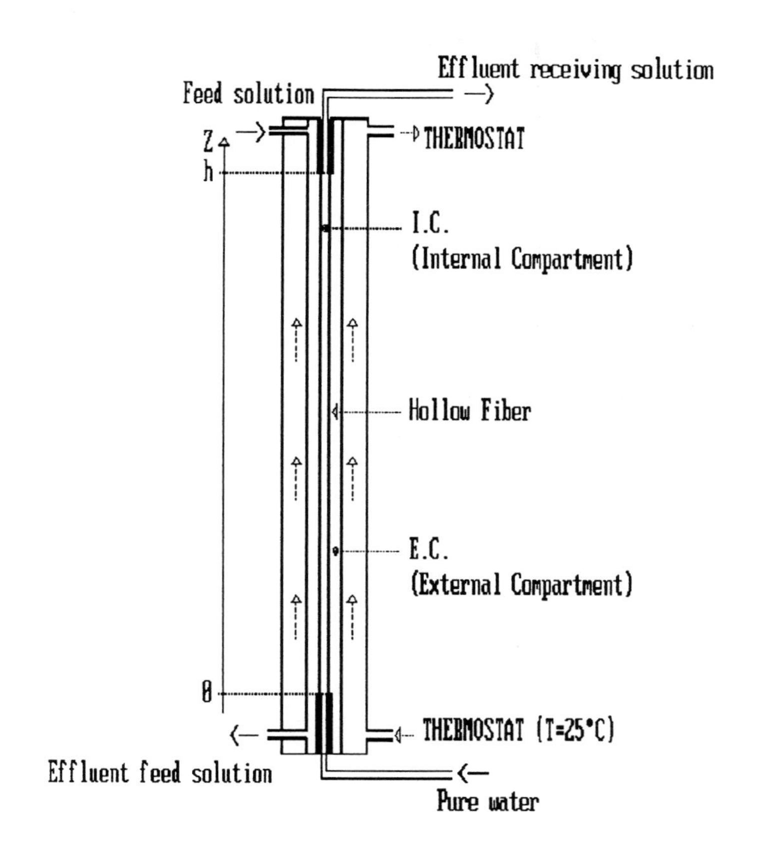

Figure 2:
Experimental device.

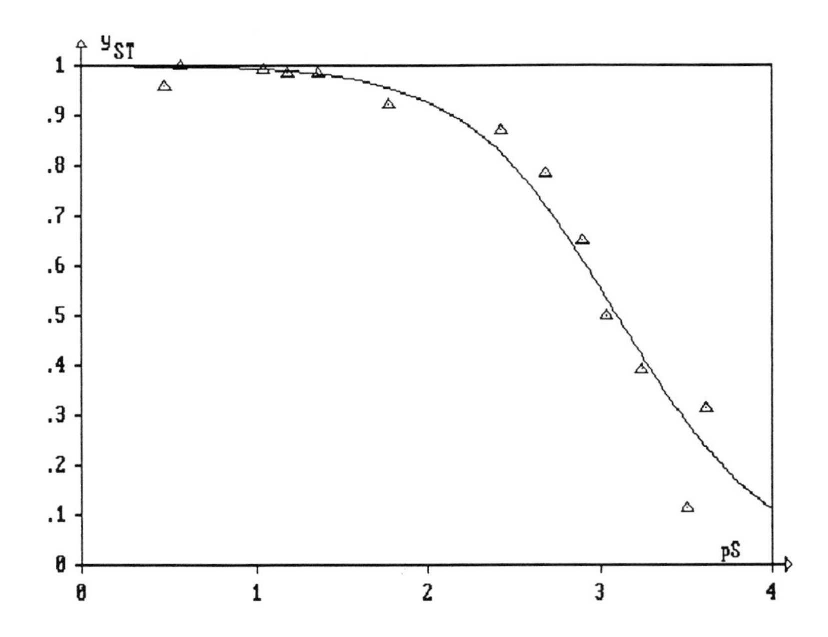

Figure 3:

Sorption isotherm of α-alanine by the Nafion fiber initially in the H^+ form. Fraction y_{ST} of the ionic sites occupied by the protonated α-alanine versus pS, = $-\log a_S$ where a_S denotes the activity of α-alanine in equilibrium solution (see eqn.11a). Curve has been computed from eqn. 11 by adjusting to 1250 the α-alanine protonation constant K^*.

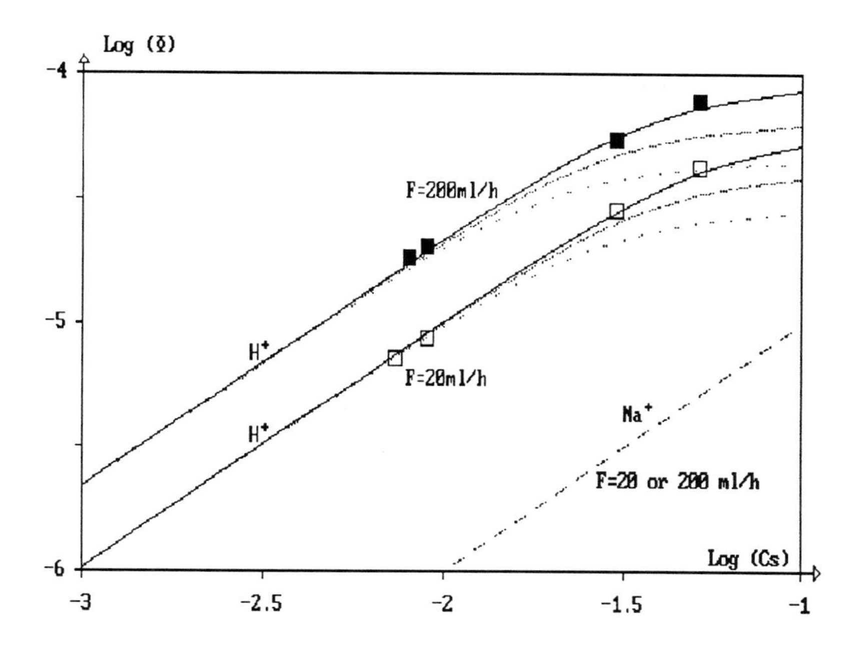

Figure 4:

Total flux Φ vs. feed concentration C_S in a log-log scale. Curves have been computed for cocurrent flow with $\overline{P_X}$ = 1 (dotted curve), 2 (dashed curve) or 4 10^{-6} $cm^2 s^{-1} m.mol\ cm^{-3}$ (full curve). The experimental points have been obtained for countercurrent flow. Ion-exchanger was initially in the reactive form (H^+). The curve, for which ion-exchanger was in the non-reactive form (Na^+), is given as reference.

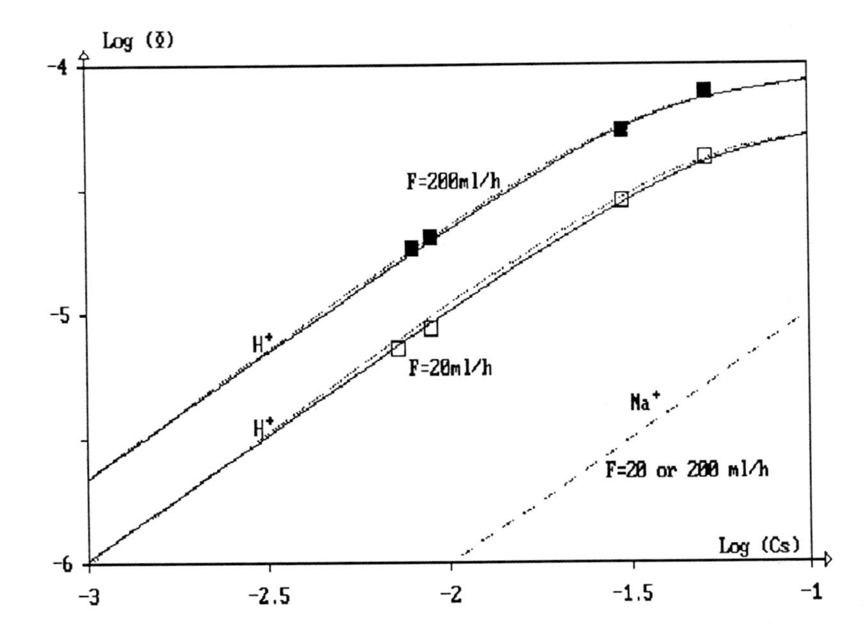

Figure 5:

Total flux Φ for cocurrent flow (full curves) and for countercurrent flow (dashed curves) with $\overline{P_X} = 4\ 10^{-6}\ cm^2 s^{-1} m.mol\ cm^{-3}$ (see fig.4).

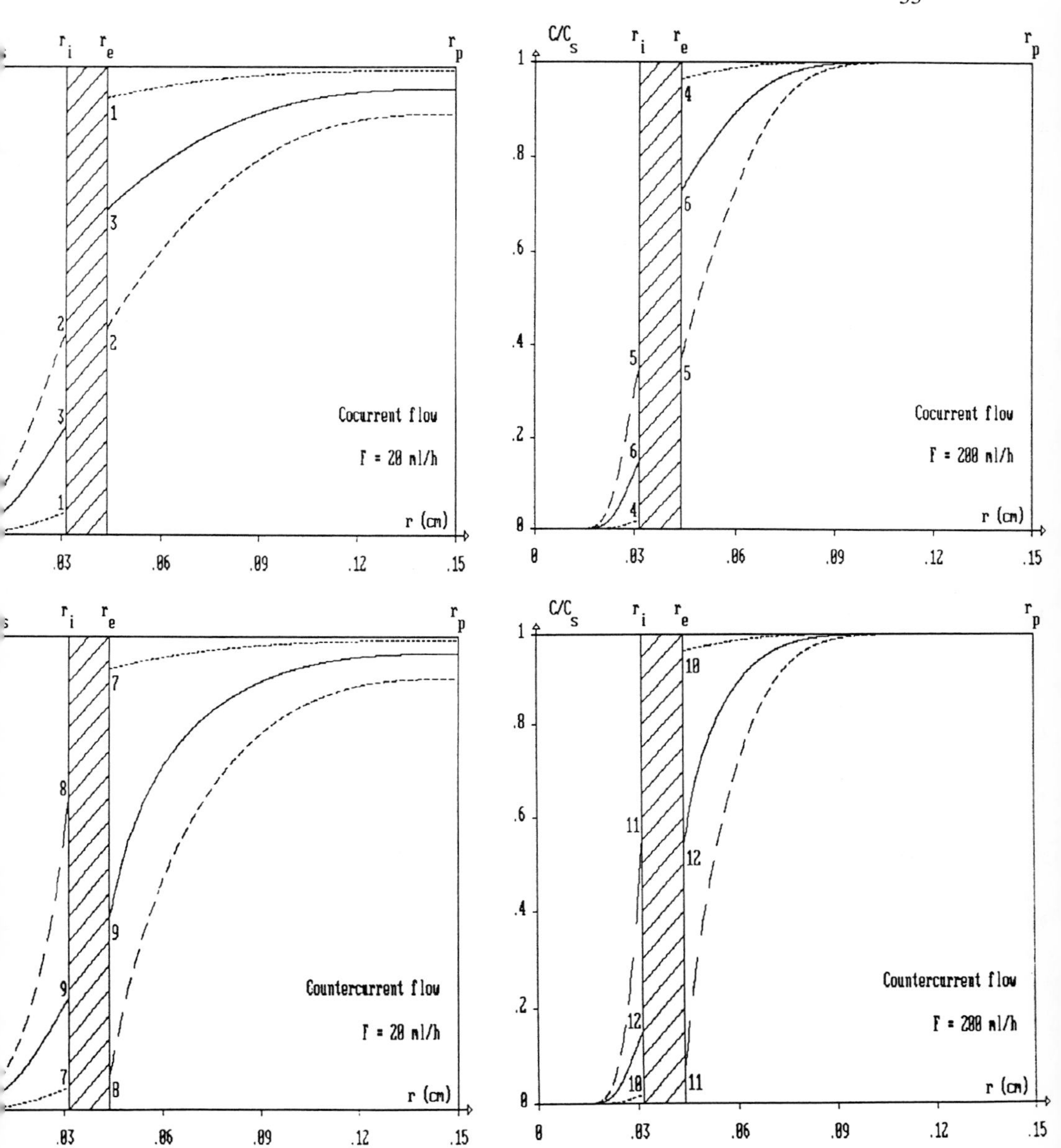

Figure 6:

Concentration polarization and computed radial profiles.

Passive diffusion whatever C_S (dotted lines and curves 1, 4, 7, 10).

Facilitated transport when C_S = 0.01 (dashed lines and curves 2, 5, 8, 11) and when C_S = 0.1 (full lines and curves 3, 6, 9, 12).

The feed solution is circulated outside the fiber at a flow rate F = 20 or 200 ml/h while the receiving solution (pure water) is circulated inside the fiber at the same flow rate.

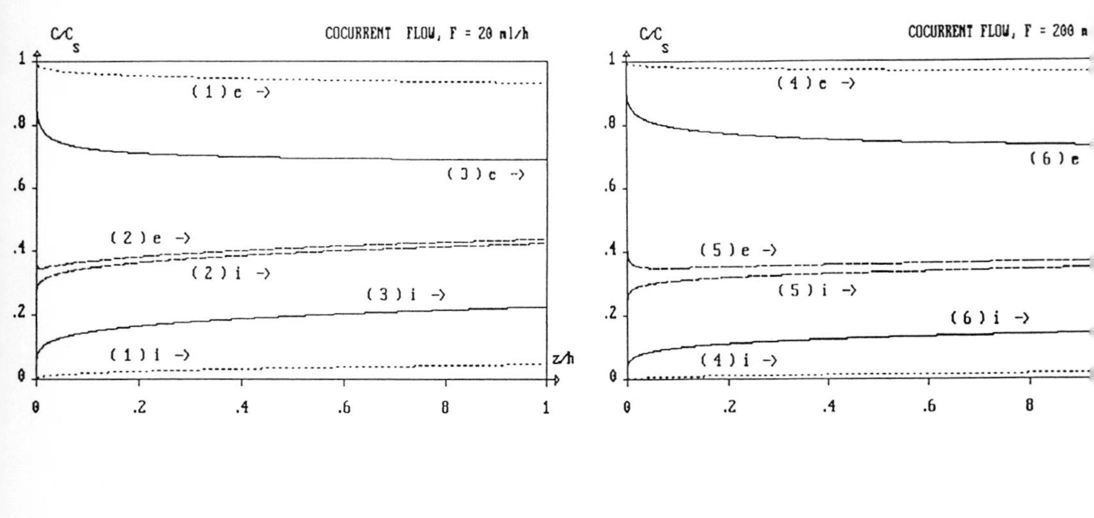

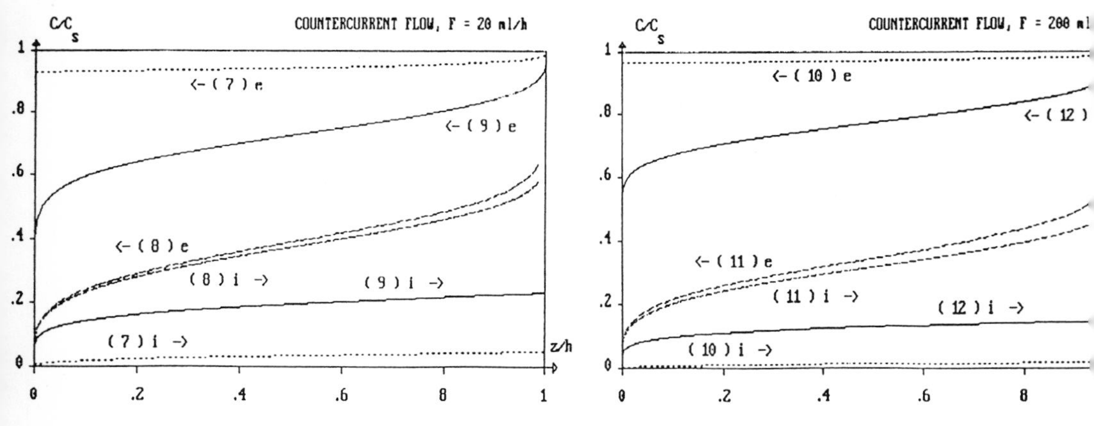

Figure 7:
Interfacial concentrations on the internal (i) or the external (e) surface of the hollow fiber. These profiles are computed along the z-axis. The arrows specify direction of flow. For lines and curve numbers see fig. 6.

THE EFFECT OF MEMBRANE FOULING AND CLEANING ON REJECTION OF LOW MOLECULAR WEIGHT TRACER IN ULTRAFILTION

N. Sanders[+] and J. Hubble
School of Chemical Engineering, University of Bath, Claverton Down, Bath BA2 7AY
(+ present address, Kraft General Foods, Banbury, Oxfordshire)

SUMMARY

Rejection of a low molecular weight chemical tracer (adenosine 5'-monophosphate, mw 347.2) by an Amicon polysulphone hollow fibre membrane (nominal molecular weight cut-off 10,000) has been measured using a spectrophotometric method in a constant flux apparatus. Fouling produced by prior exposure of the membrane to a protein solution under zero flux conditions, followed by rinsing, was found to increase tracer rejection. Cleaning of the membrane after each exposure to protein did not restore the initial rejection characteristics, although membrane permeability did not decrease substantially. The increase in rejection was reversed by prolonged soaking of the membrane in buffer. The build up of a polarised protein layer resulting from the convective deposition of protein at the membrane surface had little effect on tracer rejection, when compared to the membrane with adsorbed protein only, and was only discernible at the highest fluxes compatible with the pressure rating of the membrane.

INTRODUCTION

Solute rejection by ultrafiltration membranes is a complex phenomenon, which often cannot be adequately described by the nominal molecular weight cut-off (mwco) data as supplied by manufacturers. Frequently, molecules larger than the mwco are not completely rejected [1], whilst molecules much smaller than the mwco are partially rejected [2, 3]. The rejection of smaller molecules in ultrafiltration may be important in a number of circumstances, for instance in enzyme membrane reactors which require relatively small product molecules to pass through the membrane, whilst the macromolecular enzyme must be retained [1, 4].

Measurement of the rejection of molecules over a range of sizes has been used as a means of membrane characterisation [2, 3, 5, 6, 7]. HPLC is frequently used to measure solute concentrations for determination of rejection (for example [7]), although photometric [1], refractive index and chemical [8] techniques have also been used.

The effect of concentration polarisation of saccharides during rejection tests was assumed negligible by Hanemaaijer et al [9] due to the high diffusivity and relatively low rejection of the molecules tested. Wendt et al [10] made a theoretical investigation, using the Spiegler-Kedem equation, of the effect of membrane heteroporosity on rejection coefficients. They concluded that for real membrane characteristics, the effect of solute concentration on rejection should be relatively small.

Rejection of molecules such as dextrans has been modelled according to hard sphere theory [11]. However, deviation from the hard sphere theory can occur due to membrane pore size distribution [12] or to solute-membrane interactions, resulting in a reduction of the effective pore size [13]. Exposure of membranes to proteins has been shown to result in a reduction in pore size, as quantified by an increase in rejection of smaller solutes [2, 6]. Protein adsorption occurring during ultrafiltration can cause increased rejection even after rinsing of the membrane [8]. The rejection increase was affected by ultrafiltration pressure, suggesting that greater adsorption or pore blockage caused by increased concentration polarisation of protein may affect the rejection of smaller solutes.

Solute rejection is strongly influenced by macromolecule adsorption to membranes, but the additional effect of concentration polarisation of protein on the rejection of small solutes is less certain. Ingham et al [14] attributed rejection increases to adsorbed rather than polarised protein, but Calderaro et al [15] found that the degree of concentration polarisation of human plasma significantly affected rejection of a range of radio-labelled molecules by haemofilters.

The rejection of small solutes under conditions of adsorbed and polarised protein is important in any process requiring passage of the small molecules at the same time as retention of macromolecules. The behaviour of cleaned membranes, as well as new ones is significant in such systems, as membranes are likely to be cleaned and re-used many times. If the degree of protein polarisation is found to significantly affect rejection of small solutes, this could influence the operating regime of a membrane process.

Here, an investigation into the rejection of a low molecular weight tracer molecule (adenosine 5'-monophosphate, 5'AMP, mw 347.2) by an Amicon polysulphone hollow fibre membrane of mwco 10,000 has been undertaken. The effects of adsorbed and polarised protein on the tracer rejection are measured under a range of operating conditions, and the effect of fouling and membrane cleaning on long-term rejection characteristics is assessed.

MATERIALS AND METHODS

Tracer Rejection Experiments

Figure 1. Constant Flux ultafiltration rig. PT - pressure transducer. TT - t

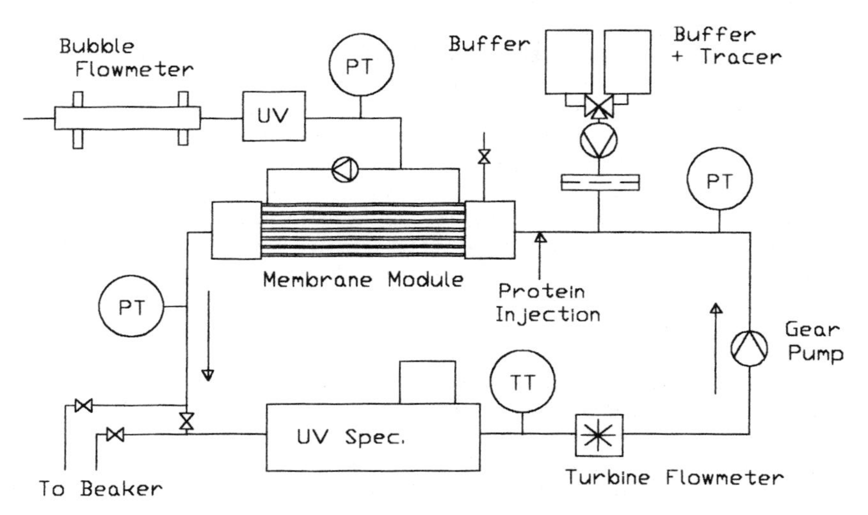

emperature probe. UV - ultraviolet absorbance measurement.

The apparatus used for this work was based on a constant flux hollow fibre ultrafiltration cell [16] and is described in Fig 1. The membrane was an Amicon H1P10-20 polysulphone hollow fibre cartridge of nominal 10,000 molecular weight cut-off (membrane area $0.08m^2$). In the constant flux cell, a closed recycle loop incorporates the fibre lumen. The liquid in the recycle loop is circulated by a gear pump (Flowgen V.015.12/2030) to provide cross flow. Trans-membrane flow (flux) is induced by a peristaltic pump (Watson Marlow 501U) feeding liquid into the recycle loop. Thus the flux remains constant at a given pump setting and so the concentration polarisation conditions remain constant. Feed solution can be selected from one of two vessels containing buffer with and without tracer.

Recycle (cross flow) rate is monitored with a turbine flowmeter (Litre Meter LM25GN), and permeate flow rate is verified using a custom - designed 'time of flight' bubble flow meter [17]. Three pressure transducers (Mediamate) allow the pressure drop along the hollow fibres, as well as the average trans-membrane pressure drop, to be determined. Temperature in the recycle loop is measured solution absorbance is measured with a UV spectrophotometer (Cecil CE 2272) fitted with a custom made high flow short path flow cell. Permeate solution absorbance is measured at 254 nm using a spectrophotometer (Shimadzu UV-120-02). Liquid in the permeate region is recirculated with a peristaltic pump (Watson Marlow 101 U/R) to avoid 'dead spaces' and concentration gradients downstream of the membrane. The outputs from all the above measurement devices are logged to a microcomputer for analysis and plotting.

The tracer selected for rejection studies was adenosine 5'- monophosphate (5'AMP) (Sigma A2252) . This was chosen on the basis of its low molecular weight of (347.2 $g.mol^{-1}$), high aqueous solubility and large optical extinction coefficient . This allows tracer concentration in both retentate and permeate to be conveniently determined by absorbance measurement. Accurate assessment of the rejection of small molecules by an ultrafiltration membrane in the presence of protein requires that there is minimal interaction between the tracer molecule and the protein. To assess this possibility a number of gel permeation experiments were conducted using a Sephadex G25 column to separate 5' AMP and BSA. Comparison of the retention volumes for each component showed that they were essentially the same when run both in isolation and as a mixture, indicating little interaction between protein and tracer.

Rejection coefficients for the cleaned membrane in the absence of protein were determined by measuring the concentrations of tracer in the retentate and permeate when equilibrium had been reached. The observed rejection coefficient is given by:

$$Robs = 1 - \frac{\text{Tracer Concentration in Permeate}}{\text{Tracer Concentration in Retentate}}$$

The effect of flux on the rejection of 5'AMP was determined by altering the feed pump setting and allowing the system to reach equilibrium at each feed flowrate.

Measurement of Tracer Rejection after Protein Adsorption

Rejection of 5'AMP was also determined for a membrane which had previously been exposed to protein (BSA). The recycle loop was opened to include a stirred beaker, to which a concentrated protein solution containing a known mass of BSA was added. Transmembrane flux was prevented at this stage by closing the permeate tube. Bulk recycle solution protein concentration was determined by UV absorbance at 280nm. At equilibrium, the mass of BSA associated with the membrane could be determined, given that the volume of the system is known. The protein solution was then rinsed out with distilled water followed by buffer, the recycle loop was closed, and tracer solution fed into the system (previous studies with this system indicate that no protein desorption occurs after rinsing). Recycle and permeate solution absorbances were monitored at 254nm, allowing the rejection of 5'AMP to be evaluated at a range of fluxes.

Measurement of Tracer Rejection with Protein Polarisation

Investigation of the rejection of 5'AMP with simultaneous concentration polarisation of protein required measurement of both BSA and 5'AMP concentrations in the same solution. This was achieved by measurement of recycle solution absorbance at both 254 and 280nm. As the spectrophotometer zero could not be adjusted during experiments, the zero readings for buffer at the two wavelengths were noted before and after each experiment, and the readings obtained during experiments were corrected for zero difference and zero drift.

The individual contributions of BSA and 5'AMP to the observed absorbance at 254 and 280nm were found not to be strictly additive, and a correction factor was required to accurately relate the concentration of the two components to the absorbance measured. The need for a correction factor indicates that there is some interaction between BSA and 5'AMP, possibly a weak association leading to a spectral shift. However, given the results of gel permeation experiments where there was little evidence that any interaction was hindering separation, it is unlikely that this would significantly affect the measured rejection of 5'AMP.

After determining the rejection of tracer by the clean membrane at a single flux, the feed pump was switched off and a small volume of concentrated protein solution containing a known mass of BSA was injected into the recycle loop via a septum. Liquid was displaced through the membrane during this operation. The feed pump was then switched on at a low setting and the recycle and permeate absorbances allowed to reach equilibrium. Recycle spectrophotometer readings at 254 and 280nm were taken, from which the concentrations of BSA and 5'AMP were calculated. Rejection, polarisation and pressure drop data at constant crossflow rate and protein loading, for arrange of fluxes, was obtained in each experiment.

Membrane Cleaning

Membrane cleaning was carried out after each experiment where the membrane was exposed to protein. 2.5l of 0.1M NaOH was backflushed through the membrane, whilst circulation of the backflushed solution through the fibres was maintained. The recycle loop was then thoroughly rinsed with distilled water, closed and about 2l of distilled water fed through the membrane via the feed pump. The recycle loop was then flushed out with buffer and buffer was filtered until the flux - pressure relationship stabilised.

The cleaning procedure described here has been found by the authors to give reproducible results for successive adsorption experiments, the adsorbed amount under zero flux conditions was measured.

RESULTS AND DISCUSSION

Rejection of Tracer by the Clean Membrane

Experiments were carried out to determine the effect of flux on the rejection of tracer at different recycle (crossflow) rates (Table 1). Crossflow had no noticeable effect on the

Expt No	Flux ml.min^{-1}	Reject.	Membrane Permeably. ml.min^{-1} psi^{-1}	Crossflow Rate ml.min^{-1}	Tracer Concn. mol.l^{-1}	Cleaning Method Before Expt.
1	7.40 14.70 25.20 40.00	0.140 0.170 0.194 0.190	2.40 2.38 2.34 2.27	110	0.001	Backflush
2	7.70 15.50 27.70 45.00	0.145 0.165 0.174 0.180	2.41 2.43 2.40 2.25	310	0.001	Not Cleaned
3	7.80 15.70 27.80 44.10	0.125 0.135 0.160 0.170	2.44 2.42 2.40 2.20	210	0.001	Not Cleaned

Table 1. Summary of tracer rejection experiments with a clean membrane

flux-rejection curves, suggesting that concentration polarisation of tracer under these conditions hardly influences rejection. This confirms the observations of Hanemaaijer et al [9], who proposed measurement of the rejection of a mixture of low molecular weight polysaccharides as a means of membrane characterisation.

Observed rejection of tracer increases with flux up to a transmembrane flowrate of about 20 ml.min^{-1}, beyond which there is little effect of flux on rejection. It seems that in this flux - independent region, additional pore blockage by tracer molecules does not occur.

Rejection of Tracer by a membrane After Exposure to Protein

Membranes were treated with BSA under zero flux conditions, followed by rinsing and determination of the flux - rejection characteristics for tracer. Adsorbed protein remaining attached to the membrane after rinsing with distilled water was found to affect the rejection characteristics (Fig 2) (Table 2).

Expt No	Flux ml.min^{-1}	Reject.	Membrane Permeably ml.min^{-1} psi^{-1}	Adsorbed Protein g.m^{-2}	Bulk Protein Concn g.l^{-1}	Tracer Concn. mol.l^{-1}
4	7.70 16.00 29.70	0.305 0.375 0.401	2.41 2.36 2.40	0.66	5.35	0.0004
5	6.75 13.60 23.80 38.50	0.280 0.370 0.410 0.410	2.70 2.62 2.62 2.56	1.64	14.42	0.0004
6	7.05 14.30 24.60 39.00	0.350 0.465 0.507 0.506	2.14 2.10 2.04 1.97	3.79	24.28	0.0004
7	6.96 14.00 24.70 39.70	0.415 0.507 0.540 0.530	2.43 2.36 2.35 2.31	0.15	1.06	0.0004
8	7.00 14.10 24.70 39.30	0.330 0.425 0.410 0.420	2.52 2.47 2.46 2.37	No protein	No protein	0.0004

Table 2. Tracer rejection experiments after protein adsorption and rinsing (Backflush cleaning)

Observed rejection was increased by the presence of adsorbed protein, as calculated from a mass balance of protein in the system, but the shape of the flux - rejection relationship was little different to that for a clean membrane. Others have found that adsorbed protein increases rejection, and the effect of the adsorbed protein has been explained in terms of reduction in pore size [2, 6].

Work on the adsorption of proteins to membranes has shown that the adsorbed amount increases with protein concentration, even at very high protein concentrations [18]. Thus the equilibrium concentration for protein adsorption might be expected to affect the adsorbed amount and hence the rejection of tracer in these experiments. Adsorbed protein layers formed by ultrafiltration at increasing pressures have been shown to exhibit increasing rejection of small solutes [8]. Higher pressures will lead to greater concentration polarisation, higher wall protein concentrations and hence increase protein adsorption. In this work, rejection was found to increase with equilibrium adsorption

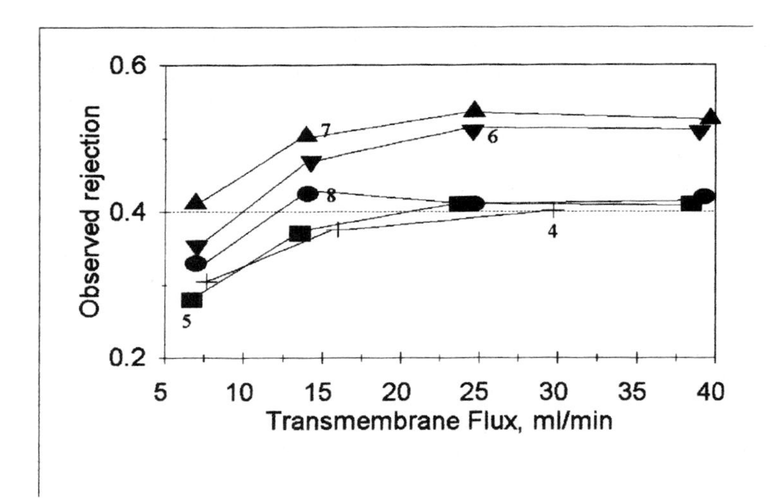

Figure 2. Effect of adsorbed protein on tracer rejection. Conditions as in Table 2, experiments carried out in the numerical order shown.

when a lower concentration of 1 $g.l^{-1}$ was used (Fig 2) (Table 2), even though the membrane was backflush cleaned after each experiment. Rejection by the unfouled membrane after backflush cleaning did not return to the values obtained before the series of fouling experiments. Thus fouling took place which considerably affected rejection and was not removed by backflush cleaning.

These results suggest that there is a progressive increase in rejection from run to run as the protein concentration is increased. As the final run, with a lower protein concentration, also shows an increase in rejection it would appear that the effects of adsorption are cumulative, and this again supports the premise that the adsorbed protein is not removed by backflushing.

It was considered possible that backflush cleaning resulted in blockage of pores by protein which had migrated into the structure of the membrane during fouling. Therefore a modified cleaning procedure was adopted, consisting of the usual backflush followed by ultrafiltration (forward flushing) of about 2l of 0.1M NaOH at high crossflow, in order to remove any protein that had been forced backwards into the pores. This cleaning procedure did not result in any reduction of the observed rejection when compared with the previous, backflush - only method (Table 3).

Expt No	Flux ml.min^{-1}	Reject.	Membrane Permeability ml.min^{-1}psi^{-1}	Adsorbed Protein g.m^{-2}	Bulk Protein concng.l^{-1}	Tracer Concn. mol.l^{-1}
9	6.90	0.330	2.52			
	14.60	0.430	2.52	No protein	No protein	0.0004
	25.40	0.470	2.46			
	41.40	0.455	2.45			
10	5.65	0.332	2.35			
	12.40	0.440	2.48	0.76	5.31	0.0004
	21.20	0.480	2.45			
	34.00	0.480	2.43			

Table 3 Tracer rejection experiments after protein adsorption and rinsing (Backflush and Forward Flush cleaning)

The 'irreversible' fouling demonstrated by these experiments also reduced the effect (on rejection) of subsequent exposure to protein (Tables 2 and 3). Such fouling seems to result in 'permanent' occupation of active sites on and in the membrane structure, reducing subsequent protein - membrane interactions and their effect on membrane characteristics. Subsequent work indicated that these 'irreversible' fouling effects may be reversible over a long period of time. The membrane, after back- and forward flush cleaning, was left in buffer for a period of four weeks. A series of experiments were then performed, exposing the membrane to BSA at a concentration of 5 g.l^{-1}, followed by rinsing and rejection measurement (Table 4).

Expt No	Flux ml.min^{-1}	Reject.	Membrane Permeably ml.min^{-1} psi^{-1}	Adsorbed Protein g.m^{-2}	Bulk Protein concn g.l^{-1}	Tracer Concn. mol.l^{-1}
11	5.40	0.072	2.97			
	11.30	0.098	2.96	0.72	5.32	0.0002
	19.80	0.122	2.89			
	32.70	0.128	2.83			
12	5.75	0.072	2.67			
	12.20	0.100	2.71	0.66	5.35	0.0002
	20.40	0.210	2.60			
13	6.10	0.440	2.43			
	12.60	0.540	2.47			0.0002
	20.00	0.584	2.38		5.0	
	32.80	0.588	2.34			

Table 4 Tracer rejection experiments after protein adsorption and rinsing (Backflush and Forward Flush cleaning after storage)

The observed rejection of tracer had dropped considerably over the period of 'resting' the membrane, but began to rise rapidly again even though the membrane was cleaned by back- and forward flushing after each experiment.

These results support findings that desorption of 'irreversibly' bound protein may take place over a long period of time [19]. Deliberate 'resting' of membranes in order to restore their ability to pass smaller molecules is likely to be ineffective, due to the rapid rise of rejection on exposure to protein. Investigation into pre-fouling by materials which occupy the active sites for adsorption, but which have less effect on rejection, may provide a partial solution to the problem.

Effect of Concentration on Permeability and Adsorbed Amount

Membrane permeability was measured during all rejection experiments on the cleaned membrane, and during experiments on the membrane after exposure to protein under zero - flux conditions (Tables 1-4). In general, there was little correlation between membrane permeability and tracer rejection. However the experiments performed after 'resting' the membrane for four weeks showed a decrease in permeability from one experiment to the next, accompanied by a rise in tracer rejection. Therefore the cleaning procedure was unable to prevent a build-up of membrane - associated protein which affected both rejection and, to some extent, flux.

The amount of protein associated with the membrane under equilibrium adsorption conditions was calculated. However, the values calculated did not correlate with tracer rejection (Table 2,3 and 4), although the observed cumulative effect of protein exposure suggests that higher adsorbed amounts do result in lower membrane permeabilities. It is possible that the errors inherent in a mass balance based estimate of adsorbed protein are masking the true situation, and indirect evidence for a correlation between adsorbed amount and rejection has been presented by others [8]. While it is possible that the amount of membrane associated protein measured at equilibrium is not equal to the amount remaining after rinsing, protein adsorption is usually regarded as irreversible or slowly reversible [19] so it is likely that at least a proportion of the membrane associated protein will remain during rinsing.

Rejection of Tracer by a Membrane with Polarised Protein

The effect of a polarised protein layer on the observed rejection of tracer has been studied. Experiments were carried out at different crossflow rates and protein loadings, resulting in different levels of protein polarisation. Each experiment consisted of measuring the tracer rejection of the clean membrane at a transmembrane flowrate of about 23 $ml.min^{-1}$, followed by addition of a known mass of protein to the recycle loop and determination of the observed rejection at a range of fluxes (Table 5). Thus the rejection of tracer under conditions of protein polarisation could be compared with rejection by the cleaned membrane before exposure to protein.

Expt No	Flux ml.min^{-1}	Rejection	Crossflow Rate ml.min^{-1}	Protein Mass g
14	21.30 6.25 13.00 22.60 31.70	0.52 0.49 0.59 0.62 0.63	310	No protein 1.0
15	22.60 7.00 13.50 20.90 28.60	0.52 0.52 0.60 0.63 0.63	150	No protein 1.0
16	22.80 7.20 13.60 21.00 25.80	0.48 0.51 0.58 0.60 0.65	90	No protein 1.0
17	23.10 7.30 13.80 21.30 29.80	0.59 0.46 0.53 0.54 0.58	90	No protein 0.2
18	23.40 7.40 14.20 21.60 30.00	0.63 0.52 0.60 0.62 0.62	310	No protein 0.2

Table 5. Summary of tracer rejection experiments in the presence of polarised protein. Tracer concentration 0.0002 mol.l^{-1}. Backflush cleaning before each experiment.

Experiments were performed at a fixed protein loading (1g of BSA in the system), and a range of crossflow rates. The presence of polarised protein increased the rejection of tracer when compared with the cleaned membrane, in agreement with the work of others [14, 15]. At the higher crossflow rates the shape of the flux - rejection curves is little different to those obtained with adsorbed protein only. However at the lowest crossflow rate, there is an indication of an increase in observed rejection at high flux, where protein polarisation is greatest. This could signal incipient gel formation. Experiments using a lower protein loading (0.2g BSA in the system) indicated that tracer rejection in the presence of polarised protein was the same or less than with the cleaned membrane (Fig 3).

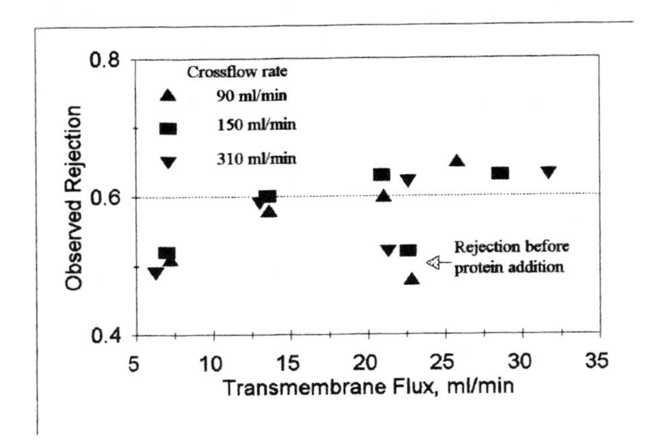

Figure 3 Effect of polarised protein on tracer rejection. Expt 17. Crossflow rate 90 ml.min^{-1}, shear rate 490 s^{-1} + Expt 18. Crossflow rate 310 ml.min^{-1}, shear rate 1685 s^{-1}.

Lower levels of protein adsorption due to the lower bulk protein concentration are likely to have reduced the effect of protein on rejection. There is also the possibility of time-dependant changes in rejection which may have been partly responsible for this behaviour. Nevertheless, the form of the curves is similar to those obtained with a protein loading of 1g, and the results indicate that adsorbed protein provides the major contribution to rejection at moderate levels of protein polarisation. At the highest fluxes obtainable in this system, there is evidence that polarised protein begins to affect the rejection of tracer. The variation of the "no protein" results shown in table 5 reflect the variations in cleaning efficiency despite a similar cleaning protocol being used throughout. Residual adsorbed protein remaining after cleaning makes comparisons difficult but it is clear that polarisation has little direct effect on rejection.

Differences between the rejection characteristics of a clean membrane, and those during polarisation, are smaller in the case of the lower protein loading. Lower bulk protein concentrations will result in lower adsorption, which seems to reduce the effect of adsorbed protein on rejection of tracer.

CONCLUSIONS

The rejection of a low molecular weight tracer has been studied for a clean membrane, the same membrane fouled by zero - flux protein adsorption, and the membrane under conditions of protein polarisation. Rejection of the tracer (molecular weight about 350) by the cleaned, (but not new) membrane of nominal molecular weight cut-off 10,000 was about 0.15, whilst rejection in the presence of polarised protein was as high as 0.65. This level of rejection of a small molecule would be significant in an enzyme membrane reactor, which may rely on low molecular weight products passing through the membrane.

Repeated exposure of the membrane to protein caused an increase in observed rejection, even though membrane was cleaned after each exposure by back- and forward - flushing with 0.1M NaOH, demonstrating that irreversible fouling took place. The amount of protein adsorbed to the membrane did not strongly affect rejection, therefore it seems that rejection is determined by small amounts of tightly - bound protein and not by the bulk of the adsorbed protein. Evidence for slow desorption of this tightly - bound material is provided by a significant drop in rejection of tracer after soaking ('resting') the cleaned membrane in buffer for four weeks. The results obtained here are consistent with those obtained from electron microscope studies which showed that denatured BSA could enter the pores of membranes with low molecular weight cut offs, and that this material could influence rejection. It is interesting to relate these findings to work conducted on the role of serum albumins in modifying the permeability of biological capillaries, where it has been shown that albumin interactions with the inner surface of the capillary are essential for its ultrafiltration properties [21].

Polarised protein was found to have little effect on rejection, except at low crossflow rates and high fluxes where protein polarisation is greatest. Under conditions of limiting flux or gel permeation, gel layers may cause increased rejection of smaller molecules. Gel polarisation cannot be achieved in this apparatus, due to the pressure rating of the membrane.

ACKNOWLEDGEMENTS

The authors wish to acknowledge the financial support of SERC grant No. GR/E/05568. NS acknowledges the support of a SERC studentship.

REFERENCES

1 Jandel L, Schulte B, Bückmann AF, Wandrey C, "Quantitative description of the rejection of polymeric catalysts by ultrafiltration membranes", J Membr. Sci, 7 (1980) 185-201.
2 Hanemaaijer JH, Robbertsen T, Van den Boomgaard Th, Gunnink JW, "Fouling of ultrafiltration membranes. The role of protein adsorption and salt precipitation", J Membr. Sci, 40 (1989) 199-217.
3 Nakao SI, Kimura S, "Analysis of solutes rejection in ultrafiltration", J Chem Eng, Japan, 14 (1981) 32-37.

4 Cheryan M, Deeslie WD, "Soy protein hydrolysis in membrane reactors", J Am Oil chem Soc, 60 (1983) 1112-1115.
5 Nobrega R, de Balmann H, Aimar P, Sanchez V, "Transfer of dextran through ultrafiltration membranes: a study of rejection data analyzed by gel permeation chromatography", J Membr. Sci, 45 (1989) 17-36.
6 Zeman LJ, "Adsorption effects in rejection of macromolecules by ultrafiltration membranes", J Membr. Sci, 15 (1983) 213-230.
7 Jonsson D, Christensen PM, "Separation characteristics of ultrafiltration membranes, Membranes and membrane processes", ed. Drioli E, Nakagaki M (1984) 179-190
8 Dorson WJ, Cotter DJ, Pizziconi VB, "Ultrafiltration of protein molecules through deposited protein layers", Trans.amer. Soc. Artif. Int. Organs. 21 (1975) 132-137
9 Hanemaaijer JH, Robbertsen T, Van den Boomgaard Th, Olieman C, Both P, Schmidt DG "Characterisation of clean and fouled ultrafiltration membranes", Desalination, 68 (1988) 93-108.
10 Wendt RP, Mason EA, Bresler EH, "Effect of heteroporosity on membrane rejection coefficients", J Membr. Sci. 8 (1981)
69-90
11 Zeman L, Wales M, "Steric rejection of polymeric solutes by membranes with uniform pore size distribution", Sepn. Sci. Technol. 16 (1981) 275-290
12 Jonsson G "Transport phenomena in ultrafiltration: membrane selectivity and boundary layer phenomena", Pure Appl. Chem. 58(12) (1986) 1647-1655
13 Long TD, Jacobs DL, Anderson JL, "Configurational effects on membrane rejection", J Membr. Sci. 9(1981) 13-27
14 Ingham KC, Busby TF, Sahlestrom Y, Castino F. "Separation of macromolecules by ultrafiltration: influence of protein adsorption, protein-protein interactions, and concentration polarisation", in Ultrafiltration membranes and applications, (ed Cooper AR) (1980) 141-158.
15 Calderaro V, Memoli B, Andreucci VE, Drioli E, Albanese O, "Influence of concentration polarisation in post-dilutional hemofiltration of human plasma", Artif. Organs. 4 (1980) 317-321
16 Turker M, Hubble J "Membrane fouling in a constant flux ultrafiltration cell", J Membr. Sci. 15 (1987) 213-230.
17 Bishop JT, Sanders N, "Bubble flowmeter for measurement of low permeate flows in ultrafiltration", Biotech. Techniques, 3 (1989) 101-106
18 Matthiasson E, Hallstrom B, Sivik B, "Adsorption phenomena in fouling of ultrafiltration membranes", in Engineering science in the food industry, (ed. McKenna BM) (1989), Elsevier Applied Science, London and New York pp 139-149.
19 Norde W, "Adsorption of proteins from solution at the solid-liquid interface", Adv. Colloid Interface Sci. 25 (1986) 267-340
20 Sheldon, J.M., Reed, I,M., Hawes, C.R. "The fine structure of ultrafiltration membranes II: Protein fouled membranes", J. Membrane Sci. 62 (1991),87-102.
21 Michel, C.C. "Ultrafiltration in living capillaries" in Effective membrane processes:benefits and opportunities, Turner, M.K. ed Elsevier Applied Science, London and New York, pp230-239.

Separation of Penicillin G with Hollow Fiber Contained Liquid Membrane System

Z.F. Yang, D. Rindfleisch, T. Scheper, and K. Schügerl*
Institut für Technische Chemie, Universität Hannover, Callinstr. 3, 3000 Hannover, F. R. Germany

ABSTRACT

In the present work Hollow Fiber Contained Liquid Membrane (HFCLM, LIQUI-CEL™, E400/1.0), in which extraction and stripping proceed at the same time, has been successfully employed to separate penicillin G from aqueous solution. Amberlite LA-2 + tributyl phosphate (TBP) were used as complexing agents for facilitated transport of penicillin G from an aqueous solution. 15-50 g/L $NaHCO_3$ were used as stripping agent. The recovery extent of penicillin G reach up to 90% in three stages in series at the following conditions: the flow rates of feed and stripping are about $0.5 cm^3/min.$, the equilibrium pH values of extraction and stripping are 4.0-4.4 and 8.1-8.5, respectively, feed and stripping flow countercurrently.

INTRODUCTION

In order to improve the extraction of penicillin G with butyl acetate as solvent, reactive extraction by secondary amines was employed (Reschke and Schügerl, 1984a&b&c). It was possible to carry out the extraction at pH 5 and the stripping at pH 7.5. In this range penicillin G is stable, thus the losses could be reduced to less than 1 %.
However, the amines are probably difficult to be used as industrial solvent for extraction of penicillin G because of the difficulties of stripping and solvent recovery (Yang, 1989, Yang et al., 1992). Therefore, the extraction of penicillin G with neutral phosphorus esters and dioctyl sulfoxide in n-butyl acetate was investigated and the possible bonding structures of the extracted species were proposed (Yang et al., 1991, 1992) The traditional industrial process for the

* To whom correspondence should be adressed.

recovery of penicillin G from the filtrate of fermentation broth usually consists of three steps: extraction, stripping, and reextraction in a centrifugal contactor with demulsifiers, which is expensive and limits the recovery extent of penicillin G.

Majumdar et al. (1988, 1989) recently proposed a new liquid membrane structure using hydrophobic microporous hollow fibers called Hollow Fiber Contained Liquid Membrane (HFCLM). This HFCLM appears to be able to overcome most of the shortcomings of aqueous SLM-s (supported liquid membrane) in gas separation (Majumdar et al., 1988, 1989) and in solute separation from aqueous solutions without any reaction (Sengupta et al., 1988a,b), and in citric acid recovery with chemical reaction (Basu and Sirkar, 1991).

The HFCLM permeator was built out of two sets of identical hydrophobic microporous hollow fibers well mixed throughout the permeator but separated at the ends. The aqueous feed solution (containing penicillin G) was pumped through one fiber set. In the other set the strip solution was flowing in the lumen. The liquid membrane in the interstices of the two sets of fibers on the permeator shell side contained a mobile complexing agent (Amberlite LA-2+TBP). The interface between the organic liquid membrane and the feed aqueous solution is located on the inside diameter of the feed fiber set since the organic membrane wets the hydrophobic fibers pores. At this interface the complexing agent reacts selectively with the solute (penicillin G). The complex then diffuses across the organic liquid membrane and dissociates at the aqueous-organic interface of the hydrophobic strip fiber, releasing the solute (penicillin G).
The recovery of penicillin G with Amberlite LA-2 in n-butyl acetate as membrane phase in HFCLM permeator has been already investigated (Schmidt, 1991). Because of the higher solubility of n-butyl acetate in water(about 2%), this membrane phase is not stable for a long time. The aim of this work is to investigate the possibility of Amberlite LA-2+TBP-kerosene as the membrane phase for the recovery of penicillin G, and the effects of feed and strip flowrate and the concentration of buffer solution on the mass transfer and the recovery of penicillin G.

EXPERIMENTAL MATERIALS AND PROCEDURE

All chemicals were used as received. Tributyl phosphate (TBP) (A.R., 99%, Riedel-de Haën AG), Amberlite, LA-2 (lauryl trialkylmethyl amine, molecular weight, 375, Rohm & Haas), butyl acetate, kerosene, sulfuric acid, sodium hydrogen carbonate, sodium citrate (A.R.and C.P., Riedel-de Haën AG), potassium penicillin G salt (activity: 1580 I.E./mg, Hoechst AG).
The fiber used is LIQUI-CEL™, E400/1.0 (Hoechst AG). The experimental setup is shown schematically in Figure 1. The membrane phase pressure (shell pressure), the inlet, and outlet pressures of the feed and strip phases were monitored using dial gauges. The aqueous phase pressure in the lumen must be kept higher than that of the solvent phase in the shell to prevent the entrainment

of solvent into the aqueous phases. The flow rates were controlled using regulating valves. For all experiments with hydrophobic HFCLM systems the pressure conditions of the feed side, strip side, and the membrane liquid are 0.30-0.65 bar, 0.30-0.65 bar, and 0.12-0.40 bar, respectively.
The penicillin G in aqueous solutions was analysed with Perkin-Elmer Polarimeter at λ = 365 nm at 20 °C. Sodium citrate has no influence on the analysis of penicillin G.

RESULTS AND DISCUSSION

Effect of different solvents
Aliphatic amines, especially Amberlite LA-2, in polar or aromatic diluents can effectively extract penicillin G even at pH5-6 (Reschke and Schügerl, 1984a-c). However, dissolved in aliphatic hydrocarbons they can not extract penicillin G (Yang et al., 1989, 1992b). A membrane phase used in HFCLM system should extract penicillin G effectively, be insoluble in water, and have low toxicity. According to earlier works (Yang, 1989, Yang et al., 1991), Amberlite LA-2+TBP in kerosene may be the best suitable system.

When 20 % TBP-kerosene is used as the membrane solvent, the extraction and recovery of penicillin G are low. The extraction and recovery of penicillin G will be obviously increased with the addition of Amberlite LA-2 in the TBP-kerosene But if Amberlite LA-2 is more than 10 (v/v)%, the recovery of penicillin G will decrease due to the difficulty in stripping (Fig. 2,3). This is due to two opposing factors which act on the extraction and stripping equilibrium of penicillin G and the mass transfer coefficient K when the Amberlite LA-2 concentration is varied. An increase in LA-2 concentration is followed by an increase in the ability for complex formation in the membrane, leading to an enhanced driving force for the mass transfer. But this enhancement of the mass transfer driving force will reach a limit because of the limit of the mass transfer area in the permeator. On the other hand, as LA-2 concentration is increased, the viscosity of the membrane liquid increases since long-chain amines are extremely viscous. This increase of viscosity reduces the diffusion coefficient of the complex and the amine. Moreover, higher amine concentration inhibits the stripping (Reschke and Schügerl, 1984a-c, Yang et al., 1991a). These effects decrease the mass transfer and enhancement of the recovery due to a higher driving force. As penicillin G concentration in the feed is 5 g/L, 5 % LA-2 gives better results than 10 % LA-2 for the recovery of penicillin G (Fig. 4). For a feed concentration of penicillin G of 20 g/L, 10 % LA-2 is more suitable than 15 % LA-2 (Fig. 5,6).

The variation of TBP has similar effect on the mass transfer and the recovery of penicillin G to the variation of the LA-2 concentration. For 10 % LA-2 the suitable TBP concentration is about 20 (v/v)% (Fig. 7). The higher TBP concentration increases the viscosity of the membrane and inhibits the reaction activity of LA-2 because of the interaction of LA-2 and TBP.

Buffer concentration

The variation of sodium citrate concentration has significant effect on the penicillin G recovery (Fig. 8,9).

There are also two opposing factors acting on the extraction and stripping equilibrium as the buffer concentration is varied. An increase in the buffer concentration may improve the extraction equilibrium of penicillin G due to the coextraction of citric acid and the salting effect of citrate ion resulting in enhanced extraction for penicillin G. On the other hand, as the buffer concentration increases, the recovery of penicillin G will be inhibited by the competitive extraction and stripping of citric acid. For the investigated system 0.1 M sodium citrate as buffer is suitable.

Penicillin G concentration and flow rate

The recovery extent of penicillin G decreases with the increase of penicillin G concentration (Fig. 10). The extraction and recovery of penicillin G decrease with the increase of the aqueous flow rate (Fig. 11). It is because of the limit of mass transfer rate and area in the HFCLM permeator. The steady mass transfer can be reached in shorter time with higher penicillin G concentration in the feed and higher aqueous flow rate.

Comparison of butyl acetate and TBP-kerosene

Earlier works (Reschke and Schügerl, 1984a-d, Yang, 1989, Yang et al., 1991, 1992) show that 3-10 % LA-2 in butyl acetate or TBP-kerosene can effectively extract penicillin G at pH 5-6. But with these solvents as the membrane phase in the HFCLM permeator, the extraction is not so effective. As the diluent of LA-2 20 % TBP-kerosene has almost the same (or better) effect as butyl acetate (Fig. 12). The penicillin G in the membrane phase is accumulated due to the slow stripping which inhibits the extraction of penicillin G, but improves the stripping with the higher penicillin G concentration difference, i.e. the higher mass transfer driving force, between the membrane phase and the strip phase. Therefore, the difficult stripping in the extraction with amines as carrier can be improved in the HFCLM system.

Recovering penicillin G in a three series stages system

The concentration of penicillin G in the filtrate of fermentation broth is at least 20 g/L. Therefore, 20 g/L penicillin G was used as the initial feed solution. 0.1 M sodium citrate was used as buffer solution, the aqueous flow rates are about 0.5 cm^3/min., extraction pH 4-4.5, 50 g/L $NaHCO_3$ as stripping phase, stripping pH 8.1-8.5. More than 90 % penicillin G could be recovered in three HFCLM stages (feed and strip flow countercurrently in the HFCLM) (Table 1). The recovery could be further enhanced at lower feed and stripping flow rate (Fig. 11). Moreover, the steady state was not reached in the operation. The recovery could be higher at the steady state. Hence, the satisfactory recovery of penicillin G could be obtained with 10 % LA-2+20 % TBP-kerosene as membrane phase in

HFCLM permeator. This process could improve the recovery extent of penicillin G (higher extraction pH values and fewer recovery stages) and reduce the operating cost (without centrifuge and demulsifiers).

Table 1. Recovering penicillin G from aqueous feed with LA-2 + 20 % TBP-kerosene in HFCLM permeator

Countercurrent Feed/Strip Flow, 50 g/L $NaHCO_3$ as strip phase, 24 °C, feed penicillin G in 0.1 M sodium citrate solution, $Q_F = Q_S = 0.50\ cm^3/min.$, $pH_F = 4.0$, $pH_S = 8.4$

No.	LA-2 v/v%	Time hr.	C_{in}^F g/L	C_{out}^F g/L	C_{out}^S g/L	P %	P, Total %
1	10	8	20	9.46	8.21	41.4	41.4
2	10	11	12	3.78	6.53	53.4	73.44
3	5	9	5	1.15	3.43	68.0	90.44

Acknowledgement

I (Dr. Zhifa Yang) gratefully acknowledge the Alexander von Humboldt Foundation for the Research Fellowship and the considerate help.

Notation

E	extraction extent, %
Es	stripping extent, %
P	recovery extent, %, $(Q_S{*}C_{out}^S)/(Q_F{*}C_{in}^F)$
Q	flow rate, cm^3/min

Subscripts

F	feed side
S	strip side
in	inlet
out	outlet

Literature Cited

Basu, R. and Sirkar, K.K., "Hollow Fiber Contained Liquid Membrane Separation of Citric Acid," AIChE J., 37(3), 383(1991).
Majumdar, S., Guha, A.K., and Sirkar, K.K., "A New Liquid Membrane Technique for Gas Separation," AIChE J., 34, 1135(1988).
Majumdar, S. et al., "A Two-Dimensional Analysis of Membrane Thickness in A Hollow Fiber Contained Liquid Membrane Permeator," J. Memb. Sci., 43, 259(1989).

Reschke M. and Schügerl K., Chem./Biochem. Eng. J., 28, B1-B9 (1984a).
Reschke M. and Schügerl K., Chem./Biochem. Eng. J., 28, B11-B20 (1984b).
Reschke M. and Schügerl K., Chem./Biochem. Eng. J., 29, B25-B29 (1984c).
Schmidt, A., "Extraktion von Bioprodukten mittels trägergestützten Flüssigmembranen," Diplomarbeit, Universität Hannover, 1991.
Sengupta, A., Basu, R., and Sirkar. K.K., "Separation of Solutes from Aqueous Solutions by Contained Liquid Membranes," AIChE J., 34, 1698(1988a).
Sengupta, A. et al., "Separation of Liquid Solutions by Contained Liquid Membranes," Sep. Sci. and Technol., 23(12,13), 1735(1988b).
Yang, Z.F., "Studies on New Solvent Systems for Extraction of Penicillin G", Doctoral Dissertation, Institute of chemical Metallurgy, Academia Sinica, Beijing, 1989.
Yang Z.F. et al., Industrial Chemistry & Engineering (China), 6, (1991a) 55-61.
Yang Z.F., Yu S.Q., and Chen C.Y., J. Chem. Technol. & Biotechnol., vol. 53, 97-103(1992a).
Yang Z.F. et al., , J. Chem. Ind. & Eng.(China, in English), 7(1), 83-92(1992b).
Yang Z.F., Yu S.Q., and Chen C.Y., J. Chem. Ind. & Eng. (China,), 6(6), 726-731(1991b).

Fig. 1. Flowsheet of the Experimental Apparatus

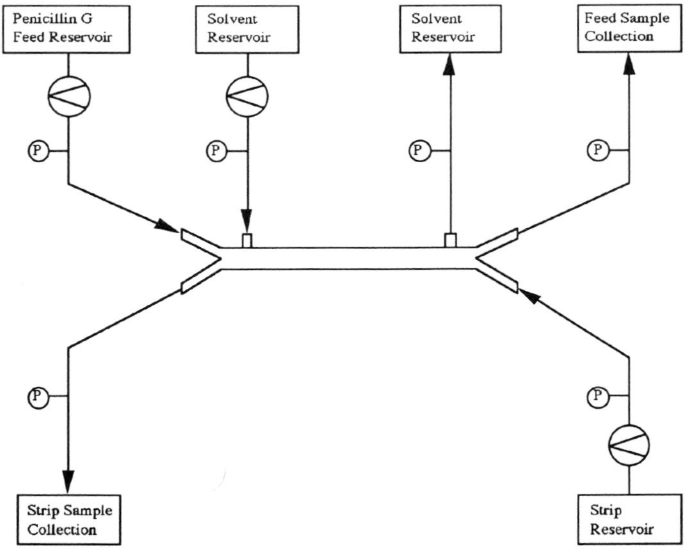

Fig.2. Effect of the conc. of LA-2 in 20%TBP-kero. on extraction of penicillin G in HFCLM permeator

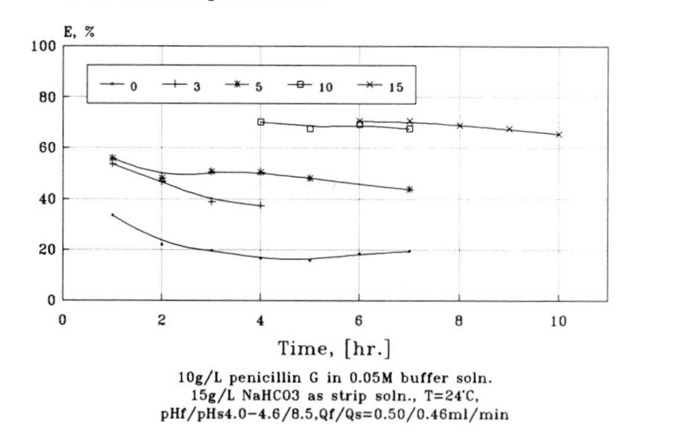

10g/L penicillin G in 0.05M buffer soln.
15g/L NaHCO3 as strip soln., T=24°C,
pHf/pHs4.0-4.6/8.5,Qf/Qs=0.50/0.46ml/min

Fig.3. Effect of the conc. of LA-2 in 20%TBP-kero. on recovery of penicillin G in HFCLM permeator

Time, [hr.]

10g/L penicillin G in 0.05M buffer soln.
15g/L NaHCO3 as strip soln., T=24°C,
pHf/pHs4.0-4.6/8.5,Qf/Qs=0.50/0.46mL/min

Fig.4. Effect of the conc. of LA-2 in 20%TBP-kero. on extraction & recovery of penicillin G in HFCLM permeator

Time, [hr].

5g/L penicillin G in 0.1M buffer soln.
50g/L NaHCO3 as strip soln., T=24°C,
pHf/pHs=4.0/8.3,Qf/Qs=0.50/0.46mL/min

Fig.5. Effect of the conc. of LA-2 in 20%TBP-kero. on extraction & stripping of penicillin G in HFCLM permeator

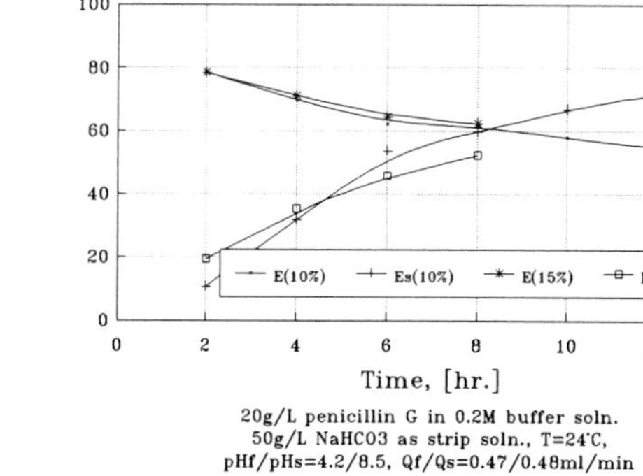
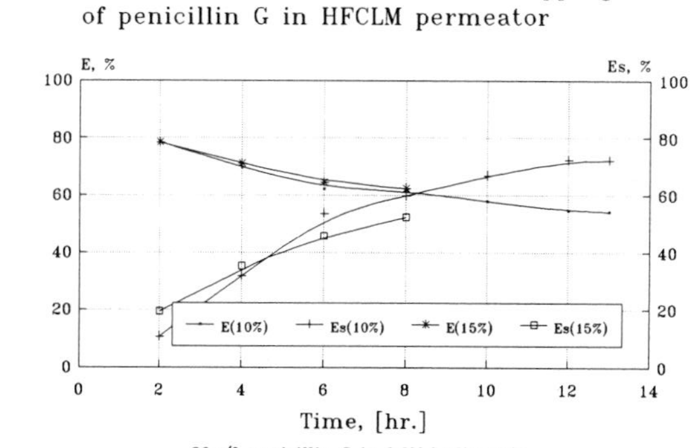

20g/L penicillin G in 0.2M buffer soln.
50g/L NaHCO3 as strip soln., T=24°C,
pHf/pHs=4.2/8.5, Qf/Qs=0.47/0.48ml/min

Fig.6. Effect of the conc. of LA-2 in 20%TBP-kero. on the recovery of penicillin G in HFCLM permeator

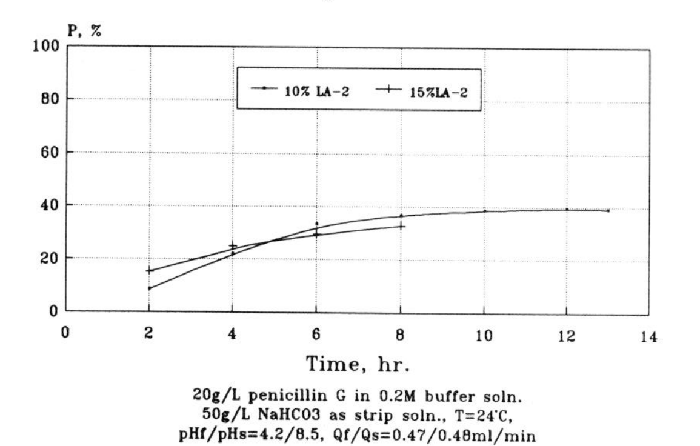

20g/L penicillin G in 0.2M buffer soln.
50g/L NaHCO3 as strip soln., T=24°C,
pHf/pHs=4.2/8.5, Qf/Qs=0.47/0.48ml/min

Fig.7. Effect of the conc. of TBP in 10%LA-2-kero. on extraction of penicillin G in HFCLM permeator

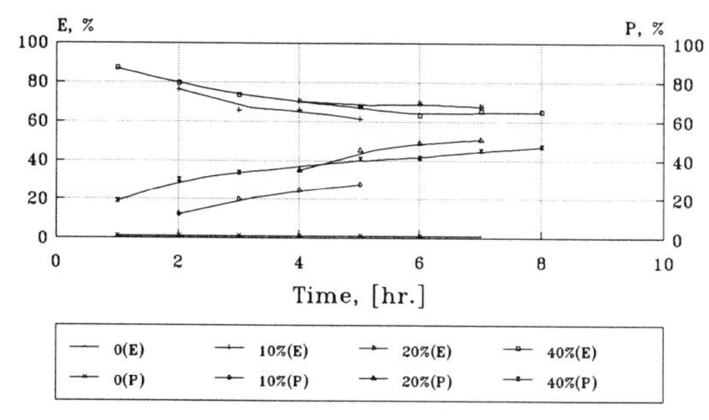

10g/L penicillin G in 0.05M buffer soln.
15g/L NaHCO3 as strip soln., T=24°C,
pHf/pHs=4.3/8.3, Qf/Qs=0.49/0.45ml/min

Fig. 8. Effect of sodium citrate conc. on extraction of penicillin G in HFCLM Permeator

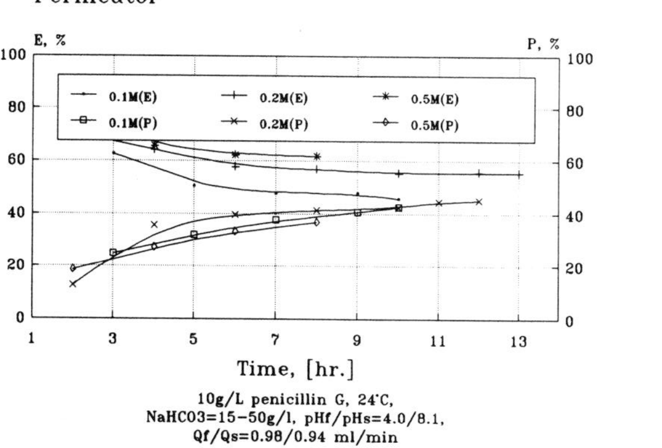

10g/L penicillin G, 24°C,
NaHCO3=15-50g/l, pHf/pHs=4.0/8.1,
Qf/Qs=0.98/0.94 ml/min

Fig. 9. Effect of sodium citrate conc. on extraction of penicillin G in HFCLM Permeator

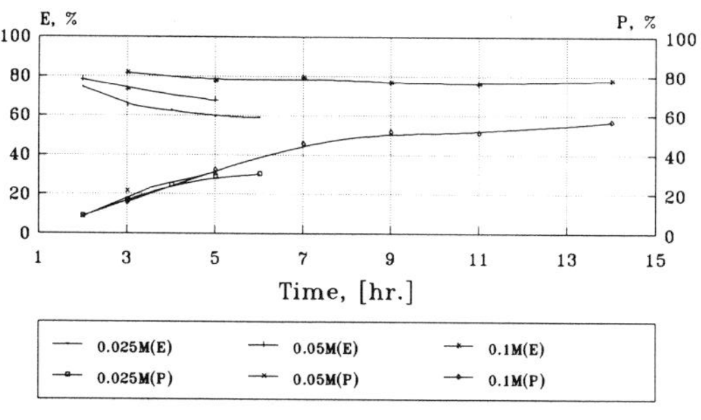

5g/L penicillin G, 24°C,
15g/L NaHCO3 as strip soln., pHf=4.0-4.6
pHs=8.3-8.6, Qf=0.50, Qs=0.44ml/L

Fig.10. Effect of the conc. of penicill-in G on the recovery with 10%LA-2+20% TBP-kero. as solvent in HFCLM permeator

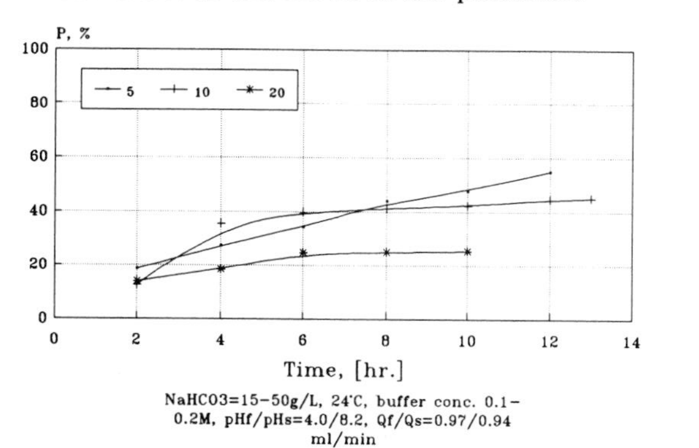

NaHCO3=15-50g/L, 24°C, buffer conc. 0.1-0.2M, pHf/pHs=4.0/8.2, Qf/Qs=0.97/0.94 ml/min

Fig.11. Effect of the aqueous flow rate on extraction & recovery of penicillin G in HFCLM permeator

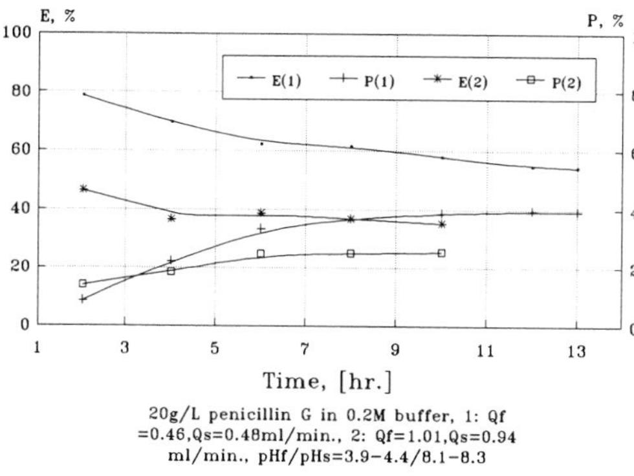

20g/L penicillin G in 0.2M buffer, 1: Qf =0.46,Qs=0.48ml/min., 2: Qf=1.01,Qs=0.94 ml/min., pHf/pHs=3.9-4.4/8.1-8.3

Fig.12. Effect of different solvents on extraction of penicillin G with 10%LA-2 in HFCLM permeator

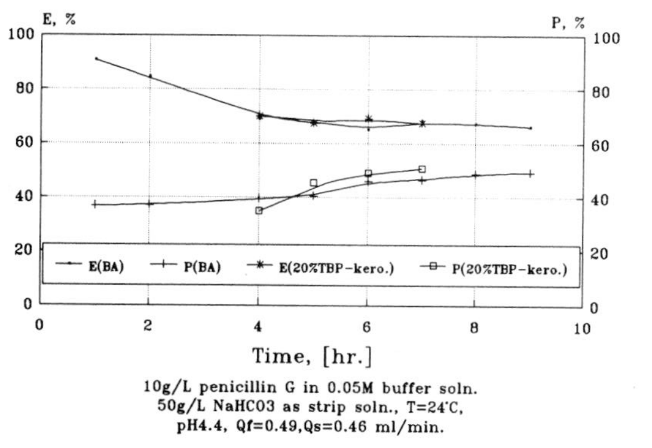

10g/L penicillin G in 0.05M buffer soln. 50g/L NaHCO3 as strip soln., T=24°C, pH4.4, Qf=0.49,Qs=0.46 ml/min.

NOVEL HOLLOW FIBRE UF MEMBRANES FOR PHARMACEUTICAL AND BIOPROCESSING APPLICATIONS.

A C J ORCHARD AND L J BATES
(PALL PROCESS FILTRATION LTD)

INTRODUCTION

The purpose of this paper is to describe a unique range of hollow fibre ultrafiltration modules. Polysulphone (PS) membrane versions with 6000 molecular weight cut-off rating (MWCO), which are steam-sterilisable, have been used in recent years in the production of purified water (PFW) but there is little published data for other applications. Furthermore,there has been very little published information on modules with different molecular weight cut-offs and in a different membrane material polyacrylonitrile (PAN), which are also available. These hollow fibre modules are also of interest in a much broader range of applications in the pharmaceutical and bioprocessing industries. Typical applications include purification of bulk products and media, removal of endotoxins or other contaminants (e.g. cell debris), and concentration of proteins.

The data presented in this paper reviews newly published claims for both membrane materials. Details presented include biological safety, cleaning and sterilisation techniques, and other test work carried out with endotoxin solutions, viruses and protein solutions in which retention characteristics, flux stability and flux recovery were investigated.

UNIQUE HOLLOW FIBRE UF MODULE

The general advantages of hollow fibre modules for UF systems in terms of high packing density (high surface area to volume ratio), clean design, simple module maintenance and low power consumption are well documented. The performance and reliability of individual hollow fibre modules, however, depend very much on a number of key factors:-

(i) Fibre/membrane characteristics
- Rejection performance
- Flux
- Strength

(ii) General module construction
- Suitability of materials
- Construction methods (eg potting of fibres)

In a unique UF module design referred to as Microza, the membrane consists of a hollow fibre which contains not only a uniform skin on the inside surface but also a similar functional uniform skin on the outer surface of the fibre. Between the two skins is a macroporous area in which a relatively thick denser porous layer exists. The structure is illustrated in figure 1 below.

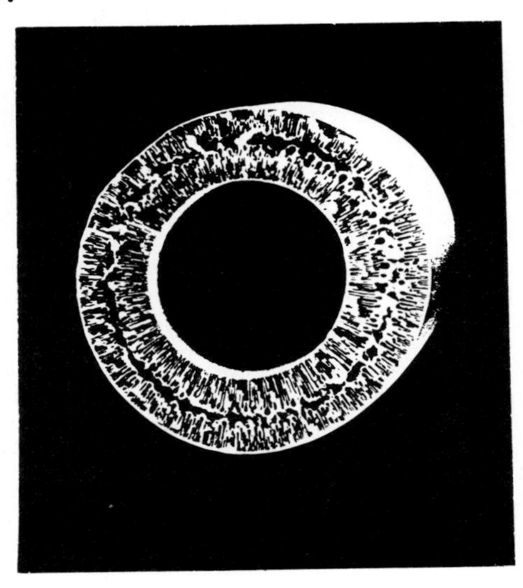

Fig 1. SEM cross-section of double-skinned hollow fibre.

The overall construction results in self-supporting fibres of very high mechanical strength and long life as demonstrated in Table 1.

BURST STRENGTH	19.7 bar (1970 kPa)
COMPRESSIVE CREEP STRENGTH	11.8 bar (1180 kPa)
COMPRESSIVE CREEP STRENGTH AFTER 2 YEARS	5.9 bar (590 kPa)

Table 1 Mechanical strength of 10,000 MWCO rated polysulphone double-skinned fibre.

The fibres are potted at both ends into specially formulated epoxy resin and the bundle is sealed into a housing. The housing contains a port at each end of the fibres allowing flow along the internal bores of the fibre, and two ports which connect to the external surface of the fibres.

According to application requirements, a number of parameters can be varied in the module construction. For example, the membrane material can be either polysulphone or polyacrylonitrile and internal fibre diameters are available from 0.6mm up to 1.4mm The potting resin formulations are also selected according to different process requirements. Biological safety requirements of the pharmaceutical industry are met by ensuring that in specific modules (VP,AP,SP types) using polysulphone and polyacrylonitrile membranes, all components meet specifications for class VI plastics at 121°C listed in the latest revision (XX11) of the U.S. Pharmacopeia. All component materials for the same modules are also listed in relevant FDA documentation (Title 21 of the U.S. Code of Federal Regulations, parts 170-199).

In addition, the unique double skin structure and high mechanical strength mean that permeate flow can be in either direction through the membrane. In most applications, feed flow through the Microza UF module is internal to the fibres and effective reverse filtration routines with "outside to inside" high reverse transmembrane pressure can be employed to assist stable flux rates. Furthermore, the smooth skin on the outer surface resists plugging by any contaminants on the permeate side during the reverse filtration cycle.

The unique structure also means that, when required, permeate flow can be from out-to-in on a continuous basis. This principle is used only when feed water is sufficiently clean not to require uniform cross-flow velocity over the membrane surface (e.g. high purity water for ULSI semiconductor manufacture).

REJECTION OF ENDOTOXIN AND VIRUS

A VIP-3017 was selected for these tests because validation data exists to show that this module can be steam sterilised for up to 100 cycles each of 30 minutes at 121°C. The module consists of a 6000 MWCO rated polysulphone membrane and has an initial clean water flux of at least 1000 l/h at 25°C and 1 bar (100 kPa) average transmembrane pressure. (TMP)

a) Endotoxin challenge

During the endotoxin challenge test, water containing 2.5×10^9 EU/ml of E coli purified endotoxin was circulated in the module at 20°C and 1 bar ((100 kPa) TMP. No endotoxin activity was detected in the permeate by standard test of 0.125 EU/ml test sensitivity.

The feed endotoxin level in this test was many orders of magnitude higher than observed in practice in pharmaceutical applications. Other tests with a range of endotoxins at lower concentrations and in long term challenge have confirmed similar membrane retention.

b) Virus challenge

PP7 bacteriophage was used for these tests because it is a small particle 0.025µm in size. Under similar conditions to those used above, a VIP-3017 module was challenged with water containing a virus count of 8 X 10^8 per ml. No viral particles were detected in the permeate.

TEST WITH PROTEIN SOLUTIONS

It is widely known that fouling of membranes by proteins can cause flux losses which affect the performance and economics of UF systems. It is also known that fouling effects can be a function of membrane characteristics including molecular weight cut-off rating and degree of hydrophobicity.

Therefore, for these tests we selected 3 module types:-

ACP - 1010	13,000 MWCO, polyacrylonitrile membrane
AIP - 1010	6,000 MWCO, polyacrylonitrite membrane
SIP - 1013	6,000 MWCO, polysulphone membrane

The minimum initial clean water fluxes of these modules are 10l/h, 35l/h and 40l/h at 25°C and 1 bar (100kPa) TMP respectively. These are all small pilot modules. The SIP-1013 module was included, not only because the membrane is polysulphone, but also because data from this module can be scaled up to module type VIP-3017, which is steam sterilisable. The fibres in all three module types have an internal diameter of 0.8mm.

The test solution was bovine serum albumin (BSA) because a considerable amount of information has been published on tests with this protein in UF systems. The initial tests used a solution of 0.1% w/v in sterile filtered, demineralised water. Processing conditions were 25°C at 1 bar (100kPa) TMP and velocity in the fibre lumens of 1.58 - 1.67 m/s. During the tests both retentate and filtrate were recycled to the 20 litre feed vessel.

The data are presented in figures 2, 3 and 4 below.

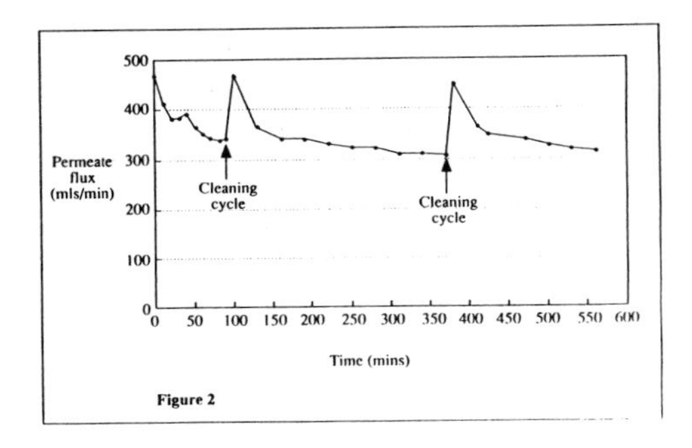

Fig 2 Flux data for ACP-1010 module

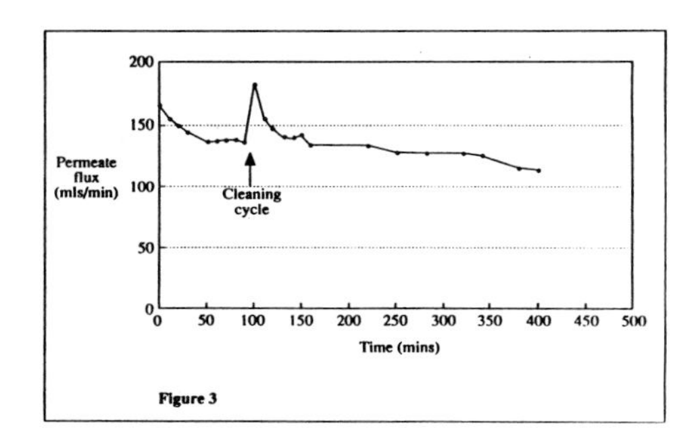

Fig 3 Flux data for AIP-1010 module

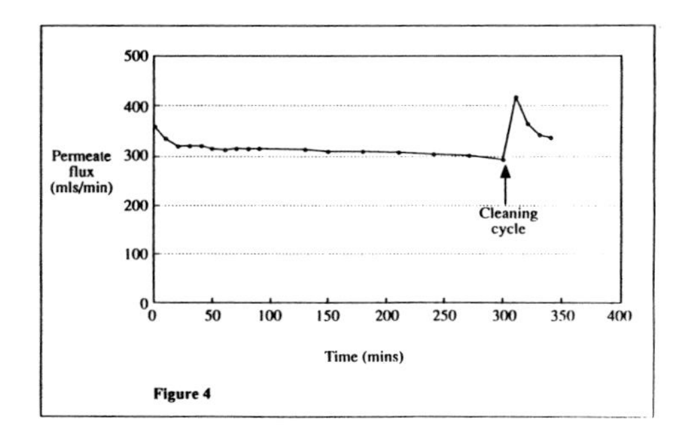

Fig 4 Flux data for SIP-1013 module

The initial results show that the flux profile for 3 membranes is quite similar. They also suggest full recovery upon cleaning with a solution containing 0.4% sodium hydroxide and 300 ppm sodium hypochlorite at 35°C in each case. Table 2 suggests that the polysulphone membrane performance (expressed as a percentage of initial water flux,) was marginally more affected by protein but the stabilised flux was still significantly higher than the polyacrylonitrile membrane of the same MWCO rating.

Module	MWCO	Membrane Material	Permeate flowrate in BSA solution		
			Stable[1]	Initial[2]	End[2]
ACP - 1010	13,000	PAN	310	75%	50%
AIP - 1010	6,000	PAN	120	90%	55%
SIP - 1013	6,000	PS	285	60%	42%

1. Typical value after 5 hours, in mls/min.
2. Expressed as a percentage of initial flux in clean water.

Table 2 Comparison of permeate flowrates.

CONCLUSIONS

Hollow-fibre membranes with a unique double skin construction offer potential benefits, especially in terms of mechanical strength and long term reliability. Test with endotoxins and a small virus demonstrated complete rejection using a steam sterilisable 6000 MWCO rated module.

No great difference in fouling characteristics between polyacrylonitrile and polysulphone membranes were seen in initial tests using 0.1% BSA solution,but the polysulphone membrane showed higher stabilised flux. However, longer term testing may be required in some applications before the optimum selection is made.

Further work will compare performances in higher protein concentrations, and in the presence of antifoams.

A Novel Membrane Reactor For Two Phase Biocatalysis

A.M. Vaidya, P.J. Halling, G. Bell
Departments of Chemical Engineering & Bioscience and Biotechnology
Strathclyde University
75, Montrose Street
Glasgow G1 1XJ
Scotland

Introduction

Biocatalysts can be used economically to carry out a number of chemical transformations of interest in industry. Amongst the reactions that constitute potential candidates for such transformations, there is a large proportion which require access to two mutually immiscible liquid phases[1,2]. This is necessary because one of the reactants and/or products is only appreciably soluble in an organic phase. A few examples of such reactions are:

- ☐ The hydrolysis of esters and triglycerides by lipases. The products of the reaction are an alcohol which passes into the aqueous phase and a fatty acid - or a mixture of fatty acids - which dissolve in the organic phase.
- ☐ The hydrolysis of menthol acetate by a microbial esterase. The products in this case are acetic acid which dissolves the aqueous phase while the menthol produced remains in the organic phase.
- ☐ The halogenation of long chain alkenes by lactoperoxidase. In this reaction the halide ions and hydrogen peroxide are supplied via the aqueous phase.

A number of reactor designs have been suggested in the published literature for carrying out a two liquid phase biocatalytic reaction. Many of these designs have been reviewed recently[3,4]. Some of the more frequently suggested ideas are:

- ☐ **Packed Beds:** This reactor design with the biocatalyst immobilized on the packing has been used by a number of workers. This type of reactor can be further classified on the basis of the reactant feed regime employed.
  - Two reactant phases are pumped simultaneously - either co-currently or countercurrently through the bed.[5,6]
  - The bed is exposed alternately to the two reactant streams[7]. In this mode of operation, the wettability of the packing material by one of the two phases is

exploited to ensure that a supply of both reactants is always present at the reaction interface.

- **Mixer-Settler Cascades:** In this type of reactor the enzyme is supplied in solution in the aqueous phase. An interface between the two phases is created by forcing the liquids through a disperser pad made of cellulose[8].
- **Membrane Reactors:** In this reactor design a membrane selectively wetted by one of the two phases is employed[9]. By applying a small positive pressure on non-wetting phase an interface between the two phases is created in the plane of the membrane.
- **Emulsion Reactors:** Such reactors consist of stirred tanks in which a large interface between the two liquid phases is created by stirring[10]. The enzyme is supplied as a solution in the aqueous phase.

These - and all other reactor designs which have been proposed so far - suffer from one, or more, major disadvantage. For instance:

1. Packed bed reactors are liable to have poor interfacial contact between the two phases when they are pumped simultaneously. When the packed bed is operated as a pulsed bed there is a gradual change in its wetting characteristics due to the generation or depletion of surface active components which frequently occur in such reactions.
2. The operation of mixer-settler cascades is likely to be hampered by the formation of stable emulsions, particularly in the presence of dissolved enzyme from the aqueous phase.
3. The maintenance of a stable interface in the plane of a membrane reactor is made extremely difficult by the generation of surface active components[11]. Further, the limited interfacial areas offered by such devices may make their operation uneconomical[12].
4. The difficult downstream processing required to recover the two liquid phases makes emulsion reactors an unattractive option for industrial use.

Evidently, a successful reactor design for two phase biocatalysis must be based on a systematic assessment of the features that such a device should provide. In the subsequent sections a reactor developed by identifying these features will be described.

Reactor Design

A successful two liquid phase reactor design must meet the following criteria:

- The reactor should offer a high area of interfacial contact between the two phases. Ideally, this should be of the same order of magnitude as offered by an emulsion system.
- Interfacial contact should be accomplished between bulk liquid phases - i.e. without emulsification - in order to avoid complicated downstream separation problems.

- The reactor should operate at, or close to, atmospheric pressure with no excessively accurate pressure control requirements.
- Enzyme immobilization - if required at all - should be simple. Ideally, it should be possible to immobilize the enzyme by physical adsorption.
- It should be possible to operate the reactor as a continuous device.

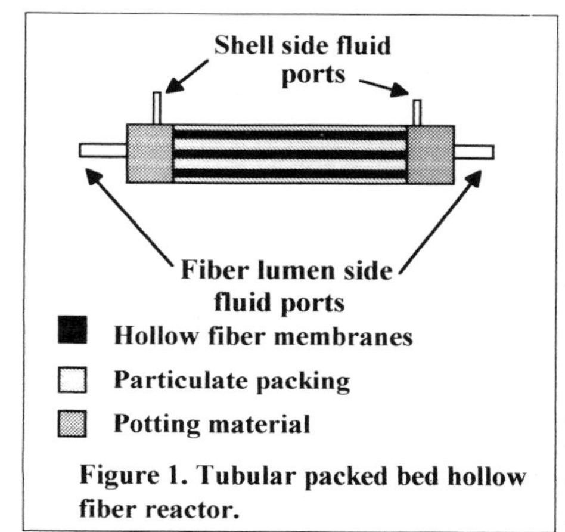

Figure 1. Tubular packed bed hollow fiber reactor.

All these criteria are satisfied by a packed bed hollow fiber reactor (**PBHR**). The reactor, which has been described in detail in a recent patent application[13] consists of microporous or ultrafiltration hollow fiber membranes potted in a tubular shell, as shown in **Figure 1.** The potting procedure employed is identical to the one used for making conventional hollow fiber dialyzers. The packing material and the membrane are selected so as to have similar wetting characteristics. By this means, it is possible to ensure a continuous diffusion path from the fiber lumen to the outside surface of the packed particles. The phase which wets the membrane and the packing is fed to the reactor through the fiber lumen and the microporosity of the packing ensures that a continuous supply of that phase is always available at the surface of the packing. The 'non-wetting' liquid phase is fed through the reactor shell and flows through the interstitial spaces in the packed bed. A two phase system with two **continuous** phases is thus formed - see **Figure 2.**

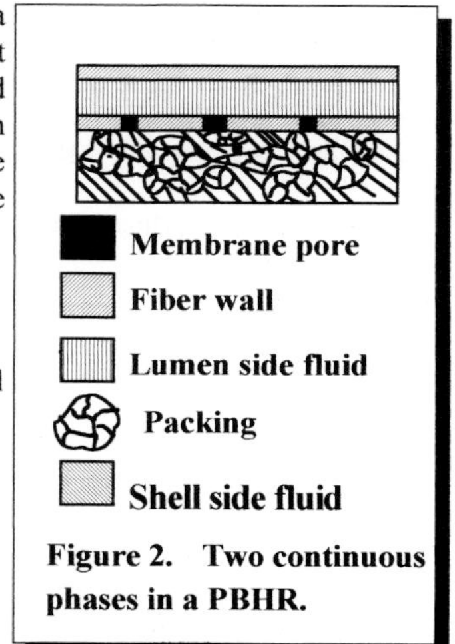

Figure 2. Two continuous phases in a PBHR.

Materials

- The details of the chemicals used in experimental work are given in **Table 1.**

- Reactors were made using two types of membrane.

  1. Cuprophan - regenerated cellulose fibers obtained from Enka Ltd. and
  2. Celgard X10 240 - stretched polypropylene - fibers obtained from Hoechst Celanese Ltd.

Chemical	Supplier	Catalog No
Lipase, *Candida rugosa*	Sigma Chemical Co, Ltd., Poole	L 1754
Gum Arabic	"	G 9752
Tributyrin	"	T 5142
Ethyl laurate	"	
Pepsin	"	P 7012
t-butyl ammonium hydroxide	Aldrich Chemical Co Ltd, Gillingham	19,187

Table 1. Chemicals used in experimental work.

All other chemicals used in the experiments were GPR grade materials purchased from Merck Ltd.

☐ The properties of the membranes are summarized below.

Celgard hollow fibers

Outside diameter: 240 μm
Wall thickness: 30 μm
Porosity: 30%
Surface pore size: 0.05 x 0.15 μm
Bulk pore size: 0.075 μm

Cuprophan hollow fibers

Outside diameter: 240 μm
Wall thickness: 20 μm
Pore size: < 2 nm

Araldite CW 1303 resin mixed with Araldite HY 1303 hardener - both supplied by Ciba Geigy Ltd., Cambridge - were used to pot the fibers into the tubular reactor shells.

☐ Two types of particulate packing were used in the reactors.

1. XAD-8, a nonionic, modified polystyrene adsorbent obtained from Sigma Chemical Company, Poole and
2. Accurel EP 100 polypropylene microstructures supplied by Akzo GmbH, West Germany.

The physical characteristics of these packings are given in **Table 2.**

Packing Property	XAD-8	Accurel EP-100
Surface area (m^2/g)	160	90
Average pore size	22.5 nm	0.1 - 1 µm
Mesh size	20 - 60 (wet)	> 65
Void volume	0.39 cc/g	0.93 cc/g

Table 2. Physical properties of reactor packings.

☐ Reactor shells were made out of stainless steel tubing as shown in **Figure 3.** The shorter of the two shell sizes was used to pot-up the Celgard fibers.

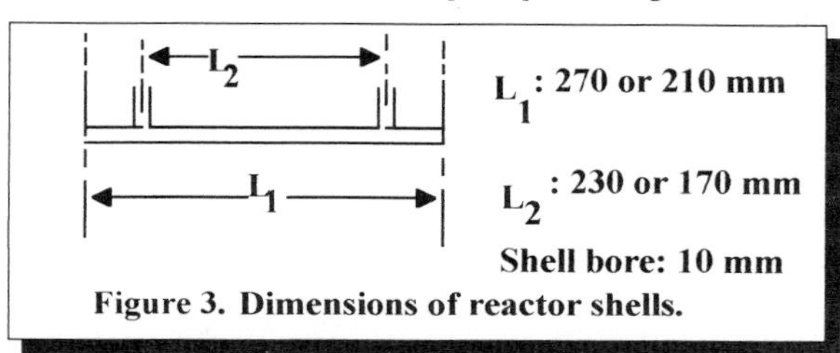

Figure 3. Dimensions of reactor shells.

Both sizes of shell had side arms 25 mm in length.

Methods

☐ **Potting of reactor shells**: Fibers were introduced into the stainless steel shell - which had previously been cleaned and dried - by drawing a fiber bundle through the shell, which had a short length (ca. 25 mm) of silicone rubber tubing attached to either end. The number of fibers was determined from their weight, and a knowledge of the linear density of the fiber. The ends of the silicone rubber tubing were sealed with Hoffman clips and the entire assembly was placed on a turntable rotating at 120 rpm. The Araldite potting compound was mixed and injected into the side arms of the shell. Flow of potting compound to the center of the shell was prevented by rotating the turntable for a period of ca.

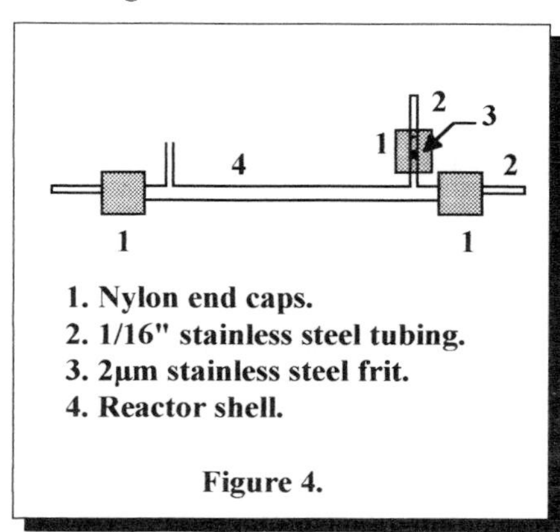

Figure 4.

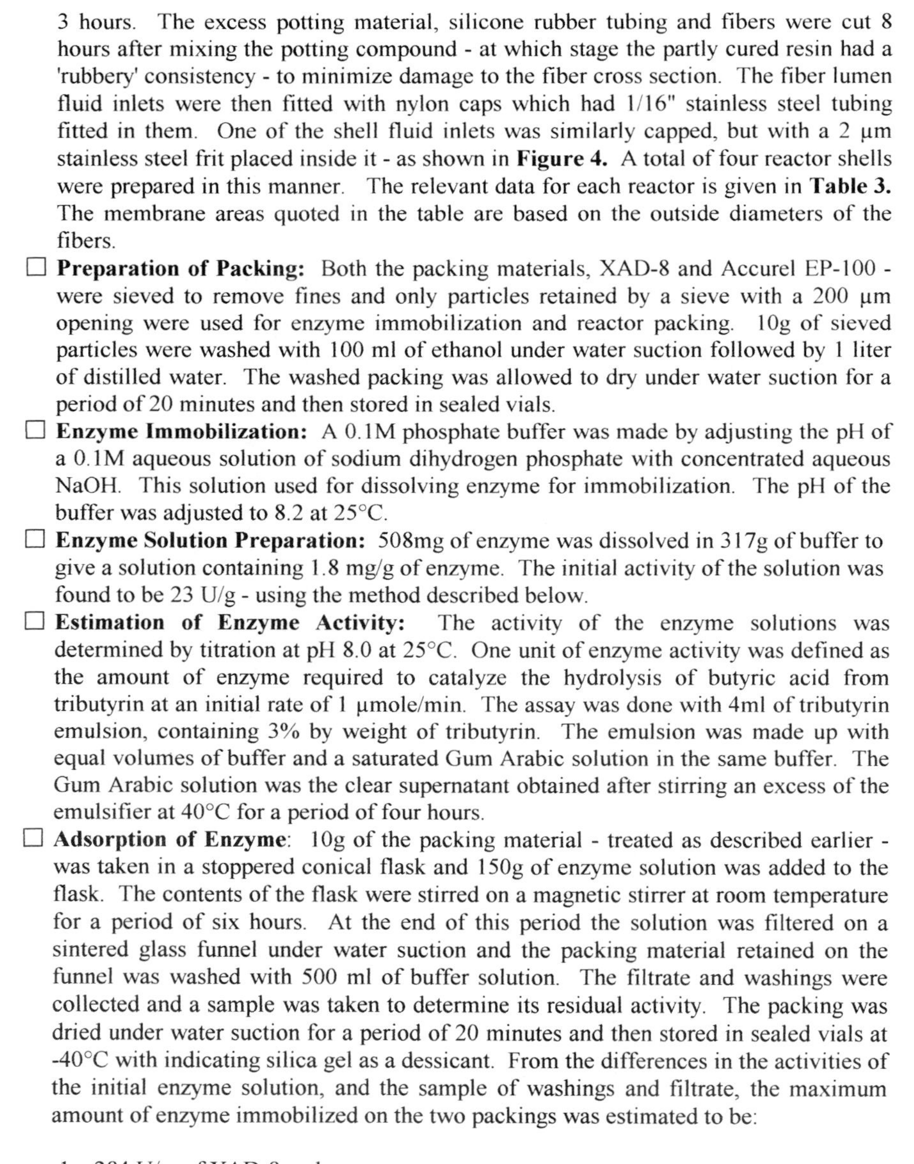

3 hours. The excess potting material, silicone rubber tubing and fibers were cut 8 hours after mixing the potting compound - at which stage the partly cured resin had a 'rubbery' consistency - to minimize damage to the fiber cross section. The fiber lumen fluid inlets were then fitted with nylon caps which had 1/16" stainless steel tubing fitted in them. One of the shell fluid inlets was similarly capped, but with a 2 μm stainless steel frit placed inside it - as shown in **Figure 4.** A total of four reactor shells were prepared in this manner. The relevant data for each reactor is given in **Table 3.** The membrane areas quoted in the table are based on the outside diameters of the fibers.

- ☐ **Preparation of Packing:** Both the packing materials, XAD-8 and Accurel EP-100 - were sieved to remove fines and only particles retained by a sieve with a 200 μm opening were used for enzyme immobilization and reactor packing. 10g of sieved particles were washed with 100 ml of ethanol under water suction followed by 1 liter of distilled water. The washed packing was allowed to dry under water suction for a period of 20 minutes and then stored in sealed vials.
- ☐ **Enzyme Immobilization:** A 0.1M phosphate buffer was made by adjusting the pH of a 0.1M aqueous solution of sodium dihydrogen phosphate with concentrated aqueous NaOH. This solution used for dissolving enzyme for immobilization. The pH of the buffer was adjusted to 8.2 at 25°C.
- ☐ **Enzyme Solution Preparation:** 508mg of enzyme was dissolved in 317g of buffer to give a solution containing 1.8 mg/g of enzyme. The initial activity of the solution was found to be 23 U/g - using the method described below.
- ☐ **Estimation of Enzyme Activity:** The activity of the enzyme solutions was determined by titration at pH 8.0 at 25°C. One unit of enzyme activity was defined as the amount of enzyme required to catalyze the hydrolysis of butyric acid from tributyrin at an initial rate of 1 μmole/min. The assay was done with 4ml of tributyrin emulsion, containing 3% by weight of tributyrin. The emulsion was made up with equal volumes of buffer and a saturated Gum Arabic solution in the same buffer. The Gum Arabic solution was the clear supernatant obtained after stirring an excess of the emulsifier at 40°C for a period of four hours.
- ☐ **Adsorption of Enzyme**: 10g of the packing material - treated as described earlier - was taken in a stoppered conical flask and 150g of enzyme solution was added to the flask. The contents of the flask were stirred on a magnetic stirrer at room temperature for a period of six hours. At the end of this period the solution was filtered on a sintered glass funnel under water suction and the packing material retained on the funnel was washed with 500 ml of buffer solution. The filtrate and washings were collected and a sample was taken to determine its residual activity. The packing was dried under water suction for a period of 20 minutes and then stored in sealed vials at -40°C with indicating silica gel as a dessicant. From the differences in the activities of the initial enzyme solution, and the sample of washings and filtrate, the maximum amount of enzyme immobilized on the two packings was estimated to be:

1. 284 U/g of XAD-8 and
2. 348 U/g of Accurel EP-100.

Expt. No	Reactor	Fiber type & Number	Membrane area cm^2	Packing type	Packing weight	Organic Phase flow rate [cc/hr]	Organic phase placement	Residence time [hrs]	Conversion	Acid production rate $moles/hr.m^2$
1	R1	Cuprophan (552)	875	Accurel EP-100	7.6g	2.36	Shell	3	72.6%	0.087
2	R2	Cuprophan (541)	856	XAD-8	7.2g	3.76	Shell	0.75	86.1%	0.168
3	R3	Celgard (480)	615	Accurel	4.1g	1.28	Lumen	3.16	89.5%	0.093
4	R4	Celgard (507)	650	XAD-8	3.9g	2.04	Lumen	2.1	30.1%	0.042
5	R2					3.2	Shell	0.875	96.5%	0.118

Table 3. Packed bed hollow fiber reactors used in experimental work. Ethyl laurate was used as the organic phase in all experiments except for experiment no **5.** in which olive oil was used. The organic phase residence times were calculated by making separate estimate of the specific void volume of the packing. In calculating the acid production rate for olive oil hydrolysis, it has been assumed that pure oleic acid was being produced.

- ☐ **Packing of Reactor Shells:** A polypropylene slurry reservoir was fitted onto the uncapped side arm of the reactor shell, prepared earlier. The reactor shell, with the reservoir attached, was clamped in an orbital shaker. A syringe pump was used to force distilled water through the fiber lumen at a rate of 200 μl/min to prevent collapse of the fibers during the packing operation. A 1% (w/w) slurry of the packing material in hexane - with enzyme immobilized on it, as described earlier - was poured into the slurry reservoir which was then pressurized to 1 bar. The entire reactor assembly, along with the slurry reservoir, was agitated by operating the orbital shaker at a speed of 120 rpm. All the reactor shells were packed in this manner. The type and quantity of packing used with each shell is summarized in **Table 3.** The slurry reservoir was replaced by a nylon end cap with a 1/16" stainless steel feed tube, as described earlier. Distilled water was flushed through the shell and lumen of the reactors for a period of 48 hours to remove all traces of hexane. This operation was carried out at 5°C to minimize deactivation of the enzyme.
- ☐ **Reactor Operation:** The four reactors described in **Table 3** were used to hydrolyze lauric acid ethyl ester. The flow rates of the two phases were adjusted so as to obtain a residence times indicated in the table. The aqueous phase flow rate was controlled to give a residence time of two hours with all the reactors. Samples of effluent organic phase were collected at three hour intervals and their acid content was estimated on a gas chromatograph after derivatization of the acid. The flow rates used are summarized in **Table 3.**
- ☐ **Regeneration of Enzyme:** Reactor R2 - used earlier for ester hydrolysis, as described above was cleaned to deactivate enzyme already present on it using the following procedure.

  1. Distilled water was passed through the packing at 40°C for a period of 48 hours at 1cc/hr.
  2. This was followed by a 1mg/g solution of pepsin in 0.1M HCl at room temperature for a 24 hour period.
  3. Distilled water was then pumped through the packing at the same flow rate for a further 100 hours at room temperature.

During each of these steps a water flow rate of 1cc/hr was maintained through the fiber lumens to avoid collapse of fibers under the pressure resulting from shell side flow.

The cleaned reactor was placed in a water bath maintained at 5°C and a 3mg/g solution of lipase was pumped through the packing at a rate of 1.5cc/hr. The enzyme solution and the depleted effluent solution were also maintained at 5°C during this operation, in order to minimize deactivation of the enzyme. This 'in-situ' immobilization was carried out for a period of 10 hours. A sample of the entire effluent was assayed for enzyme activity, as was the original enzyme solution. From these assays, it was estimated that 172U of enzyme had been adsorbed per gram of packing in the column.

This regenerated reactor was used for the lipolysis of Olive oil. As a result of the high viscosity of the oil it was found necessary to maintain a lower flow rate of 3.2 cc/hr which resulted in a higher residence time of 0.875 hours for the oil. Buffer solution was fed through the fiber lumen and the effluent organic phase was collected continuously, with stirring. A 2g sample of oil was withdrawn at intervals of 3.5 hours, diluted with THF and titrated against 01M t-butyl ammonium hydroxide. A solution of bromothymol blue in methanol was used as the end point indicator. This reactor was run for a period of 28 hours.

Results

The results of the five experiments described in the previous section are summarized in **Table 3.** The production rates quoted in the table are the maximum rates attained during the entire duration of each experiment. The variation in rates during the course of the experiments is shown in **Figure 5.** The same data is shown in **Figure 6.** recalculated on the basis of a unit of immobilized enzyme.

Discussion

The experimental results presented in this paper are, of necessity, of a preliminary nature. In this work, no attempt was made to test a wide variety of packing material-membrane combinations. Intuitively, it was felt that it would be necessary to have the same wetting characteristics for both the packing and the membrane. The four reactors for which results have been reported were made to test the validity of this assumption. The combination of wetting characteristics for the packing and hollow fibers used in them is summarized in **Table 4.**

Reactor	Experiment	Fiber type	Packing type	Expected reaction rate	Actual reaction rate
R1	1	Cuprophan Hydrophilic	Accurel Hydrophobic	Low	Low
R2	2	Cuprophan	XAD-8 Hydrophilic	High	Highest
R3	3	Celgard Hydrophobic	Accurel	High	High
R4	4	Celgard	XAD-8	Low	Lowest

Table 4. Expected and actual reaction rates for various combinations of reactor packing and membrane.

From **Figures 5 and 6,** and **Table 4.,** it is evident that the experimental results bear out this assumption. The low reaction rates obtained for reactor **R1** - which had a combination of a highly hydrophilic membrane and an extremely hydrophobic packing - can be accounted for by the poor interfacial contact that must have resulted with this combination of packing and membrane. It is slightly surprising that reactor **R4** should give the lowest reaction rate since one expects the modified polystyrene packing used to

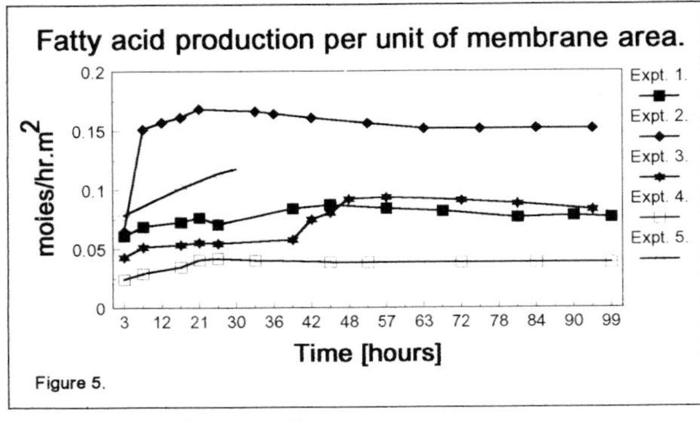

Figure 5.

have a sufficient residual hydrophobic character to permit some wetting by the organic phase. A possible reason for this apparent anomaly is the fact that the packing was water wet at the start of the experiment. This combined with the effects of shell side fluid pressure could have prevented the passage of organic liquid into the pores of the packing.

The production rates quoted in **Table 3.** are significantly higher than the figure of 0.045 moles/hr.m^2 with optimal enzyme loading, reported by Pronk and Van't Riet[10] who ran a Cuprophan membrane reactor. For instance, the reactor of Expt. 2. gave a maximum acid production rate of 0.168 moles/hr.m^2. This is higher by a factor of 3.73. The same reactor was also used after regeneration for the hydrolysis of olive oil where a fatty acid production rate of 0.118 moles/hr.m^2 was obtained - 2.62 times greater than the optimal rate obtained by Pronk and Van't Riet[10]. Goto et al.[14] have reported an acid production rate of 0.029 moles/hr.m^2 for their PTFE hollow fiber reactor in which triolein was hydrolyzed by Lipase *Candida rugosa*. This rate is 4 times smaller than the comparable figure for a PBHR.

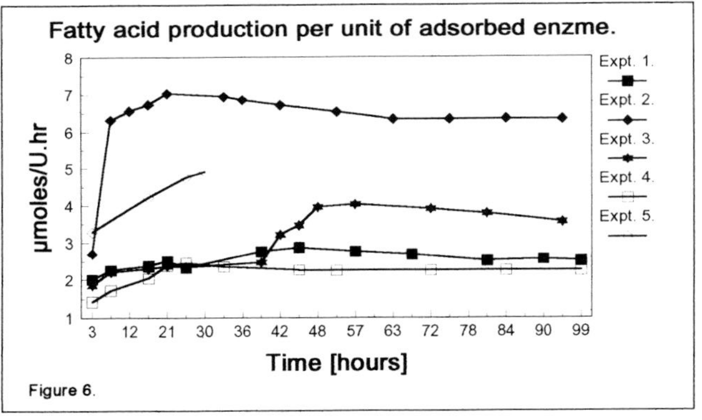

Figure 6.

Comparisons with packed bed reactors are difficult to make since much of the published literature does not explicitly state an acid production rate. From the work of Brady et al.[5] a value of 2.2 moles/hr.kg of packing - a high density polyethylene Accurel with a particle size of 2-3 mm - can be made. The corresponding figure for the reactor of Expt. 3. - in **Table 3.** - (with a different type of Accurel and another strain of Lipase) is 1.35 moles/hr.kg of packing.

It must be borne in mind that the experimental results presented in this paper were obtained without any efforts being made to optimize reactor performance. With a systematic choice of membrane and/or packing it is likely that much higher reaction rates can be obtained.

Summary

- ☐ A two phase biocatalytic reactor design which overcomes many of the disadvantages of the reactor types that have been proposed so far has been presented in this paper.
- ☐ The construction and operation of this reactor - a packed bed hollow fiber reactor (*PBHR*) - have been described.
- ☐ The superior performance of the reactor has been demonstrated by comparing the acid production rates obtained for the lipase catalyzed hydrolysis of ethyl laurate and olive oil to the results quoted in the literature for *membrane* reactors and continuous co-current packed bed reactors operating under similar conditions. The production rates were found to be higher by a factor of up to 4.
- ☐ A systematic investigation of this reactor design is planned to determine the criteria to be used in the selection of packings and fibers for use in such a device for a variety of two phase biocatalytic systems. As part of this investigation, the possibility of using a multiple plate and frame assembly in place of the current tubular design and that of using a reactor with an 'in-situ' generated packing will also be explored.

[1] Lilly, M.D., Brazier, A.J., Hocknull, M.D., Williams, A.C. and Woodley, J.M., Biological conversions involving water-soluble inorganic compounds in: Laane, C., Tramper, J. and Lilly, M.D. (Eds), Biocatalysis in Organic Media, Elsevier, Amsterdam, 1985, 3-17.

[2] Mukherjee, K.D., Lipase-catalyzed reactions for modification of Fats and other Lipids, *Biocatalysis.*, **3**, 1990, 277-293.

[3] Malcata, F.X., Reyes, H.B., Garcia, H.S., Hill, C.G. and Anderson, C.H., Immobilized lipase reactors for the modification of fats and oils - a review. *J.Am.Oil.Chem.Soc.*, **67**,1990, 890 -914.

[4] Nishimoto, T, Izumi, T. & Kubata, H., The modification of fats and oils by bioreactor process., *Yukagaku.*, **41**, 1992,960-968

[5] Brady, C., Metcalfe, L, Slaboszewski, D. & Frank, D., Lipase immobilized on a hydrophobic microporous support for the hydrolysis of fats. *J.Am.Oil.Chem.Soc.*, **65**, 1988, 917-919.

[6] Yang, D. & Rhee, J.S., Continuous hydrolysis of olive oil by immobilized lipase in organic solvent., *Biotechnol.Bioeng.*, **40,** 1992,748-52.

[7] Pat.Appl.Pend., 9204843.8, University of Strathclyde, 1992

[8] Funada, T., Hirano, J., Hashizume, R. & Tanaka, Y., Development of a bioreactor for water/oil heterogeneous system, *Yukagaku.*, **41**, 1992, 423-7.

[9] Hoq, M.M., Yamane, T., Shimizu, S, Funada, T. & Ishida, S., Continuous synthesis of glycerides by lipase in a microporous membrane bioreactor., *J.Am.Oil.Chem.Soc.*, **61,**1984,776.

[10] Pronk, W & Van't Riet, K., The interfacial behavior of lipase in free form and immobilized in a hydrophilic membrane reactor., *Biotechnol.Appl.Biochem.*, **14**,146-154, 1991.

[11] Vaidya, A.M., Bell, G. & Halling, P.J., Aqueous-organic membrane bioreactors. Part I. A guide to membrane selection., *J.Membrane.Sci.*, **71**,1992,139-149.

[12] Pronk, W., Van der Burgt, M., Boswinkel, G. & Van't Riet, K., A hybrid membrane-emulsion reactor for the enzymatic hydrolysis of lipids., *J.Am.Oil.Chem.Soc.*, **68**, 852-6.

[13] Pat.Appl.Pend., 9226820.0., University of Strathclyde, 1992

[14] Goto, M., Goto, M., Nakashio, F., Yoshizuka, K. & Katsutoshi, I., Hydrolysis of triolein by lipase in a hollow fiber reactor., *J.Membrane.Sci.,* **74,** 1992,207-214.

A COMPARISON OF STIRRED TANK AND MEMBRANE BIOREACTORS FOR AQUEOUS-ORGANIC BIPHASIC ENZYMATIC REACTIONS

P.J.CUNNAH and J.M.WOODLEY

Advanced Centre for Biochemical Engineering,
Chemical and Biochemical Engineering,
University College London,
Torrington Place,
London, WC1E 7JE, UK.

The hydrolysis of benzyl acetate by porcine pancreatic lipase, a two-liquid phase biotransformation, was investigated in a stirred tank reactor (STR) and a membrane bioreactor (MBR) as a means of comparing the two reactors for carrying out two-liquid phase biotransformations. An effective phase ratio of 0.5 and an aqueous enzyme concentration of 4g/l was used in both reactors. From experimental results highest activities and greatest conversion was achieved in the STR whereas greater catalyst stability was observed in the MBR.

INTRODUCTION

Aqueous-organic biphasic reaction systems offer several potential advantages over conventional aqueous phase media for the biotransformation of poorly water-soluble organic compounds of commercial interest [1-3]. In such systems a water immiscible organic liquid is contacted with an aqueous phase containing the biocatalyst. The organic liquid acts either as a reservoir for poorly water-soluble organic substrate(s) of the reaction or the organic liquid may actually be the substrate itself. Contact of the two liquid phases creates an interface which depending on the nature of the biocatalyst can function as a site for biocatalytic activity. Alternatively the interface can serve to facilitate the transfer of the poorly water-soluble organic substrate into the aqueous phase where biocatalysis may occur. Products generated by the reaction may also be preferentially soluble in the organic phase of the reaction medium and thus the presence of the interface can also facilitate product removal. The biocatalyst may be either an intact cell or an isolated enzyme (purified or crude), its selection being determined by it's ability to carry out the reaction of interest and its specific characteristics (eg solvent tolerance) [4].

In order to operate two-liquid phase biotransformations efficiently a number of process aspects need to be understood. Biocatalyst selection will primarily be influenced by the ability of the catalyst to carry out the reaction at a high rate while retaining operational stability. The influence on the biocatalyst of the non-conventional environment also needs to be understood. Aspects of solvent selection from a biological (influence on the biocatalyst) and an engineering (influence on reactor operation) standpoint need to be addressed. Likewise efficient biocatalysis is in part determined by selection of suitable reactors in which to exploit these catalysts. It is this last issue which is addressed in this paper.

Early investigations of two-liquid phase biotransformations were done in shaken tubes and flasks where the interface was created by shaking. Although crude, the use of these vessels gave some indication of the ability to perform these two phase reactions. However they were not readily amenable to scale-up. The next progession was to transfer these reactions into vessels which were suitable for commercial application. The stirred tank reactor (STR) is such a vessel in which a wide variety of processes in both the chemical and biological industries take place and thus a large amount of data pertaining to its operation is already available. It is a suitable vessel for the operation of two-phase reactions, agitation of the two phases bringing about the dispersion of one phase within the other thus creating the interface which facilitates the biotransformation. However more recently another device has been brought to our attention which also allows the creation of an interface between the two phases this being the membrane bioreactor (MBR).

REACTOR OPTIONS

Membranes have been applied to a range of technologies in the past. Their use for reaction engineering relies primarily on the ability of the membrane material to separate liquid and/or solid phases while at the same time selectively allowing specific components to transfer across the membrane or alternatively providing a support for the catalyst. The combination of these two membrane functions in a single unit operation makes them of potential interest for applications within the field of two-liquid phase biocatalysis.

Our interest lies in the potential of the MBR as a device in which to carry out two-liquid phase biotransformations and its potential performance compared to such reactions carried out in STRs. Early membrane bioreactors were constructed using flat sheets of porous membranes across which contact of the two phases could take place and to which biocatalyst could be attached [5,6]. More recently membrane technology has led to the development of hollow fibre membranes of various materials with differing properties. The incorporation of many fibres into a single unit generates a reactor with a large area for contact of the two phases. The particular reactor in this study is such a device. Each individual fibre is made of a porous hydrophilic material. On the lumen (internal) side of

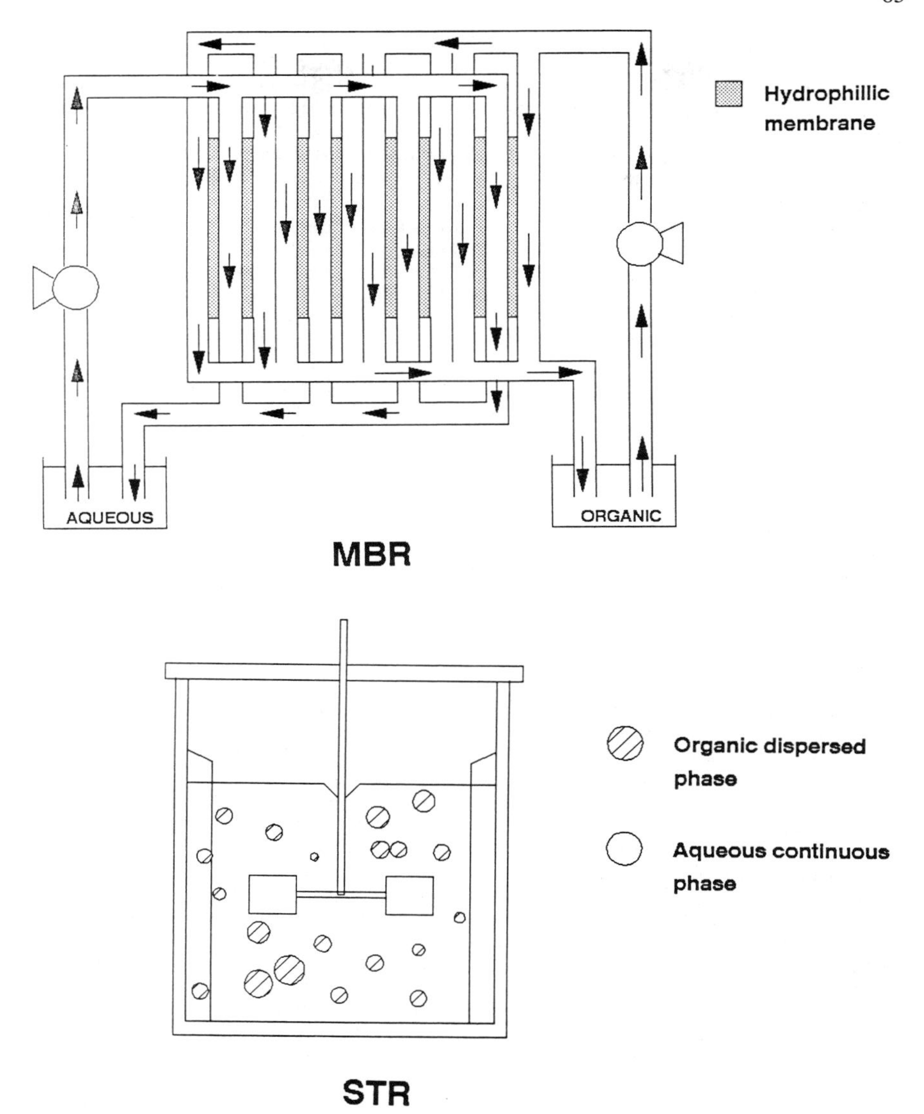

Fig 1: Diagramatic representations of the MBR and the STR for carrying out two-liquid phase biotransformations.

the fibre is a biocatalyst-impermeable layer which allows entrapment of the enzyme within the matrix of the hydrophilic material. The aqueous process stream is carried down the lumen of the fibre while the outer shell portion of the fibre carries the organic process stream. Interface is created at the shell surface of the fibre. Flow through the membrane is achieved via pumps allowing a range of flow rates to be set and independent control of the process streams.

The fundamental difference between the two reactors is in the area available for phase contact and how this is defined. In the STR, interfacial area depends upon the dispersion of one phase within the other, and this will be a function of droplet size. Thus the reactor operating variables which influence this are the phase ratio (the volume fraction of organic liquid in the reactor) and the agitation of the reactor contents. However specific interfacial area may be a function of biotransformation process time due to effects of biocatalyst denaturation and its subsequent effects on the relative rates of droplet breakage and coalescence. In contrast in the MBR the interface is a fixed absolute area for a particular module and is independent of phase ratio. The two reactors are illustrated in Fig 1.

MATERIALS AND METHODS

Biotransformation

In order to illustrate their application and make an initial comparison of the two reactors a suitable test system was chosen: the hydrolysis of benzyl acetate to benzyl alcohol by porcine pancreatic lipase (PPL). The substrate benzyl acetate is a poorly-water soluble organic liquid, (water solubility: 13.5 mM) and thus forms a second liquid phase when present at reasonable concentration. The aqueous phase is a 0.1M phosphate buffer of pH 7 and contains the lipase. The lipase is active at the interface and generates the product benzyl alcohol which is preferentially soluble in the substrate itself. A byproduct of the reaction is acetic acid which leads to a reduction in pH of the reaction liquor affecting enzyme stabilty and this was accounted for by titration with sodium hydroxide thus maintaining the pH of the aqueous phase.

Reactor Operation

Likewise to make a valid comparison of the two reactor types conditions in the STR should as much as possible resemble those in the MBR. Ideally the same interfacial area in the STR and the MBR should be attained. However this is impractical since no truly acccurate method of determining interfacial area for a two-liquid phase dispersion (as opposed to an emulsion) has been found at the present time. In order to make this initial comparison it was deemed suitable to take into account measurable parameters the most obvious of these being the enzyme concentration and liquid volumes used.

Biotransformation in the MBR

4 g of proteinaceous material having lipase activity was loaded onto the membrane. 500 ml of 0.1M phosphate buffer pH 7 was circulated at a rate of 500 ml/min through the lumen, this serving as the aqueous phase for hydrolysis. Reaction was initiated by the circulation of the substrate benzyl acetate through the shell. A pH-stat was used to maintain the pH at 7 by addition of 10M NaOH. This also served to monitor the reaction rate and amount of product formed. The temperature was maintained at 30°C by the use of jacketed vessels containing the constituent phases.

Biotransformation in the STR

Into the STR was placed the aqueous phase (0.1M phosphate buffer pH 7), followed by an identical volume of the substrate benzyl acetate, corresponding to a phase ratio 0.5 and thus mimicing the phase ratios used in the MBR experiments. Agitation was initiated at 1000 rpm , a value based on previous work with an esterase which was found to be sufficient to produce a reasonable amount of interface without compromising enzyme effectiveness [20]. Proteinaceous material corresponding to an aqueous phase concentration of 4 g/l was added to the reactor to commence reaction. Reaction monitoring and pH maintenance was by addition of 10M NaOH using a pH-stat. The temperature was maintained at 30°C using a jacketed vessel.

RESULTS

The results of these initial experiments are depicted in Fig 2. Reactions in both vessels were initally monitored over a period of 50 hours. The results for the MBR (▲) and the STR (●) show the amount of product generated, based on reactor liquid volume, as reaction time proceeds. The results confirm that in both vessels reaction occurs. However there is a substantial difference in the performance of the two reactors.

Reaction rates

Based on the preceeding results it is possible to make an assessment of the specific reaction rate at any given time during the reaction based on the amount of product being generated with time per gram of protein. Fig 3, depicts specific rates as reaction proceeds in the MBR (▲) and in

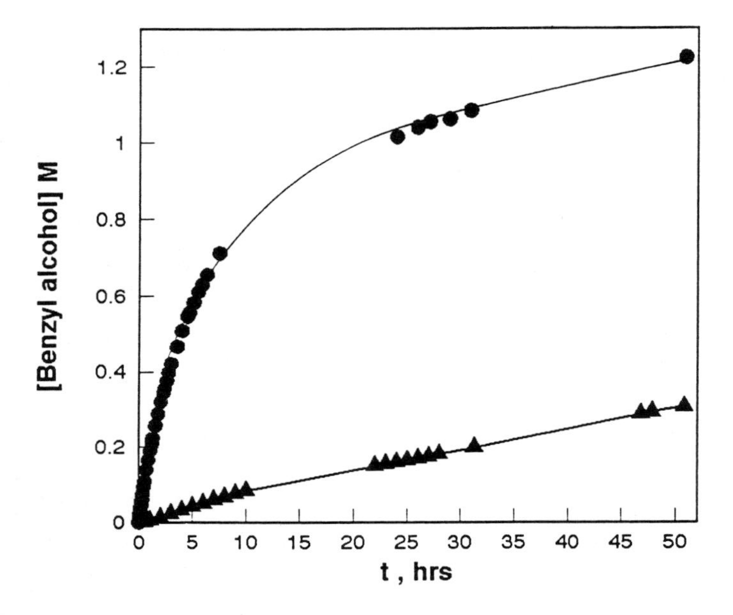

Fig 2: Benzyl acetate hydrolysis by porcine pancreatic lipase in a STR (●) and a MBR (▲).

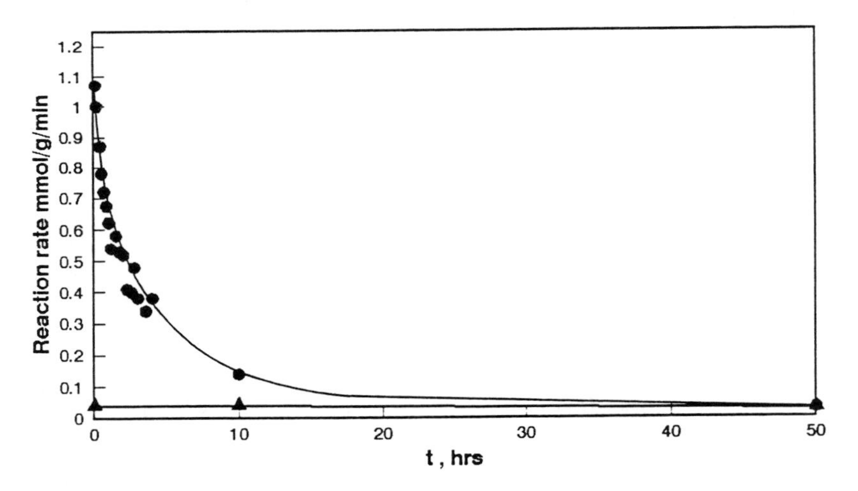

Fig 3: Reaction rate as a function of time for the hydrolysis of benzyl acetate by porcine pancreatic lipase in a STR (●) and a MBR (▲).

the STR (●). In the STR this rate is initially high but falls off dramatically within the first few hours of the reaction and then continues to decline further as the reaction proceeds. In the MBR the initial rate is observed to decrease slightly over the period of time the reaction was run. Direct comparison of the rates obtained shows that initially reaction in the STR is 12.5 times greater, displaying activity at 0.52 mM/g/min, than the initial rate attained in the MBR (0.041 mM/g/min). After 10 hours where a marked reduction in the reaction rate in the STR has been observed it is still 3.5 times greater, at 0.14 mM/g/min, than that observed for the MBR at that time, 0.041 mM/g/min. Over the next 35 hours observed rates in the STR continue to be in excess of those in the MBR until at 45 hours the rates are seen to be equal at 0.028 mM/g/min and this rate is maintained in both vessels until the end of the monitoring period at 50 hours.

Product concentrations and substrate conversion.

Over the observed period of 50 hours product was generated in both types of reactor. In the MBR product was generated at a steady rate to yield a final concentration after 50 hours of 0.32 M. In contrast in the STR 0.32 M was attained in 2.5 hours and after 50 hours of operation the concentration observed in the STR was 1.22 M. The concentration of substrate available for conversion in both reactors, based on total reactor volume, was 3.3 M. Fig 4 depicts the percentage conversion as a function of reaction time. In the MBR the conversion after 50 hours of operation is 10% (▲) while in the STR it is 37% (●) in the same time period.

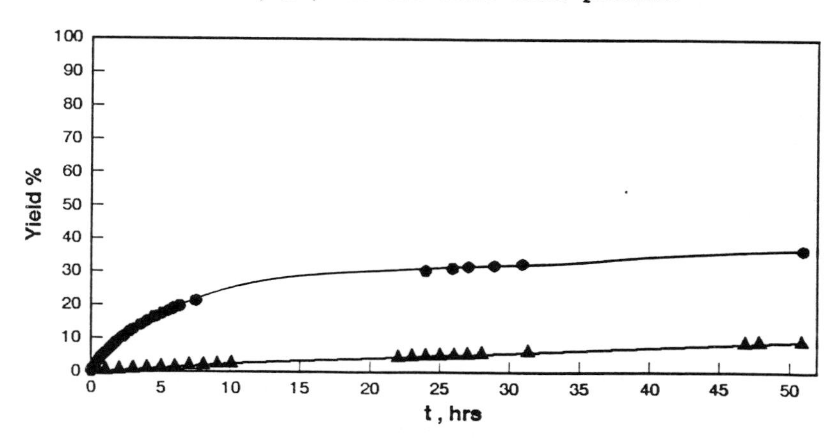

Fig 4: Reaction conversion as a function of time for a STR (●) and a MBR (▲).

DISCUSSION

Many of the current two-liquid phase biocatalytic processes use STR technology [7-9] and an understanding of the principles of operation with both bacterial and enzymic catalysts is becoming established, [10-14]. In addition much of this work with two-phase processes in STRs [15-20] has identified the key operating parameters and their role in determining the performance of these reactors.

In order to achieve a similar understanding of the processes in the MBR and to identify its potential for application, the experiments carried out and reported here were of a similar nature to facilitate an accurate comparison. The lipase used was an impure preparation and its loading onto the membrane was determined by activity assay. It is believed that the loading of an impure preparation onto the membrane reactor acts as a purifiction step, [21]. Knowing the amount of activity on and in the membrane (measured by comparing the activity of the enzyme solution before and after loading) one could provide the identical amount in the STR. In this study all available activity was supported by the membrane and this activity was in 4g of the crude protein. Hence 4g of the protein was used in the STR, making available the same amount of activity in both reactors.

A comparison of the initial activities obtained in the two reactors showed a marked difference: those in the STR being 13.5 times higher than those in the MBR. One might reasonably state that the specific interfacial area achieved in the STR was far in excess of that in the MBR thus making available far more area for greater activity. This is supported by previous studies of mass transfer in the MBR, [22], where achievable rates of transfer are far less than those obtained in a STR and this was attributed to the differences in the interfacial area. However the lower MBR activities may also be attributable to diffusional limitations within the membrane since not all enzyme supported by the membrane will necessarily be exposed to the organic/aqueous interface. McConville and co-workers, [23], drew a similar conclusion when studying the enzyme effectiveness of a lipase in a MBR to resolve Ibuprofen. The nature of the interface is also difficult to assess. Pores within the membrane may be occupied on the membrane surface with enzyme. Hence interface is created by entry of organic into the pore while the hydrophilic nature of the membrane retains the aqueous phase. However not all pores may become occupied with organic phase and the efficiency of occupation will be dependent on the relative pressures of the two process streams and interfacial tension relative to pore size. This needs further investigation. In contrast, in the STR using the dissolved enzyme all available interface can be utilised by the enzyme and there will be no diffusional limitations. Guit and co-workers, [24], evaluated an MBR to elucidate lipase kinetics for the hydrolysis of triacetin comparing these results with results obtained in a STR. Analysis of specific activity at low enzyme loading showed higher activities in the MBR. It was suggested that this was possibly a result of a large lipase fraction remaining in the aqueous phase of the STR. However at more optimal enzyme loadings the STR shows higher activity than the MBR as reported here. A different character to the interface exists within the MBR for the organic solvents used in the study of Guit et al compared to benzyl acetate. Physical phase properties will influence the pore filling characteristics and therefore the nature of the interface.

Comparing the operational stability of the enzyme in the two reactors, half life in the STR was 53 mins while that in the MBR was in excess of 50 hours. Half life in the STR is low in comparison to the MBR possibly because of the direct exposure of the free enzyme to the interface. Using a support in the STR may result in an increase in the stability but there may also be disadvantages and further study in the STR is required to resolve this issue. A protective effect on the enzyme is thus implicated in its immobilisation within the MBR and this seems to be mirrored by a number of other studies of enzymes in MBRs, [24-26], which also report long half lives. However it may be that the maintenance of such low rates as observed in the MBR may also be possible in the STR. It is not until after 50 hours of operation that the rates in the two reactors become identical. The maintenance of high rates in the STR may be made possible not only due to saturation of the interface with enzyme initially but an excess of enzyme may in the STR lead to a constant "new for old" exchange faciltating maintenance of these rates. In the MBR because of the immobilised nature of the catalyst such an exchange is not possible. When the reaction rate does drop, new enzyme cannot replace old.

The higher conversions in the STR over the MBR are a direct consequence of the higher activity. The implications of this are that factors affecting enzyme activity directly link to the productivity of the reactor. To obtain the same conversion in the STR as obtained in the MBR takes far less time eg 10% conversion in the MBR requires 50 hours of operation whereas the equivalent conversion in the STR requires 2.5 hours of operation. Furusaki and co-workers, [25], have examined a membrane enzyme reactor and concluded that in order to achieve high conversion high enzyme concentration in the membrane along with high residence time is required and they also note a reduced activity of the immobilised enzyme. Having an understanding of this relationship thus provides a useful aid to analysing the efficiency of the reactor.

Although we have only considered here the performance of the reactor based on productivity a number of other elements not yet addressed experimentally may need to be considered when analysing the relative merits of the two particular reactors. The reactor and what occurs in it will influence other operations which take place within the process as a whole. Of particular relevance are downstream operations to separate and purify products and recycle unconverted substrates and potentially catalyst. The operation of the two-phase system in the STR may create emulsions, further complicated by the presence of denatured

		STR	MBR
Reactor Design Features	Fluid-fluid interface	Dynamic	Static
	Catalyst immobilisation	Optional	By definition *
	Interfacial area	Variable	Fixed (low)
	Operating variables	Phase ratio Enzyme conc. Agitation rate	Phase ratio (Recycle amount) Enzyme conc. (on membrane) Shell and lumen flow rates
	Phase hydrodynamics	Well mixed droplet/continuous	Plug flow
	Reaction site	Interfacial/Bulk	Interfacial *
	Catalyst loading	Variable	Variable (limiting)
Observed Performance	Catalytic activity	+	-
	Catalyst stability	-	+
	Phase separation	-	+
	Mass transfer [22]	+	-

* As operated here - Poor performance + Good performance

Table 1. A summary of the reactor comparison of a STR and a MBR for carrying out two-liquid phase biotransformations.

catalyst. Further processing would thus be needed in order to remove product and recycle substrate. In contrast in the MBR the two process streams are kept separate eliminating troublesome emulsions, [27]. Further the separation of process streams in the MBR also facilitates carrying out in situ extractions, separations and enrichments, [28,29]. The operation of two phase processes in the STR is generally in batch mode whereas the MBR can be operated in a continuous mode.

CONCLUSION

Comparing the productivity of a MBR with a STR has indicated that for the lipase catalysed two-liquid phase reaction under investigation higher activities and thus higher conversion of product are achieved in a STR. A number of explanations of why this difference exists have been presented. The stability of the enzyme however is much increased by its immobilisation in the MBR and this may prove an advantage for some reactions. In the discussion a number of other areas of comparison were identified. Table 1 summarises the conclusions which we have drawn from this work and identifies some of the key reactor design features. Further investigation of this reaction in the two reactors is ongoing at this laboratory and it is believed that this study will lead to a better understanding of two-liquid phase reactions carried out in these reactors thus providing a scientific basis for reactor evaluation. Ultimately this will provide guidelines for complete process design for a specific biotransformation.

ACKNOWLEDGEMENT

UCL is the Science and Engineering Research Council's Interdisciplinary Research Centre for Biochemical Engineering and the Council's support is gratefully acknowledged.

REFERENCES

1. Lilly,M.D., Woodley,J.M., In Biocatalysis in Organic Syntheses., Tramper,J., van de Plas,H.C., Linko,P. [Eds], Elsevier, Amsterdam, 3 (1985).
2. Lilly,M.D., Brazier,A.J., Hocknull,M.D., Williams,A.C., Woodley,J.M., In Biocatalysis in Organic Media., Laane,C., Tramper,J., Lilly,M.D. [Eds], Elsevier, Amsterdam, 3 (1987)
3. Woodley,J.M., Lilly,M.D., In Biocatalysis in Non-Conventional Media., Tramper,J., Vermue,M.H., Beeftink,H.H., von Stoker,V. [Eds], Elsevier, Amsterdam, 147 (1992).
4. Woodley,J.M., Harrop,A.J., Lilly,M.D., In Enzyme Engineering 10, Annals of the New York Academy of Sciences, Okada,H., Tanaka,A., Blanch,H.W. [Eds], 613, 191 (1990).
5. Hoq,M.M., Yamane,T., Shimizu,S., J. Amer. Oil and Chem. Soc., 61, 776 (1984).
6. Hoq,M.M., Yamane,T., Shimizu,S., J. Amer. Oil and Chem. Soc., 62, 1016 (1985).
7. Umemura,T., Hirohara,H., In Biocatalysis in Agricultural Biotechnology., Whitaker,J.R., Sonnet,P.E. [Eds], Amer. Chem. Soc., Washington, 371 (1989).
8. Schutt,H., Schmidt-Kastner,G., Arens,A., Preiss,M., Biotechnol. Bioeng., 27, 420 (1985)
9. Evans,T.W., Kominek,L.A., Wolf,H.J., Henderson,S.L., Eur. patent appl. No 0127 294 A1.
10. Furuhashi,K., Shintani,M., Takagi,M., Appl. Microbiol. Biotechnol., 23, 218 (1986).
11. Mukataka,S., Kobayashi,T., Takahash,J., J. Ferment. Technol., 63, 461 (1985).
12. Brookes,I.K., Lilly,M.D., Drozd,J.W., Enz. Microb. Technol., 8, 53 (1986).
13. Carrea,G., Riva,S., Bovara,R., Pasta,p., Enz. Microb. Technol., 10, 333 (1988).

14. Brazier,A.J., Lilly,M.D., Herbert, A.B., Enz. Microb. Technol., 12, 90 (1990).
15. Williams,A.C., Woodley,J.M., Ellis,P.A., Lilly,M.D., In Biocatalysis in Organic Media., Laane,C., Tramper,J., Lilly,M.D. [Eds], Elsevier, Amsterdam., 399 (1987).
16. Williams,A.C., Woodley,J.M., Ellis,P.A., Narendranathan,T.J., Lilly,M.D., Enz. Microb. Technol., 12, 260 (1990).
17. Woodley,J.M., In Biocatalysis., Abramovicz,D.A. [Ed], Van Nostrand Reinhold, New York, 157 (1990)
18. Woodley,J.M., Brazier,A.J., Lilly,M.D., Biotechnol. Bioeng., 37, (1991).
19. Lilly,M.D., Dervakos,G.A., Woodley,J.M., In Opportunities in Biotransformations., Copping,L.G., Martin,R.E. , Pickett,J.A., Bucke,c., Bunch,A.W. [Eds], Elsevier, London, 5 (1990).
20. Woodley,J.M., Cunnah,P.J., Lilly,M.D., Biocatalysis., 5, 1 (1991).
21. Xavier,F., Hill,C.G., Amundson,C.H., Biotech. Bioeng., 38, 853 (1991).
22. Cunnah,P.J., Woodley,J.M., In Proceedings of the 1992 IChemE Research Event., Institution of Chemical Engineers,Rugby.
23. McConville,F.X., Lopez,J.L., Wald,S.A., In Biocatalysis.,Abramovicz,D.A. [Ed], Van Nostrand Reinhold, New York, 167 (1990).
24. Guit,R.P.M., Kloosterman,M., Meindersma,G.W., Mayer,M., Meijer,E.M., Biotechnol. Bioeng., 38, 727 (1991)
25. Furusaki,S., Nozawa,T., Nomura,S., Biopr. Eng., 5, 73 (1990).
26. Pronk,W., Kerkhof,P.J.A.M., van Helden,C., van't Reit,K., Biotech. Bioeng., 32, 512 (1988).
27. Stanley,T.J., Quinn,J.A., Chem. Eng. Sci., 42, 2313 (1987).
28. Bratzler,R.L., In Proceedings Biotech 87., Online Publications, Pinner, 23 (1987).
29. Quinn,J.A., Matson,S.L., Lopez,J.L., In Enzyme Engineering 10 Annals of the New York Academy of Sciences., Okada,H., Tanaka,A., Blanch,H.W. [Eds], 613, 155 (1989).

FOOD

CONTINUOUS HYDROLYSIS OF CASSAVA FLOUR STARCH IN AN ENZYMATIC MEMBRANE REACTOR

Rual Lopez-Ulibarri* & George M. Hall

Dep. of Chemical Engineering, Loughborough University of Technology, Leicestershire, LE11 3TU, England.

ABSTRACT

Cassava gelatinised by extrusion was continuously hydrolysed with glucoamylase (amyloglucosidase) in an dead-end stirred membrane reactor. Starch conversions of 97-98% were achieved with an enzyme concentration of 300 Sigma AG units per gram of cassava flour and a retention time of 150 minutes. The best performance of the reactor was achieved using a substrate concentration of 40 mg/ml in acetate buffer. Amicon YM5 and YM10 membranes with a nominal molecular weight cut-off of 5,000 and 10,000 respectively were used in the reactor. A very small product rejection was found and basically pure glucose was obtained in the permeate. The bioreactor presented a good performance for both enzyme activity and stabilisation up to 30-32 hrs of continuous hydrolysis.

Key words: Enzymatic membrane reactor, saccharification, cassava, glucoamylase, glucose syrup.

* To whom all correspondence should be addressed.

INTRODUCTION

The industrial preparation of glucose syrups involves maize starch saccharification to oligo-saccharides using a special thermostable α-amylase and a second hydrolysis to glucose using glucoamylase. Starch hydrolyzates with a high dextrose equivalent (DE) are extensively used in the food industry and as source of fermentable sugars (Patil, 1991). Glucose syrups with a concentration of 96-98% (high-glucose syrups, HGS), are used for the production of crystalline D-glucose or as starting material for the production of high-fructose syrup (HFS) (Fogarty, 1983). Traditionally, the final hydrolysis to glucose has been carried out in batch mode which results in: a) an inefficient use of the enzyme since it is not recovered and reused, b) additional costs for glucose separation from other sugars c) variation in the final product between batches and d) long reaction times (24-96 hrs) depending on the desired level of glucose in the product (Sims and Cheryan, 1992). The continuous production of glucose syrup from liquefied maize starch has been studied using traditional immobilization techniques of glucoamylase but the required conversion yield has not been reached yet and the immobilization costs are too high (Madgavkar, 1976, Freire & Sant'Anna, 1990). The use of ultrafiltration equipment as an Enzymatic Membrane Reactor (EMR) for the saccharification of starches or cellulose has been proposed in the past with different levels of success due mainly to membrane fouling and enzyme deactivation in long term experiments (Butterworth, *et al.*, 1970, Ghose and Kostick, 1970, Closset, *et al.*, 1974, Hägerdal, *et al.*, 1980, Darnoko, *et al.*, 1989)

Cassava (*Manihot esculenta* Crantz) is one of the ten most important tropical crops. It is used as starch provider for human consumption and animal feeding in Africa, Asia and Latin America. The potential yield of cassava is high and, by 1987, production is reported to have risen to nearly 135 million tons (Bertram, 1990). Cassava flour is a good potential substrate for the production of HGS, since it has several commercial advantages: a high content of good quality starch (85-90%, of dry matter); because low technology is required to produce it, cassava is a source of starch cheaper than maize and it contains much less protein and minerals than maize or potato flour (Cock, 1985). Additionally, cassava starch has a lower gelatinisation temperature, higher amylopectin content and higher amylose solubility than maize starch (Balogolapan, *et al*, 1988). Moreover, the potential use of cassava flour as a substrate for the production of glucose syrups in an EMR avoids additional costs due to the starch extraction process and final product purification needed in the production of cassava starch.

The aim of this work was to examine the utilization of cassava flour as source of starch in a continuous EMR which would reduce the production cost of high glucose syrups.

MATERIALS AND METHODS

Substrate

Frozen peeled pieces (food grade) of cassava root (*Manihot esculenta* Crantz) were obtained in a commercial preparation (Prosanca) produced and packed by Proexpo, S.A., San José, Costa Rica. This material was defrosted, chopped, dried, ground and sieved. The cassava flour obtained was then extruded with a diluted phosphate buffer (1% v/v, pH 6.5) in a Brabender Food Extruder with a three sectioned barrel, water cooled feed and operated with a Do-Corder "E" torque rheometer Model 330 (Brabender Extruder technical information). The operating parameters used for the extrusion were: Barrel temperature (starting from the feeding section): 110˚, 140˚ and 130 ˚C, screw speed: 250 rpm and barrel output (die head) diameter: 4.0 mm. These processing conditions produced a cassava flour extrudate expansion close to the optimum (Badrie and Mellowes, 1992).

Table 1. *Composition of the extruded peeled cassava.*

Component	*%*
Dry matter	*92.37*
Starch	*84.22*
Reducing Sugar	*1.11*
Fibre (crude)	*1.95*
Protein	*2.06*
Fat	*1.16*
Ash	*1.87*

The extruded cassava obtained was then dried at 65 °C in a convection oven for 24 hrs., ground and sieved (final particle size <= 300 μm). The composition of the extruded cassava flour obtained is presented in Table 1. This material was suspended to the concentration required and used as substrate in the enzymatic membrane reactor.

Enzyme

Amyloglucosidase (Glucoamylase; Exo-1, 4 α-glucosidase; EC 3.2.1.3) from *Aspergillus niger*, was purchased in liquid form from Sigma Chemical Co. (Dorset, England). The activity reported was 6,100 AG units/ml. Unit definition: one amyloglucosidase unit (AG) will liberate 1.0 mg of glucose from starch in 3 min at pH 4.5 and 55 °C. The enzyme was used as received.

Batch Reactor.

An spherical glass reactor with a nominal volume of 500 ml was used as batch reactor for preliminary runs. In these experiments, 250 ml of the desired concentration of cassava in acetate buffer (pH 4.5) were heated up, with continuous internal agitation, in a water bath. When the temperature desired was reached, the glucoamylase was added. Aliquots were taken at regular periods and analysed for sugars (reducing, glucose and oligosaccharides) and total solids.

Ultrafiltration Membrane Reactor

Description. A "dead-end cell" ultrafiltration equipment from Amicon (Model 402) was modified and used as Continuous Stirred Tank Membrane Reactor (CSTMR) (Cheryan and Mehaia, 1986). Figure 1 is an schematic representation of the system used. Modifications made to the ultrafiltration cell allowed both internal temperature monitoring and sampling. The bioreactor had a maximum process volume of 400 ml and holds ultrafiltration flat-sheet membranes, 76 mm diameter, with an effective permeation area of 41.8 cm^2 (Amicon technical publication). The EMR was immersed in a water bath to obtain the desired temperature. A pressurized container was used to feed either buffer or substrate to the reactor. In order to obtain the necessary differential pressure for permeate collection, the whole system was pressurized with nitrogen. A small rheometer was installed to measure the permeate flow.
Membranes. YM5 and YM10 Amicon ultrafiltration flat-sheet membranes (made from regenerated cellulose) with a nominal cut-off of 5,000 and 10,000 MW respectively were used for substrate, enzyme and product permeation-rejection tests and selected for saccharification of cassava flour.
Operation. At the beginning of each experiment, 250 ml of the desired concentration of substrate in acetate buffer (pH 4.5 +/- 0.1) was put into the EMR and immersed in the water bath. While being continuously stirred and when the internal temperature reached the process temperature (55 °C +/- 0.2 °C), the required amount of enzyme was added and the system was closed. After 1 hr. of reaction, with valves 3 and 4 closed, pressure was applied in both containers. When equilibrium was achieved (15-20 sec), valve 2 was closed and valve 3 was opened to operate the system in continuous mode. Internal samples were obtained, when necessary, closing valve 3 and opening valve 4 and analysed for

obtained, when necessary, closing valve 3 and opening valve 4 and analysed for total solids, and identification and measurement of sugars. After sampling, valve 2 was opened again to restore the pressure in the membrane reactor, then closed again and valve 3 was opened to continue the process. The pressure applied was modified in order to obtain the required permeation flow and, therefore, the retention time. Preliminary experiments were made to establish the operational pressure necessary to obtain the desired permeate flow. In this study, a constant permeate flow of 1.66 ml/min was used for the hydrolysis process (See details in discussion).
Permeate was collected in a graduated cylinder and analysed for sugars (glucose and oligosaccharides) and total solids. At the end of each experiment, the reactor was emptied, washed and rinsed with distilled water. The membrane was rinsed several times with running distilled water and then reinstalled in the clean membrane reactor to operate with distilled water at 25-30 psi until the original permeate flux was achieved.

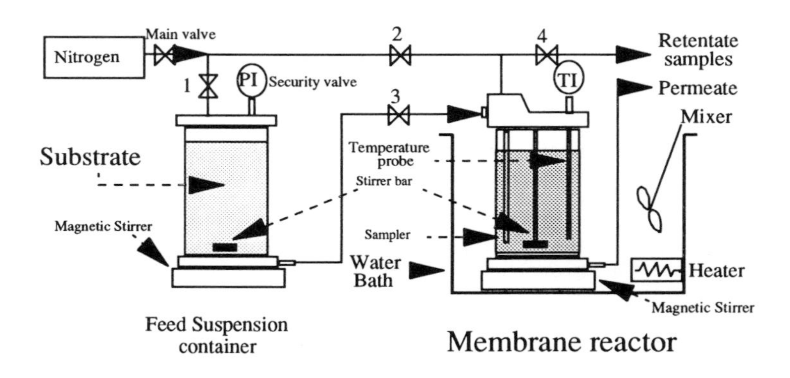

***Figure 1**. Schematic representation of the enzymatic membrane reactor. See text for details.*

Analytical Methods.

Dry matter, starch, carbohydrates, fibre, protein, fat and ash were determinated using the Oficial Methods of Analysis (AOAC, 1990). Glucose and other sugars were determinated by HPLC using a Zorbax NH_2 column and a refractive index detector (Waters Assoc., Milford, Massachusetts). The mobile phase used was acetronitrile: water (72:28) and the dilution flow was 1.5 ml/min. Soluble potato starch, glucose, maltose and other oligosaccharides for standard preparations were obtained from Sigma Chemical Co. (Dorset, England).

RESULTS AND DISCUSSION

Preliminary batch experiments were made to establish the more suitable environmental conditions for the hydrolysis of extruded cassava with glucoamylase. Peeled cassava was selected as starting material instead of whole cassava roots in order to obtain a final material rich in carbohydrates and little inert material (i.e. fibre and protein) as possible. Besides, the presence of fibre can decrease the amylolytic activity in saccharification of cassava (De Menezes *et al*, 1978). The optimal conditions for cassava flour were found as those reported for the enzyme manufacturer, using soluble potato starch (pH 4.5 and 55 °C). Glucoamylase activity for extruded cassava flour was 94.6% +/- 0.97% of that obtained using soluble potato starch. A higher glucoamylase activity was found at pH 4.5 and 60 °C, but the stability at that temperature was very poor, compared with that obtained at 55 °C, as it can be seen in Figure 2.

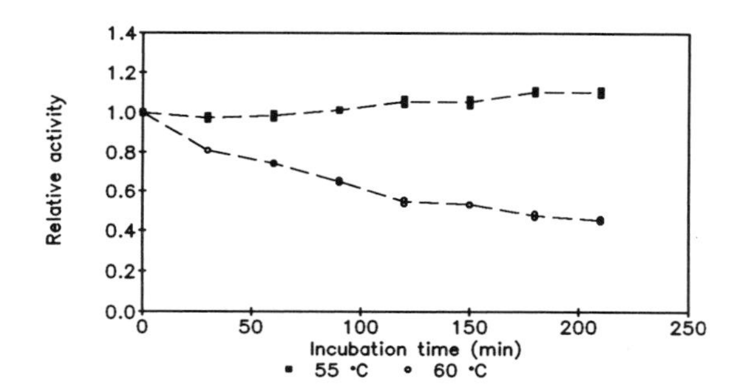

Figure 2. Effect of incubation time in relative glucoamylase activity at 55 °C and 60 °C.

In order to establish an adequate enzyme concentration for the continuous saccharification of cassava (CSC), experiments were performed in the batch bioreactor. The effect of glucoamylase concentration on the hydrolysis of cassava using a 1% (w/v) substrate concentration is shown in Figure 3. All enzyme concentrations are expressed in glucoamylase (AG) Sigma units per gram of substrate (dry basis). Note that the use of just 5 AG units g^{-1} produced almost half of the maximum hydrolysis achieved but, on the other hand, with 400 AG units g^{-1} the hydrolysis achieved is not substantially different from that using 300 AG units g^{-1}. To guarantee no enzyme limitation and a fast initial hydrolysis, the enzyme concentration of 300 AG units g^{-1} was selected for the CSC process. This also enhances the production of a highly pure glucose permeate (Darnoko *et al*, 1989). Starch conversion to glucose (%) was calculated as follows:

$$\text{Conversion} = ([G] \cdot 0.9/[S]) \cdot 100$$

where:

[G] = Concentration of glucose
[S] = Concentration of starch

The maximum conversion using the selected parameters was reached in about 150 minutes and this was selected as the retention time for the CSC in the EMR. The permeate flow rate to achieve this was 1.66 ml min^{-1} and it is calculated as process volume of the reactor (250 ml) divided by the retention time (150 min).

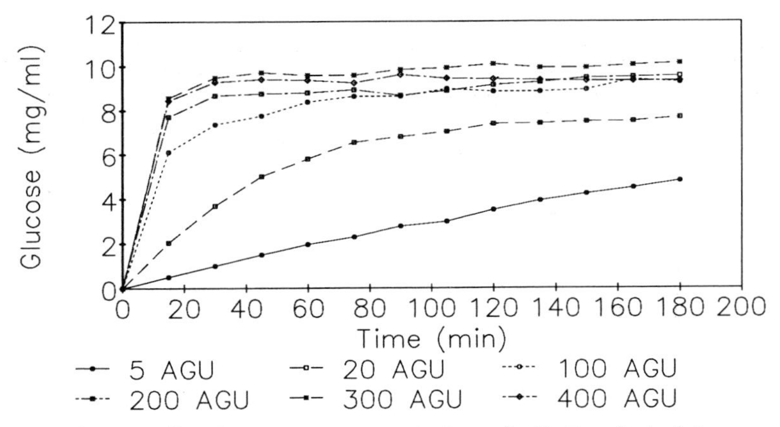

Figure 3. Effect of enzyme concentration in the saccharification of extruded cassava flour in a batch reactor. Enzyme concentrations are given in AG Units per gram of cassava (dry basis). [S] = 10 mg/ml. Conditions: pH 4.5 and T=55°C.

Hydrolysis of cassava flour in the EMR.

In hydrolysis trials in the EMR, both YM5 and YM10 membranes gave 100% rejection of both substrate and enzyme. No oligosaccharides were detected on permeate samples using both membranes. The chromatography analysis of permeate samples demostrated that 99.5% of the solids content was glucose. However, YM5 membrane presented a slightly higher glucose rejection than that obtained with the YM10 membrane.
Glucose rejection (%) was calculated as follows:

$$R = [1 - (G_p/G_r)] \cdot 100$$

where:

G_p = Concentration of glucose in permeate
G_r = Concentration of glucose in retentate

Figure 4 illustrates the use of these membranes on glucose rejection (%). It was observed that in order to obtain the selected permeate flow, a higher process pressure had to be applied to trials with the YM5 membrane than those using the YM10 membrane.

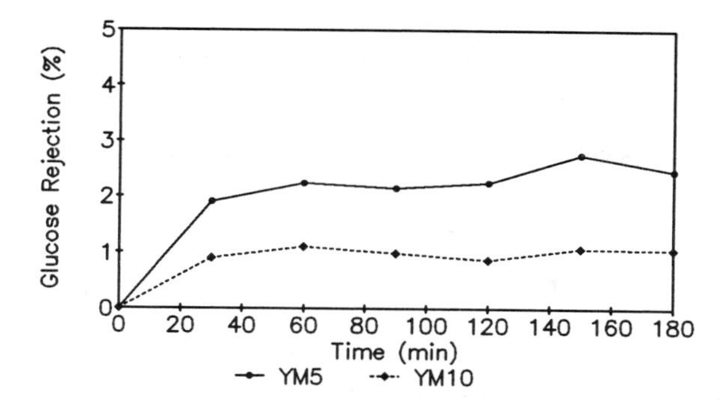

Figure 4. Glucose rejection for YM5 and YM10 membranes for contimuous hydrolysis of cassava flour starch. [S] = 10 mg/ml, [E] = 300 AG Units g^{-1} of cassava flour. Same conditions than Figure 3.

It was also noticed that the configuration using the YM5 membrane was less stable and the pressure had to be corrected more often in order to maintain the permeate flow. This was probably due to the combination of a smaller pore size and the accumulation of insoluble material. Because of its better performance, the YM10 membrane was selected and used in the subsequent experiments.

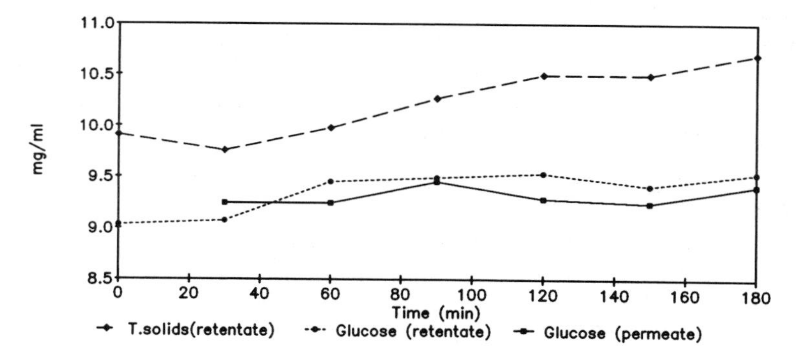

Figure 5. Continuous hydrolysis of cassava flour in the EMR. Membrane: YM10, [S]= 10/mg/ml, [E]= 300 AGU g^{-1} of cassava flour; pH 4.5 and T=55°C.

Figure 5 shows typical results of CSC obtained with the YM10 membrane using a 1% substrate suspension and 300 AG Units g^{-1} of glucoamylase. It was found that disturbances created by internal sampling made steady state difficult to achieve. However, it was useful to establish the general pattern (behaviour) for the CSC on partial starch conversion and permeation-rejection analysis. Since internal sampling caused a relatively long recovery of the steady state, no internal samplings were made for long trial runs.

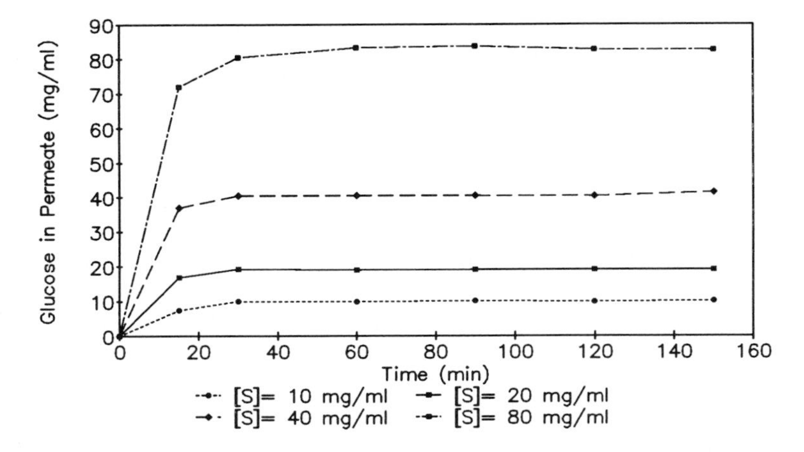

Figure 6. Effect of substrate concentration for the continuous hydrolysis of cassava flour in the EMR. Membrane: YM10. [E]= 300 AGU g^{-1} Conditions: pH 4.5 and 55 °C. Permeation flow: 1.66 ml min^{-1}.

In order to establish the most suitable substrate concentration for the CSC, experiments using 10, 20, 40 and 80 mg/ml of cassava flour (dry basis) and 300 AG Units g^{-1} of cassava were performed and the results are shown in Figure 6. As was expected, results up to one exchange of reactor volume showed no enzyme limitation during this process. In all cases, a high and stable glucoamylase activity was observed and no influence of substrate concentration on the purity of the product was detected. This was probably due to the effect of no enzyme limitation pointed out above. On the other hand, drastic rheological changes occurred for the higher concentration, reflected by stirring problems and membrane fouling. This is normal since using more concentrated substrate, the accumulation of insoluble materials is greater.

Table 2. *Performance of the EMR with different substrate concentrations[1].*

Substrate concentration (mg/ml)	*Relative productivity[2]*	*Glucose rejection (%)*
10	*3.37*	*1.05*
20	*3.19*	*1.90*
40	*3.40*	*2.22*
80	*3.37*	*2.31*

1 Membrane YM10; retention time: 150 min; Experimental conditions: T= 55 °C, pH 4.5.
2 Total grams of glucose in permeate per glucoamylase unit (AGU) after one volume replacement.

However, using a substrate concentration of 40 mg/ml no drastic rheological changes were detected and the relative productivity in terms of product obtained per unit of enzyme and general performance are adequate (See Table 2). Darnoko *et al.* (1989) using cassava starch (tapioca) in an EMR for the production of glucose syrups in similar conditions, found that increasing the substrate concentration had a negative influence on both starch hydrolysis and glucose purity in permeate. The use of cassava flour as substrate for hydrolysis in an EMR offers a better performance since none of these effects were detected.

Experiments were carried out to establish the long-term operational stability of the system cassava flour-EMR. Since an excessive accumulation of insoluble material could produce adverse rheological changes in the bioreactor for CSC experiments, long term trials were expected to have both stirring and membrane fouling problems in a short period. However, the EMR showed a very good stability and reproducibility in experiments with a maximum substrate concentration of 40 mg/ml and up to approx. 13 volume replacements (30-32 hours), as it can be seen in Figure 7. Trials using higher substrate concentrations presented faster membrane fouling which could not be overcome using a higher process pressure. Darnoko *et al.* (1989) found that cassava starch (tapioca) as substrate in a EMR resulted in a good conversion (96.6%) for short periods (up to 9 hrs) and low substrate concentrations (1%) . In this study, the EMR presented a starch conversion of approx. 98% and very low enzyme inactivation was detected. In spite of the problems presented, the productivity (grams of glucose produced per unit of glucoamylase used) of the EMR compared with that of the batch reactor was much higher (approx. 13 times).

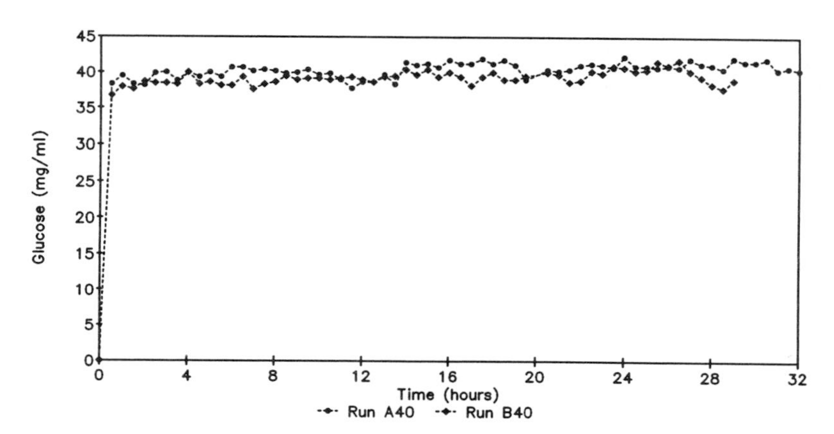

Figure 7. Long-term operational estability of the membrane reactor. [S]= 40 mg/ml. [E]= 300 AGU g^{-1} of cassava flour. Membrane: YM10. Conditions were the same as Figure 6.

As was pointed out before, the accumulation of insoluble material in the reactor for long operational periods caused a remarkable increase in the viscosity of the retentate (clearly observed, but not measured), a decline in the permeate flow due most probably to membrane fouling and, eventually, the necessity to stop the process. De Menezes, *et al.* (1978) found that the use of cellulolytic enzymes in the saccarification of cassava can reduce the viscosity of the product in batch experiments. This could be used in the EMR to increase the reaction time. Besides, the use of cellulases could enhance the hydrolysis of cassava flour to glucose.

CONCLUSION

Cassava flour can be continuously hydrolysed in an enzymatic membrane reactor (EMR) obtaining a high glucose syrup. The stirred dead-end cell used as an EMR was useful for laboratory experimentation. The best performance of the reactor for the continuous saccharification of cassava flour extrudate was obtained using a YM10 Amicon membrane (MW cut-off 10,000), resulting in a 100% of rejection for both substrate and enzyme and very low rejection of the product (glucose). The starch conversion depends on the substrate/enzyme ratio. The best substrate and enzyme concentrations for the CSC in this study were 40 mg/ml (dry basis) and 300 AG Sigma units g^{-1} of cassava flour, respectively. The best parameters for the CSC with glucoamylase are 55 °C, pH 4.5 and a retention time of 150 min. A true steady state was achieved with a good enzyme estability and no enzyme inactivation for runs up to 13 volume replacements. Although the productivity of the EMR against the batch reactor is much higher, the accumulation of insoluble materials increased viscosity and, therefore, produced stirring and membrane fouling problems, being the main cause to stop the process. In order to establish the feasibility of this process at a higher level, further research involving the use of other type of membranes such as hollow fibres is necessary.

ACKNOWLEDGMENTS

The authors thank Mr. P. Glover for technical support on the extrusion process at the Department of Applied Biochemistry and Food Science of the University of Nottingham, Sutton Bonington. This research was partially supported by the National Council of Science and Technology (CONACyT), Mexico (Grant No. 54167).

REFERENCES

AOAC. 1990. Official Methods of Analysis. Association of Official Analytical Chemists, 15th edition, Kenneth Helrich Ed.

Badrie, N. and Mellowes, W.A. 1992. Cassava Starch or Amylose Effects on Characteristics of Cassava (*Manihot esculenta* Crantz) Extrudate. *J. Food Science*, **57**, No.1, 103-107.

Bagalolapan, C., Padmaja, G., Nanda, S.K. & Moorthy, S.N. 1988. Cassava in Food, Feed and Industry. Boca Raton, FL: CRC Press Inc.

Bertram, R.B. 1990. Cassava. In *Agricultural Biotechnology: Opportunities for International Development*, G.J. Persley ed. Biotechnol. in Agric. Series No.2, C.A.B. International, Wallingford, Oxon, U.K., 241-261.

Butterworth, T.A., Wang, D.I.C. and Sinskey, A.J. 1970. Application of Ultrafiltration for Enzyme Retention During Continuous Enzymatic Reaction, *Biotechol. Bioeng.*, **12**, 615-631.

Cheryan, M. and Mehaia, M.A. 1986. Membrane Bioreactors. In *Membrane Separations in Biotechnology*, W.C. McGregor ed., Bioprocess Technology Series, **1**, Marcel Dekker Inc. Ed. New York and Basel. 255-301.

Closset, .P., Cobb, J.T. and Shah, Y.T. 1974. Study of Performance of a Tubular Membrane Reactor for an Enzyme Catalyzed Reaction. *Biotechnol. Bioeng.*, 16, 345-360.

Cock, J.H. 1985. Cassava: New Potential for a Neglected Crop. Westview Press, Boulder, Colorado and London.

Darnoko, D., Cheryan, M. and Artz, W.E. 1989. Saccharification of cassava starch in an ultrafiltration reactor. *Enzyme Microb. Technol.*, **11**, March, 154-159.

De Menezes, T.J.B., Arakaki, T., DeLamo, P.R. and Sales, A.M. 1978. Fungal Cellulases as an Aid for the Saccharification of Cassava. *Biotechnol. Bioeng.*, **20**, 555-565.

Freire, D.M.G. & Sant'Anna, G.L., Jr. 1990. Hydrolysis of Starch with Immobilized Glucoamylase. *Appl. Biochem. and Biotech.*, **26**, No. 1, 23-34.

Fogarty, W.M., 1983. Microbial Amylases. In *Microbial Enzymes and Biotechnology*, W.M. Fogarty ed., Applied Science Publishers LTD, Essex, England.

Ghose, T.K. and Kostick, J.A. 1970. A Model for Continuous Enzymatic Saccharification of Cellulose with Simultaneous Removal of Glucose Syrup. *Biotechnol. Bioeng.*, **12**, 921-946.

Hägerdal, B., López-Leiva, M. and Mattiasson, B. 1980. Membrane Technology Applied to Bioconversion of Macromolecular Substrates and Upgrading of Products, *Desalination*, **35**, 365-373.

Madgavkar, A.M. 1976. A Study of Ultrafiltration Membrane Reactors in Starch Hydrolysis by Solubilized and Immobilized Glucoamylase. PhD Thesis, University of Pittsburgh.

Sims, K.A. and Cheryan, M. 1992. Continuous Production of Glucose Syrup in an Ultrafiltration Reactor. *J. Food Science*, **57**, No.1, 163-166.

Hydrolysis of Butteroil by Immobilized Lipase in a Hollow Fiber Membrane Reactor: Optimization and Economic Considerations

F. X. Malcata[1] and C. G. Hill, Jr.[2]
Department of Chemical Engineering
University of Wisconsin
1415 Johnson Drive
Madison, WI USA 53706

SUMMARY

A lipase from *Aspergillus niger* immobilized by physical adsorption in a hollow fiber membrane reactor can be used to bring about the controlled hydrolysis of the glycerides of melted butterfat. McIlvane buffer and butteroil are fed continuously to the lumen and shell sides of the reactor, respectively. The effects of reactor space time, time elapsed since immobilization of the enzyme, pH, and temperature on the effluent concentrations of both total free fatty acids and ten specific free fatty acids have been reported in previous papers. The rate expressions which provide the best fit of the kinetic data for the hydrolysis reactions are formally equivalent to multisubstrate Michaelis-Menten mechanisms involving inhibition by product species. The kinetics of the enzyme deactivation phenomena are best modeled by two parallel first-order processes, one of which leads to an inactive form of the enzyme and the other of which leads to a less active form.

These rate expressions were utilized in a plug flow model of reactor performance to determine optimum operating conditions. The best compromise between high hydrolysis rates and low rates of enzyme deactivation is achieved by operation at ca. 40°C and pH 7. Computer simulations based on the use of these operating conditions and a requirement for constant effluent composition permit one to determine the requisite schedule for the feedstock flow rate and the optimum time of operation before regeneration with fresh lipase.

An assessment of the merits of the hollow fiber membrane reactor process relative to the conventional free enzyme process indicates that the membrane reactor provides a very attractive alternative to traditional batch reactor technology.

[1] Now with Escola Superior de Biotecnologia, Universidade Catolica Portuguesa, Rua Dr. Antonio Bernardino de Almeida, 4200 Porto, Portugal
[2] Author to whom correspondence should be addressed

INTRODUCTION

In recent years membrane processes have been employed with increasing frequency to replace or complement conventional industrial methods for separation, concentration, and purification. Engineers have also explored additional alternative applications of microporous membranes which capitalize on the high surface area to volume ratios of these materials. An example of such an application is the use of membranes as supports for immobilization of enzymes. The focus of the present paper is consideration of factors relevant to optimization of the performance of an immobilized lipase being used to bring about the hydrolysis of butteroil in a hollow fiber reactor. The impetus for interest in this problem is the fact that the current surplus of milkfat production poses a major problem for the dairy industry in the United States. Literature relevant to the use of immobilized lipase reactors for modification of fats and oils has recently been reviewed by Malcata *et al.* (1).

During the past two decades increasing consumer interest in the relationship of diet to good health has led to concerns about the amount of milkfat an individual should consume. The resultant increases in demand for low fat dairy products and the concomitant declines in demand for whole milk, butter, cream, etc., have created the necessity for finding new markets for milkfat.

The triglycerides of butterfat contain an unusually wide variety of fatty acid residues, among which is a large proportion of short- and medium-chain residues. Hydrolysis of these triglycerides can impart sensations of richness, creaminess, buttery flavor, and a variety of cheese aromas. This hydrolysis can be effected at relatively high rates by lipases. The resultant lipolyzed flavor bases have an expanding number of uses in confections, bakery goods, snack foods, condiments, salad dressings, and pet foods. Controlled hydrolysis of butterfat may thus provide a partial solution to the problem of surplus butterfat, provided that it becomes an economically viable technique on a commercial scale.

In the presence of water, lipases are able to catalyze the cleavage of the ester bonds of glycerides, thus releasing free fatty acids. Traditionally, the manufacture of lipolyzed products has consisted of a cumbersome and expensive multistep batch process. By contrast, we have focused on a continuous process for the production of a lipolyzed butteroil with the philosophy that the use of an immobilized lipase reactor should provide both better process economics and opportunities for better control of both the process and product quality. In this work an sn-1,3-specific lipase from *Aspergillus niger* was employed. Hence for these studies only two of the three ester bonds in the triglyceride are susceptible to attack. Table 1 contains a summary of the advantages of a continuous immobilized enzyme process relative to the traditional batch process.

Table 1

Advantages of the Continuous Immobilized Enzyme Process Relative to the Conventional Free Enzyme Batch Process

High productivity per unit of enzyme employed
Low costs for raw materials.
Operating temperature and pH of buffer may be manipulated to permit development of different flavor notes
Reductions in enzyme activity may be compensated for by changes in flow rate (space time)
No need for use of emulsifiers or vigorous stirring
No contamination of product with residual lipase
No need for downstream thermal treatment
Ease of regeneration of reactor with fresh lipase
Crude preparations may be used as source of lipase

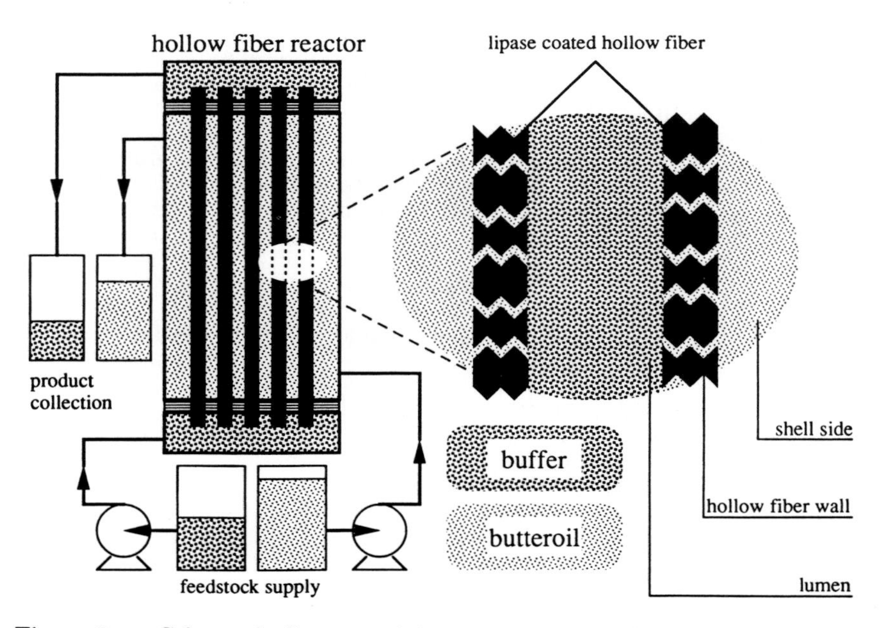

Figure 1: Schematic diagram of the bioreactor, with magnified view of the vicinity of the hollow fiber. Reproduced with permission from (4), © 1992, John Wiley & Sons.

BACKGROUND

Figure 1 is a schematic representation of the equipment used in the experimental component of the research. Details concerning the experimental protocols, analytical procedures, etc., are provided in the earlier papers in this series (2-7). The lipase was immobilized by adsorption on a bundle of polypropylene hollow fibers. These fibers were encased in a shell to form a reactor configuration resembling a shell and tube heat exchanger. McIlvane buffer was pumped through the lumens of these fibers, while melted butterfat was pumped cocurrently through the shell side. Nonlinear regression methods were used to fit both uniresponse data (3) (overall extent of hydrolysis) and multiresponse data (4) (release of ten different types of fatty acid residues) to nested rate expressions derived from a Ping-Pong Bi Bi enzymatic mechanism, coupled with three nested rate expressions for deactivation of the enzyme. Temperature (5,7) and pH (6,7) effects were also modeled using both multiresponse and uniresponse data.

Because the enzyme deactivates within a time frame which is comparable in magnitude to the space time of the reactor, it is essential to account for the kinetics of the deactivation process. In the case of the enzyme in question, it turns out that the best statistical fit of the data is provided by a model which assumes that two deactivation processes are occurring in parallel.

For purposes of optimization and economic analysis the time dependent, plug flow model of the performance of the reactor given in Table 2 is appropriate for consideration. The plot shown in Figure 2

Table 2

Equations for the Time Dependent Plug Flow Reactor Model

$$v_{oil}\frac{\partial[G]}{\partial z}+\frac{\partial[G]}{\partial t}=-a_t\{t\}r_{hyd}\{[G]\}$$

where

$$a_t\{t\}=\theta_1(\mathbf{exp}[-\theta_2 t])+(1-\theta_1)$$

and

$$r_{hyd}\{[G]\}=\frac{\theta_3[G]}{1-\theta_4[G]}$$

Initial condition:

$$\text{at } t=0 \text{ for } 0\le z\le L, [G]=[G]_0$$

Boundary condition

$$\text{at } z=0 \text{ for } t\ge 0, [G]=[G]_0$$

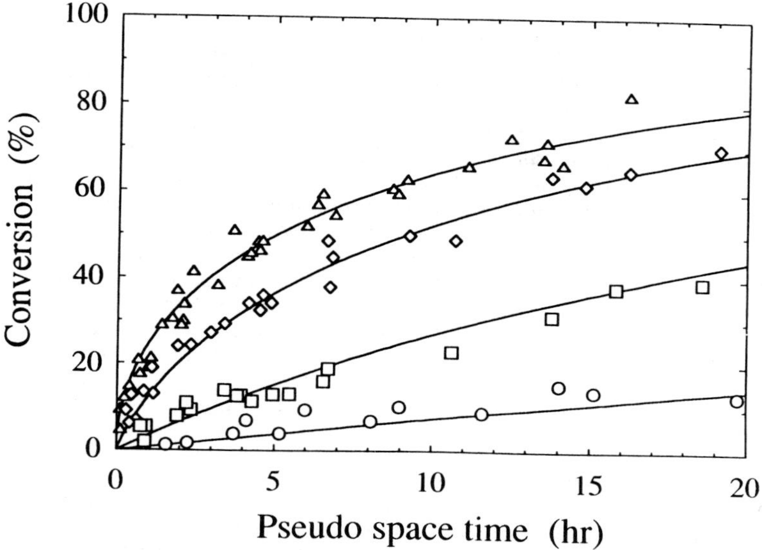

Figure 2: Percent conversion of glyceride bonds susceptible to lipase action versus the pseudo space time. The solid lines correspond to the best fit in each case, whereas the symbols represent the experimental data obtained at 40°C (Δ), 50°C (◊), 55°C (□), and 60°C (O). Reproduced with permission from (5), © 1992, John Wiley & Sons.

---

indicates the quality of the fit provided by this model. The pseudo space time employed as the abscissa in this plot is the space time (the ratio of the reactor volume on the shell side to the volumetric flow rate of butteroil) that would give the indicated conversion if no deactivation of the enzyme occurs. For the model of interest, the actual space time (τ), the pseudo space time (τ^*) and the time elapsed since immobilization of the enzyme are related by the following equation (3):

$$\tau^* = \frac{\theta_1}{\theta_2}\left[e^{-\theta_2(t-\tau)} - e^{-\theta_2 t}\right] + (1-\theta_1)\tau \qquad (1)$$

where θ_1 and θ_2 are rate constants characteristic of the process by which the enzyme loses activity. This process is characterized by two parallel first order terms, one of which leads to an inactive form of the enzyme, and the other of which leads to a form with significantly reduced activity.

DESCRIPTION OF THE PROCESS

The premise on which the process design and the optimization analysis is based is that the process for continuously producing hydrolyzed butterfat will be installed as an addition to an existing dairy plant. The heart of the process is a hollow fiber reactor of the type employed in the experimental work (3-7), but much larger in size. The lipase is immobilized by spontaneous physical adsorption on these fibers. The feed to the reactor is assumed to be the effluent oil stream from a centrifuge which produces butterfat and skim milk from whole milk. Periodic regeneration of the lipase activity in the reactor is accomplished by a protocol consisting of (i) removing the residual fat and desorbing the lipase with a mixture of hexane and ethanol, (ii) rinsing with water, (iii) rinsing with appropriate buffer, (iv) rinsing with the buffered lipase solution, and (v) finally rinsing with plain buffer.

OPTIMIZATION OF OPERATING CONDITIONS

There are several factors whose impacts must be assessed if one is to optimize both the performance of the reactor and the efficacy of the process. These factors include not only such operating variables as the temperature and pH, but also such process parameters as the length of time between initiation and termination of a run, i.e., the length of time a reactor of this type can be operated before it becomes necessary to remove deactivated lipase from the hollow fibers and recharge the system with fresh lipase. This issue is the primary focus of the present paper.

Selection of the appropriate pH at which to operate the hollow fiber membrane reactor is a simple task, given the availability of data reported elsewhere (6) which indicates that at 40°C there is a maximum in the rate at approximately pH 7.0, but that substantial activity is observed over a fairly broad pH range (from 4.0 to 7.5).

The question of selecting the optimum operating temperature involves not only the sensitivity of the hydrolysis and deactivation reactions to changes in temperature, but also the tendency of low-melting constituents of the butterfat to crystallize if the temperature of the oil phase is too low. This factor places a lower limit of ca. 30°C on the operating temperature. Conversion versus pseudo space time data reported elsewhere (5) indicate that for operation at pH 7.0, the temperature which maximizes the conversion at a given value of the pseudo space time is ca. 40°C.

Because of these results, efforts to develop an appropriate operating schedule were based on the use of kinetic parameters characteristic of operation at 40°C and pH 7.0.

In terms of the reactor space time based on the oil phase (τ) at the reactor outlet for clock times (t) greater than the space time (i.e., after the initial startup transient is over) the equation describing the variation of the total effluent concentration of free fatty acids [A] with time takes the form

$$\left[\mathbf{ln}\left[\frac{[G]_0}{[G]_0-[A]}\right]\right]-\theta_4[A]=\theta_3\left[\frac{\theta_1}{\theta_2}[\mathbf{exp}[-\theta_2 t]][[\mathbf{exp}[\theta_2\tau]]-1]+[1-\theta_1]\tau\right] \quad \text{for } t \geq \tau \qquad (2)$$

where $[G]_0$ represents the total concentration of ester bonds in the glycerides constituting the feed stream and the θ_i are lumped parameters which are related to the rate constants, Michaelis constants and inhibition constants for the hydrolysis reaction as well as the rate constants for deactivation of the enzyme. The variation of the effluent concentration of free fatty acids as a function of the time elapsed after immobilization of lipase, t, and the space time of the reactor, τ, is depicted as a three-dimensional plot in Figure 3.

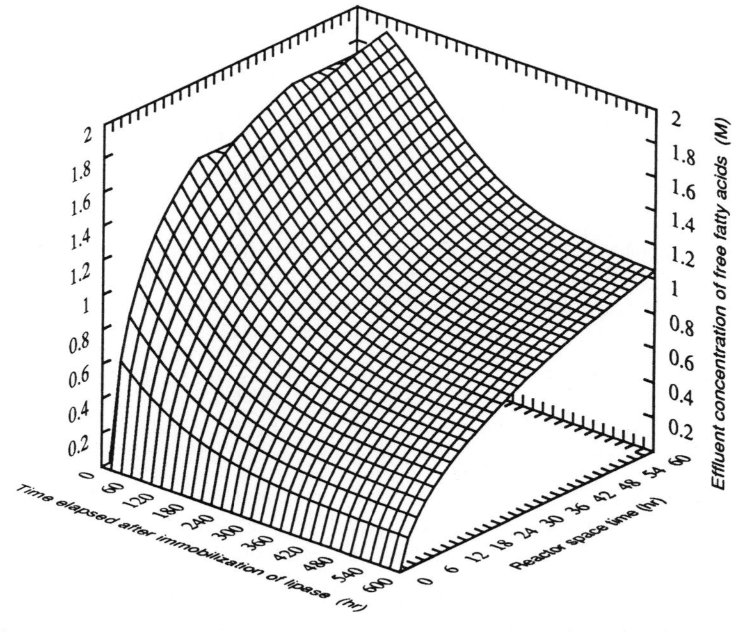

Figure 3: Variation of the effluent concentration of fatty acids, [A], as a function of time elapsed after immobilization of the lipase, t, and the reactor space time, τ.

This figure is based on numerical values corresponding to the use of an sn-1,3 lipase from *Aspergillus niger* at 40°C and pH 7.0. These operating conditions correspond to conditions which give an acceptable free fatty acid profile in the effluent oil stream. For a constant space time, the total concentration of free fatty acids in the effluent stream changes with time along an iso-τ line.

Two basic approaches to selecting an operating strategy for the reactor are to: (i) strive to obtain a constant level of hydrolysis (given by the desired final concentration of free fatty acids) or (ii) operate at a constant flow rate. In the second mode of operation, the composition of the effluent stream changes continuously. One may then proceed to obtain higher levels of hydrolysis for a portion of the feedstock which can then be mixed with unhydrolyzed feedstock to yield the desired level of free fatty acids. Although this second possibility permits one to operate at lower flow rates, it would require an extra mixing tank downstream for standardization of the product. The curvature of the three dimensional surface shown in Figure 3 is such that the average space time per unit of final lipolyzed product is minimized when all the fatty acids are produced by a reactor operating at the desired effluent conversion level. In general, it is not economic to mix material at different stages of conversion.

Changes in the activity of the immobilized lipase with time due to deactivation and isomerization processes prevent one from obtaining a constant extent of hydrolysis in the reactor unless other operating parameters are changed. Since changing the operating pH or temperature may lead to instability of the physically adsorbed lipase and since our analysis is based on operation at the temperature and pH which give the maximum rate (40°C, pH 7.0), shifts in these operating parameters can not be used to offset declines in activity. The remaining option is to decrease the flow rate at the concomitant expense of decreasing the productivity of the reactor. Earlier work has indicated that it is also undesirable to change the overall extent of hydrolysis because this will change the free fatty acid profile of the effluent oil stream. Such a change may present some disadvantages because it is known that each type of free fatty acid is responsible for different flavor key notes. Hence, the remainder of the analysis it is assumed that the reactor will be operated at a constant level of hydrolysis, i.e., that [A] is held constant.

After the startup transient has occurred (i.e., for $t \geq \tau$), the relation between the space time necessary to achieve a specified outlet concentration, $[A]_f$, and the elapsed time since immobilization can be determined from the following rearrangement of Equation 2:

$$t = \frac{1}{\theta_2}\ln\left[\frac{\theta_1\theta_3\left[\left[\exp[\theta_2\tau]\right]-1\right]}{\theta_2\left[\left[\ln\left[[G]_0/\left[[G]_0-[A]_f\right]\right]\right]-\theta_4[A]_f-\theta_3[1-\theta_1]\tau\right]}\right] \quad (3)$$

This equation can be employed to determine how the flow rate of oil (or the reactor space time) should be varied to maintain the desired effluent composition. However, it should be recognized that Equation 3 and its precursor are based on steady state operating conditions. Nonetheless, it is assumed that it can be used to determine how the feed flow rate should be varied as time elapses. The underlying rationale for this assumption is that pseudo steady state conditions prevail in the reactor because the necessary change in the flow rate during a time interval equal to one reactor space time is very small relative to the flow rate itself. Consequently, the rate at which conditions are changing is itself very, very small.

Equation 3 describes the intersection of the [A] - t - τ three-dimensional surface with horizontal planes of constant conversion corresponding to a specified value of the effluent concentration of fatty acids. The loci of $\tau\{t\}$ are represented in Figure 4 for four different extents of

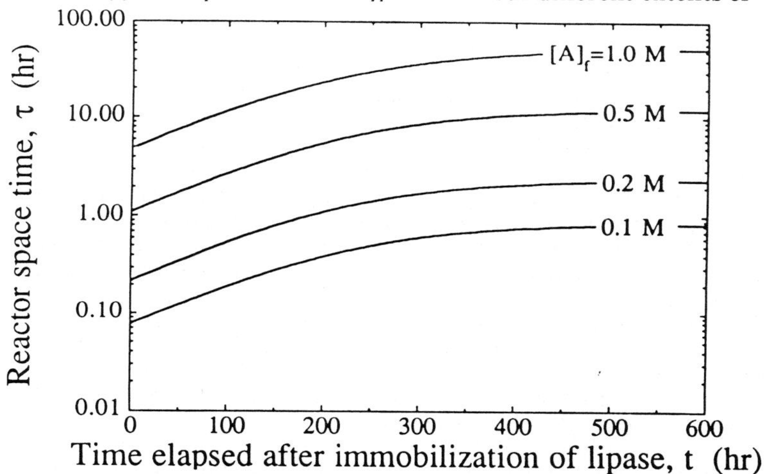

Figure 4: Required variation of reactor space time with time elapsed since immobilization for operation at a constant extent of hydrolysis.

hydrolysis. Inspection of this figure indicates that τ is a continuous function of t. For the range of variables investigated, curves representing plots of $\log\{\tau\}$ vs. t for fixed values of the specified effluent concentration are essentially parallel to one another.

Once the form of the τ vs. t relation has been determined for a fixed value of $[A]_f$, one is in a position to determine the time at which the operating reactor should be taken off line to permit cleaning and regeneration with fresh lipase. This time, t_{op}, is measured relative to the time at which the lipase was immobilized. A convenient criterion for determining this time is maximization of an objective function defined as the average overall rate of release of free fatty acids, $n_{tot}/(t_{op} + \Delta t_{reg})$, where n_{tot} is the total number of moles of free fatty acids released over the time interval $[t_{tr}, t_{op}]$, Δt_{reg} is the constant time required for cleaning up the reactor and regenerating the immobilized lipase activity, and t_{tr} is the time associated with the initial transient behavior of the reactor after startup. Hence, n_{tot} neglects the production of A during the initial transient for $0 \leq t \leq t_{tr}$. For a plug flow reactor t_{tr} can also be viewed as the residence time of the fluid element entering the reactor at t = 0. This optimization criterion is mathematically equivalent to the following relation

$$\frac{\partial}{\partial t_{op}}\left[\frac{n_{tot}}{t_{op} + \Delta t_{reg}}\right] \equiv \frac{\partial}{\partial t_{op}}\left[\frac{\int_{t_{tr}}^{t_{op}} Q[A]_f \, dt}{t_{op} + \Delta t_{reg}}\right] = 0 \qquad (4)$$

where Q is the volumetric flow rate of the butteroil through the reactor. It should be emphasized that t_{tr} and t_{op} are measured relative to a common origin, i.e., the time at which immobilization of the lipase is completed and the reactor has been filled with butteroil. Thus t_{op} is necessarily larger than t_{tr}. The quantity Δt_{reg} is a time interval (irrespective of the time origin selected). As noted above, $[A]_f$ is assumed to remain constant as time proceeds. Furthermore, it is assumed that Equation 3 is appropriate for use in calculating $\tau = \tau\{t\}$ so that the necessary schedule for the oil flow rate $Q = Q\{t\}$ can be employed. Consequently, for a fixed value of the reactor volume occupied by oil, Equation 4 can be written as

$$\frac{\partial}{\partial t_{op}}\left[\frac{\int_{t_{tr}}^{t_{op}} \frac{dt}{\tau\{t\}}}{t_{op} + \Delta t_{reg}}\right] = 0 \qquad (5)$$

Differentiation of this expression using Leibnitz's rule for differentiation of an integral gives

$$\frac{t_{op}+\Delta t_{reg}}{\tau\{t=t_{op}\}}=\int_{t_{tr}}^{t_{op}}\frac{dt}{\tau[t]} \qquad (6)$$

Use of Equation 3 to make a change of variables from t to τ gives

$$\frac{t_{op}+\Delta t_{reg}}{\tau_{op}}=\int_{\tau_{tr}}^{\tau_{op}}\frac{d\tau}{\tau(1-\mathbf{exp}[-\theta_2\tau])}+\cdots$$
$$\int_{\tau_{tr}}^{\tau_{op}}\frac{\theta_3[1-\theta_1]d\tau}{\theta_2\tau\left[\left[\mathbf{ln}\left[[G]_0/\left[[G]_0-[A]_f\right]\right]\right]-\theta_4[A]_f-\theta_3[1-\theta_1]\tau\right]} \qquad (7)$$

where $\tau_{op}\equiv\tau\{t=t_{op}\}$ and $\tau_{tr}\equiv\tau\{t=t_{tr}\}$ can be determined using Equation 3. Evaluation of the integral using a McLaurin series (terminated after the quadratic term) for the exponential term gives

$$\frac{t_{op}+\Delta t_{reg}}{\tau_{op}}=\frac{1}{\theta_2}\left[\frac{1}{\tau_{tr}}-\frac{1}{\tau_{op}}\right]+\frac{1}{2}\left[\left[\mathbf{ln}\left[\frac{\tau_{op}}{\tau_{tr}}\right]\right]+\left[\mathbf{ln}\left[\frac{2-\theta_2\tau_{tr}}{2-\theta_2\tau_{op}}\right]\right]\right]+\cdots$$
$$\cdots\frac{\theta_3[1-\theta_1]}{\Xi\left\{[A]_f\right\}\theta_2}\left[\left[\mathbf{ln}\left[\frac{\tau_{op}}{\tau_{tr}}\right]\right]+\left[\mathbf{ln}\left[\frac{\left[[A]_f\right]-\theta_3[1-\theta_1]\tau_{tr}}{\left[[A]_f\right]-\theta_3[1-\theta_1]\tau_{op}}\right]\right]\right] \qquad (8)$$

where $\Xi\left\{[A]_f\right\}$ is defined as

$$\Xi\left\{[A]_f\right\}\equiv\left[\mathbf{ln}\frac{[G]_0}{[G]_0-[A]_f}\right]-\theta_4[A]_f \qquad (9)$$

where t_{op} is given by

$$t_{op}=\frac{1}{\theta_2}\mathbf{ln}\left[\frac{\theta_1\theta_3\left[\left[\mathbf{exp}[\theta_2\tau_{op}]\right]-1\right]}{\theta_2\left[\left[\mathbf{ln}\left[[G]_0/\left[[G]_0-[A]_f\right]\right]\right]-\theta_4[A]_f-\theta_3[1-\theta_1]\tau_{op}\right]}\right] \qquad (10)$$

and where t_{tr} can be obtained from the relation

$$\left[\mathbf{ln}\frac{[G]_0}{[G]_0-[A]_f}\right]-\theta_4[A]_f=\frac{\theta_1\theta_3}{\theta_2}[1-\mathbf{exp}[-\theta_2\tau_{tr}]]+\theta_3[1-\theta_1]\tau_{tr} \tag{11}$$

The value of Δt_{reg} is assumed to be a constant. The functional dependence of t_{op} and t_{tr} on the variables under the control of the operator (i.e., t_{op} and $[A]_f$) is implicit in our equations if one recognizes that $\tau_{tr} = \tau_{tr}\{[A]_f\}$ and $t_{op} = t_{op}\{\tau_{op},[A]_f\}$. Solution of the above equations for t_{op} given $[A]_f$ and Δt_{reg} can be easily accomplished via an interval-halving method. Values of t_{op} vs. Δt_{reg} for different values of $[A]_f$ obtained in this way are plotted in Figure 5.

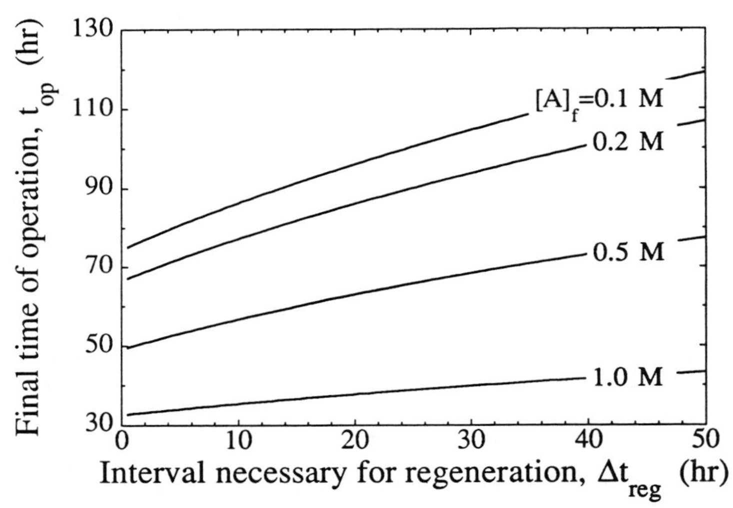

Figure 5. Final time of operation of the reactor after each regeneration, t_{op}, as a function of the time interval required for regeneration, Δt_{reg}.

The error associated with truncation of the McLaurin series after the quadratic term is less than the third order term, i.e, the maximum error is 3.5% for the range of commercial interest, i.e., for $[A]_f \leq 1.0$ M and $t \leq 600$ hr. One could, of course, reduce this error by resorting to more complex numerical integration procedures.

PROCESS ECONOMICS

It is of interest to assess the economics associated with a single production facility with a capacity of 500,000 lb/yr located downstream from a dairy plant. This production rate corresponds to less than 10% of the butterfat produced by a typical dairy plant (i.e., a plant able to produce 250,000 gal per day of skim milk containing 2% fat from an equivalent volume of whole milk containing an average of 3.6% fat). For the problem of greatest interest here, the regeneration time period is 12 hr, the optimum regeneration time, t_{op}, is ca. 79 hr and the final concentration of free fatty acids is 0.2 M. For these conditions the required reactor volume on the shell side is 18.0 dm^3. These values were selected on the basis of results obtained with our bench scale reactor. Under these circumstances, $\tau_{op} = 0.450$ hr. It is also assumed that the lipolysis reactor operates 300 days per year and 24 hr per day. The cost of the reactor would be $4,370. The capacity of the single reactor can also be expressed in terms of the requirement for the total exposed surface area of membrane via the ratio of the exposed surface area, A_R, to the void volume of reactor occupied by the oil phase, V_R. A typical value for this ratio is 49 cm^2/cm^3. Hence the required membrane area is ca. 88 m^2. Within any single operating cycle the space time of the reactor must increase from $\tau = 0.215$ hr to $\tau = 0.450$ hr between the time the regenerated reactor is first available for use and the time that the activity of the enzyme becomes insufficient for further use. Consequently, the associated rates at which butteroil should be fed to the reactor range from ca. 84 dm^3/hr (at time t_{tr}) to 40 dm^3/hr (at time t_{op}).

Experience in our laboratory indicates that for the level of conversion sought (i.e., $[A]_f = 0.2$ M) the ratio of the volumetric flow rate of buffer to that of melted butterfat can be as low as 0.025 without encountering problems due to significant accumulation of butyric acid in the aqueous phase. The total amount of marketable lipolyzed butterfat produced by a single reactor before regeneration would then be ca. 3,600 dm^3 (ca. 6,300 lb).

In preparing the cost estimate it is assumed that the process for producing hydrolyzed butterfat will be installed as an addition to an existing dairy plant, and that the optimum operating conditions determined above will be used for this process. The focus is on the difference in costs between a conventional batch process and the immobilized enzyme technology. The relevant cost elements are listed in Table 3 on a per pound basis.

As expected, the costs arising from addition of emulsifier, homogenization and stirring of the oil, and thermal processing of the lipolyzed butterfat are nil for the new process. On the other hand, the costs of addition of glycerol, and use of ethanol and hexane during the regeneration step when spent enzyme is removed and replaced with fresh lipase are nil for the traditional process. It is apparent from examination of Table 3 that the major contribution to the final cost of the lipolyzed butterfat is the cost of butterfat itself. The added costs associated with controlled lipolysis are 24.79 ¢/lb using the traditional batch process, but only 1.06 ¢/lb using the proposed continuous process. Hence, the incremental costs associated with the continuous method are more than 20 times smaller than those associated with use of the traditional counterpart. This dramatic difference arises primarily from the cost of the relatively large amount of lipase required by the traditional process.

Table 3

Relative Costs* of Raw Materials for the Immobilized Enzyme (IE) and Free Enzyme Processes

Raw Material	IE Process	Traditional Process
Butteroil	60.00	60.00
Lipase (APF-12)	0.30	23.48
Emulsifier	0.00	0.94
Buffer	0.58	0.33
Solvent	0.17	0.00
Total	61.05	84.75

* All cost entries are expressed in cents (USA) per pound of lipolyzed butteroil product.

The productivity of the immobilized enzyme system, i.e., the amount of lypolyzed butteroil produced per unit of enzyme consumed, is about 2 orders of magnitude greater than that of the soluble enzyme counterpart.

CONCLUSIONS

The reactor simulation and optimization study reported in this paper was employed as a basis for a preliminary assessment of the economic viability of a continuous process for the production of lipolyzed butteroil. The use of lipase in an immobilized form leads to a much greater productivity than the traditional batch process based on the use of the free enzyme. The associated economic and technical advantages lead one to the conclusion that the continuous process represents a very attractive alternative to conventional batch reactor technology which merits more detailed study.

NOMENCLATURE

a_t	activity of the immobilized lipase – time dependent
[A]	concentration of free fatty acids
$[A]_f$	effluent concentration of free fatty acids
[G]	concentration of glyceride bonds
$[G]_0$	concentration of glyceride bonds in the feed stream
n_{tot}	total number of moles of free fatty acids released
Q	volumetric flow rate of butteroil
r_{hyd}	rate of hydrolysis
t	time elapsed after immobilization of lipase
t_{op}	time when the next regeneration cycle is initiated
t_{tr}	time necessary for the fluid element entering at t = 0 to pass through the reactor; i.e., the time associated with the startup transient
v_{oil}	axial velocity of the oil phase
z	distance from the reactor inlet

Greek

Δt_{reg}	time necessary for the regeneration cycle
θ_1,θ_2	kinetic parameters associated with the enzyme deactivation process
θ_3,θ_4	kinetic parameters associated with the rate expression for the hydrolysis reaction
τ	reactor space time
τ_{op}	reactor space time corresponding to the time at which the next regeneration cycle is initiated
τ_{tr}	reactor space time corresponding to t_{tr}
τ^*	pseudo space time

ACKNOWLEDGMENTS

This investigation was supported in part by the College of Agriculture and Life Sciences and the Center for Dairy Research, University of Wisconsin-Madison, through funding provided by the National Dairy Promotion and Research Board. Financial support for F.X.M. was provided by JNICT (INVOTAN Fellowship, Portugal) and the Institute of Food Technologists (IFT Graduate Fellowship and John V. Luck/General Mills Fellowship).

LITERATURE CITATIONS

1. Malcata, F. X., Reyes, H. R., Garcia, H. S., Hill, C. G. and Amundson, C. H. Immobilized lipase reactors for modification of fats and oils - a review. *J. Am. Oil Chem. Soc.*, **67**, 890-910 (1990).

2. Malcata, F. X., Hill, C. G. and Amundson, C. H. Hydrolysis of butteroil by immobilized lipase using a hollow-fiber reactor: Part I. Adsorption studies. *Biotechnol. Bioeng.*, **39**, 647-657 (1992).

3. Malcata, F. X., Hill, C. G. and Amundson, C. H. Hydrolysis of butteroil by immobilized lipase using a hollow-fiber reactor: Part II. Uniresponse kinetic studies. *Biotechnol. Bioeng.*, **39**, 984-1001 (1992).

4. Malcata, F. X., Hill, C. G. and Amundson, C. H. Hydrolysis of butteroil by immobilized lipase using a hollow-fiber reactor: Part III. Multiresponse kinetic studies. *Biotechnol. Bioeng.*, **39**, 1002-1012 (1992).

5. Malcata, F. X., Hill, C. G. and Amundson, C. H. Hydrolysis of butteroil by immobilized lipase using a hollow-fiber reactor: Part IV. Effects of temperature. *Biotechnol. Bioeng.*, **39**, 1097-1111 (1992).

6. Malcata, F. X., Hill, C. G. and Amundson, C. H. Hydrolysis of butteroil by immobilized lipase using a hollow-fiber reactor: Part V. Effects of pH. *Biocatalysis* (accepted for publication).

7. Malcata, F. X., Hill, C. G. and Amundson, C. H. Hydrolysis of butteroil by immobilized lipase using a hollow-fiber reactor: Part VI. Multiresponse analyses of temperature and pH effects. *Biocatalysis* (submitted for publication).

RECYCLE REVERSE OSMOSIS SYSTEMS
SOME PERFORMANCE GUIDES

P T Cardew

North West Water Group PLC
PO Box 8, The Heath, Runcorn, Cheshire WA7 4QD, UK

SYNOPSIS

In order to achieve high concentration factors or high water utilisation with reverse osmosis systems, elements are usually configured in a tapered series of stages. However, many potential reverse osmosis applications such as that found in the food industry or in effluent treatment are too small to consider staging. Without staging low flow conditions will occur in the membrane elements and create high polarisation and fouling conditions. For those concerned with high water utilisation this problem can be overcome by partially recycling the reject stream to the feed pump. In this way high water utilisation can be achieved while maintaining a reasonable cross-flow through the element. In effluent applications recycle of the reject allows high concentration factors to be achieved, which is frequently a requirement for the next stage of processing. However, such benefits cannot be achieved without penalties in the economics and the performance of a reverse osmosis system. Simple equations to characterise the performance limits of a recycle system are derived, and can be used both as a guide to what is achievable, and as an aid to interpretation of measured data.

1 INTRODUCTION

For economical and sometimes environmental reasons there is a need to recover a large proportion of the feed water or to concentrate the feed in reverse osmosis applications. In large systems the separations unit will contain several stages, with each stage having less elements than the previous [1,2]. This is done to maintain the cross-flow across each element. If the cross-flow is allowed to fall this would increase polarisation, which reduces rejection and increases the propensity for fouling. This effect is compounded by the fact that the concentration increases through the reverse osmosis train, and in the end it is water chemistry that limits the total system recovery.

On small applications, which require high recovery the additional complexity of staging is not cost effective. Instead high recovery is achieved by recirculating some of the reject stream, usually to the feed side of the pump (Figure 1). This has the draw back of having a higher pump energy requirement, but the advantage that less elements and associated pressure vessels are required; hence a saving in capital at the expense of operating costs. Another disadvantage of the recycle system is that the element sees the harshest environment, in that the feed concentrations to the element will be much higher than that to the system.

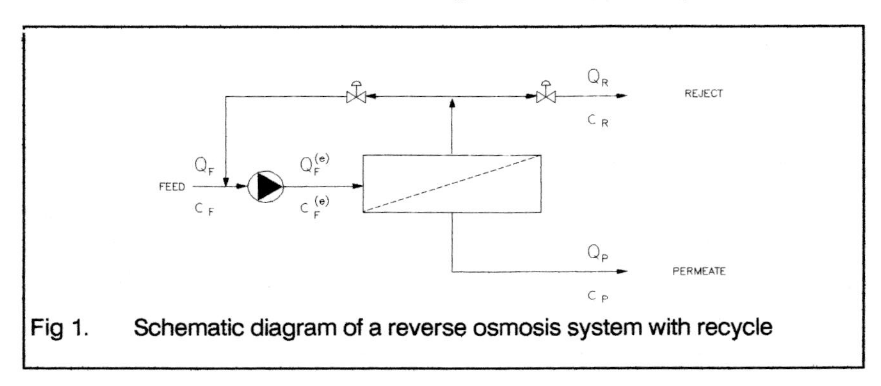

Fig 1. Schematic diagram of a reverse osmosis system with recycle

In small commercial operations monitoring is minimal and focused on system characterisation; rejection properties are frequently determined from the conductivity of the reject and permeate streams, and similarly the flow properties. The feed rate to the element is usually taken to be that determined by the pump performance curve.

The overall system performance is dictated by the element properties and the recirculation rate. In order to see how the system will perform a model is required to describe the relationship between the element and system performance. Such a model can also be used to deduce how well an element is performing from system performance data.

Three basic characteristics which are often used to describe the performance and operation of a reverse osmosis system are the concentration factor, the water recovery, and the rejection. These parameters can be applied at both the system and the element level, but it should be noted that these three factors are not independent (see section 3.2).

	System	Element
Concentration Factor	CF	CF_e
Recovery	Y	Y_e
Rejection	R	R_e

In the case of a single element with no recycle the two sets of terms are equivalent. When the system recovery is significantly higher than the element recovery then the system rejection can be substantially different to the element rejection. This is illustrated in Figure 2 for a system operating with an element recovery of 20 %, but at system recoveries varying from 20 to 80 %.

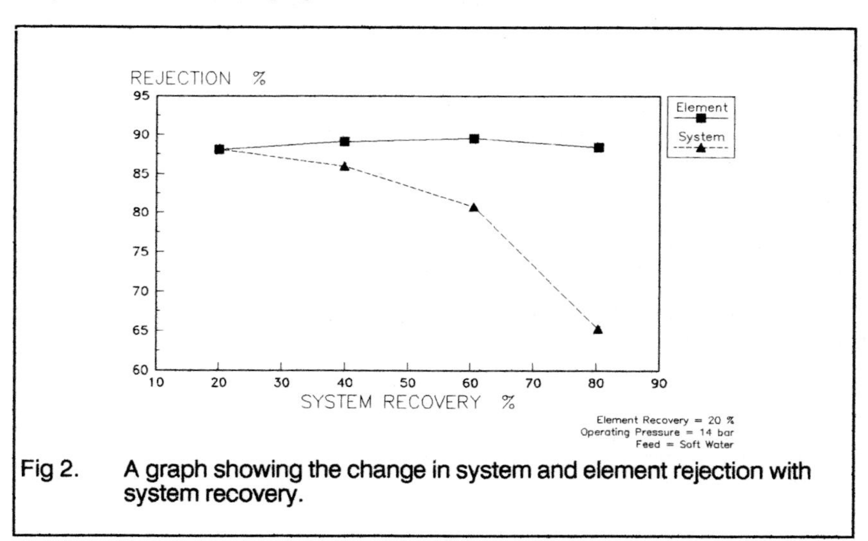

Fig 2. A graph showing the change in system and element rejection with system recovery.

The system rejection falls with increases in the level of recycle because of increases in the concentration of the feed to the element. Consequentially, if the element rejection is independent of concentration, the permeate concentration will rise, and thus a lower system rejection is recorded. To put numbers to this when the system recovery reaches 80 % recovery the concentration of the feed to the element is about 3 to 4 times that being fed to the system.

In recycle systems the system parameters are linked to the element parameters via the level of recycle which is characterised by the recycle ratio, Θ; the recycle ratio is defined as the ratio of water recycled to that rejected from the system.

$$\Theta \equiv \frac{\text{Internal Recycle Flow Rate}}{\text{Reject Flow Rate from System}}$$

The relationship between the system and element performances is obtained through writing mass balance equations for the water and the solute. From these mass balances some simple relationships can be established.

2 RECOVERY & ENERGY UTILISATION

From a mass balance on the water a simple relationship between the element and system recovery can be established involving the re-cycle ratio.

$$Y_e = \frac{Y}{1+(1-Y)\Theta}$$

This equation is readily inverted to give the converse relationship which shows how the system recovery changes with recycle ratio.

$$Y = Y_e\left(\frac{1+\Theta}{1+Y_e\Theta}\right)$$

Figure 3 shows how the system recovery varies with recycle ratios for a range of element recoveries

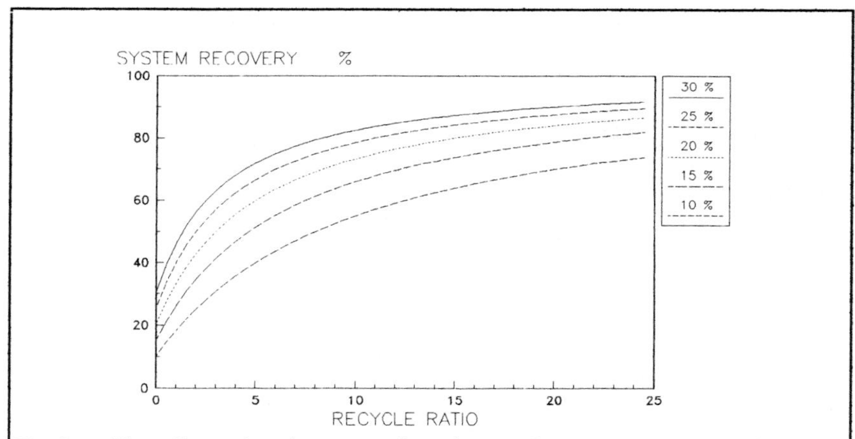

Fig 3. The effect of various recycle ratios on the system recovery for a set element recovery

Thus, to achieve high system recoveries for reasonable element recoveries, high recycle ratios need to be employed, and this brings an economic burden.

High recycle ratios lead to large pump energy costs. The proportionate increase in energy cost to that of no recycle, assuming pump efficiency is constant, is given by

$$\frac{E - E_o}{E_o} = \Theta * (1 - Y) = \left(\frac{Y - Y_e}{Y_e}\right)$$

Thus if Θ was 10 and Y = 0.5 the energy consumption will be five times as much as a once through system. Figure 3 indicates how the energy increase varies with recycle ratio.

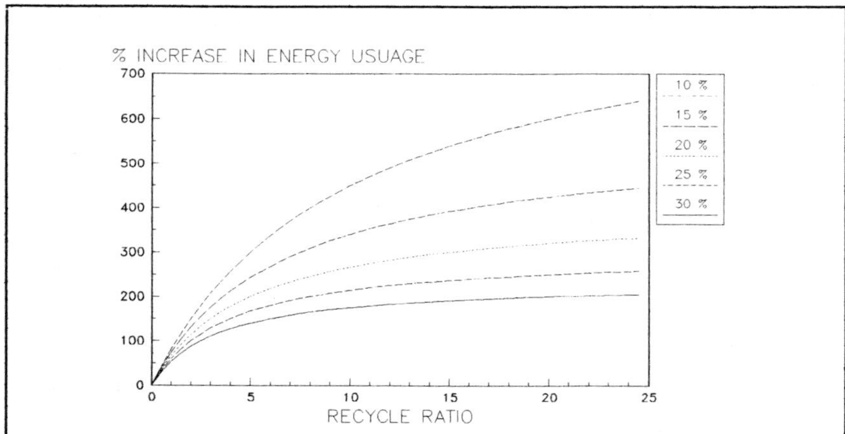

Fig 4. The effect of recycle ratio on the increase in energy achieve a system recovery of 50 % for a number of element recoveries ranging from 10 to 30 %.

As the recycle ratio increases the % increase in energy usage tends to an asymptotic value, which is

$$\lim_{\theta \to \infty}\left(\frac{E - E_o}{E_o}\right) = \frac{1}{Y_e} - 1$$

This indicates that if the element is operated at low recovery the energy increase can be large for example at 10 % element recovery the increase in usage could approach 10 fold while operating at 30 % the maximum increase in energy is 3 fold. Thus, to minimize energy one operates at as high an element recovery as is compatible with fouling and polarisation issues.

3 RELATIONSHIP BETWEEN ELEMENT AND SYSTEM PERFORMANCE

3.1 Introduction

Given a fixed operating pressure the performance of a recycle system can be changed in two basic ways.

Firstly, the amount of water being rejected from the system can be turned down forcing more liquid to recycle. One outcome of this is that the overall solute concentration being fed to the element increases, since the amount of fresh dilute solute being added will decrease. As a consequence the permeate concentration will rise unless the rejection happens to show a positive increase with concentration, which is unusual. Since the flow rate to the system has been reduced the system recovery has increased. The element recovery will not change significantly unless there are significant osmotic pressure effects altering the water permeation rate. This operation represents the limit of total system recovery (ie 1) at fixed element recovery. In practice there are limits to which this can be achieved due to fouling, precipitation or osmotic effects.

Secondly, keeping the recycle flow rate constant, the pump rate can be increased. Consequentially the reject concentration will fall, and hence the feed concentration to the element will fall to that of the feed. The higher pump rates reduce polarisation, lowers the concentration factor, and reduces the fouling potential. This operation represents the limit of zero element recovery at fixed system recovery. In practice there are limits to the level to which the pump rate can be increased due to pressure drop down the element.

The rejecting properties of an element can change slightly with element recovery due to concentration factor, polarisation, and any concentration dependence of rejection. Changes in system recovery may also have an effect by way of changing the feed concentrations, which can significantly effect the osmotic pressure of the feed, and also alter the membrane rejecting properties if concentration dependent. However, over the range of conditions usually encountered these effects are small and will not be considered here.

In the following sections the relationship between the system and element performance factors will be explored in order to assist design, and interpretation of data. Using mass balance equations a set of equations can be obtained which relate the system performance to that of the element.

3.2 Rejection

In this section two problems will be looked at. The first is a design question as to what the system will perform at given the element performance. The second question is an operational one which addresses the question as to whether an element is performing to expectations.

With regard to the first question a mass balance of the water and the solute gives the following relationship between the element rejection to that of the system rejection.

$$R = R_e\left(\frac{1 - Y}{1 - Y + (1 - R_e)(Y - Y_e)}\right)$$

This formula shows that the system rejection falls rapidly with increasing system recovery. This is illustrated in Figure 5 which shows a calculated rejection for the case shown in Figure 2.

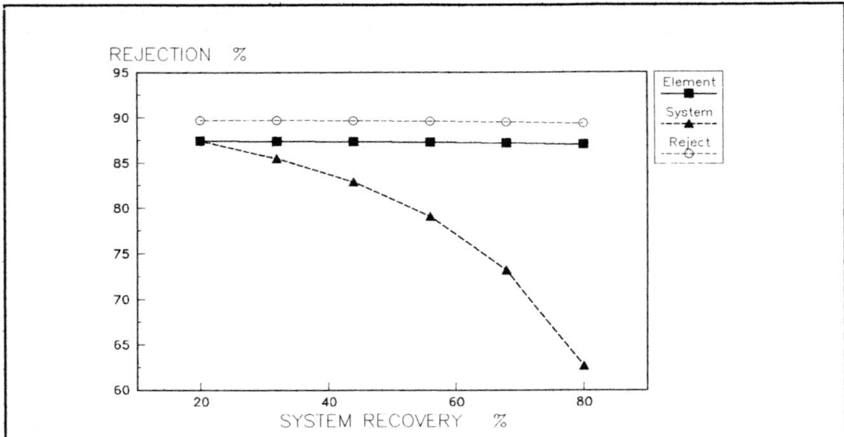

Fig 5. The effect of increasing system recovery on rejection for system operating at 20 % element recovery.

Comparison shows that the there is a close resemblance between the experiment and theory. Physically, the explanation for this behaviour is that as the system recovery is increased more reject is being returned to the feed. Consequently, the concentration of the feed to the element rises. If the element rejection is treated as a constant, then the permeate concentration will rise. Hence the system rejection will fall relative to that of the element

The above equation also shows that differences between the system and element rejection will be smaller the better the rejection. However, if the data is analysed from a transmittance point of view the relative effect is the same to first order in the transmittance.

In order to estimate the element rejection for a unit in operation the inverse of the above relationship is required.

$$R_e = R\left(\frac{1-Y_e}{(1-Y_e)-(1-R)(Y-Y_e)}\right)$$

Even if the element recovery is not known the above formula does provide an upper bound to the element rejection.

$$R < R_e < \frac{R}{1-(1-R)Y}$$

As already mentioned commercial systems frequently only monitor the reject and the permeate, and thus rejections are frequently quoted in terms of the reject concentration.

$$R_r \equiv 1 - \frac{c_P}{c_R}$$

This rejection overestimates the element rejection, as can be seen in figure 5. However, from a mass balance a simple relationship can be established between the reject based rejection and system rejection

$$R_r = R\left(\frac{1}{1-(1-R)Y}\right)$$

Comparing this formula with the bounds set on the element rejection it can be seen that the reject rejection is, not surprisingly, the upper bound for the element rejection. From figure 5 it can be seen that the reject based rejection, unlike the system rejection, closely parallels the element rejection over a wide range of system recoveries. To obtain the system rejection from the reject based rejection the above formula needs to be inverted.

$$R = R_r\left(\frac{1-Y}{1-R_r Y}\right)$$

3.3 Concentration Factor

A mass balance on the system provides a simple relationship between the concentration factor, recovery and rejection

$$CF = 1 + R\left(\frac{Y}{1-Y}\right)$$

This relationship provides a simple upper bound on the concentration factor that can be achieved:

$$CF < \frac{1}{1-Y}$$

with equality being achieved only in the ideal case of a membrane with total rejection of the solute.

A similar relationship to that given above exists at the element level. In the case of no recycle it can be established [3] that

$$CF_e = \left(\frac{1}{1 - Y_e}\right)^r$$

where r is the membrane rejection. This formula accounts for the concentration rise within the element. The basic assumption that underlies it is that the rejection is a constant over the range of concentrations of interest.

Both the above formulae indicate that in the limit of total recovery the concentration factor becomes infinite. This physically impossible limit follows from the assumption that the rejection is independent of concentration. For systems with recycle a tighter bound on the concentration factor that can be achieved indeed it will be shown that in the limit of total recovery a finite value is obtained for the concentration factor.

Using the relationship between system and element rejections given in section 3.2 in the above formula for concentration factor gives

$$CF = \frac{1 - Y_e(1 - R_e)}{1 - Y + (1 - R_e)(Y - Y_e)}.$$

The difference between the value predicted by this equation and the earlier formulas is illustrated in Figure 6.

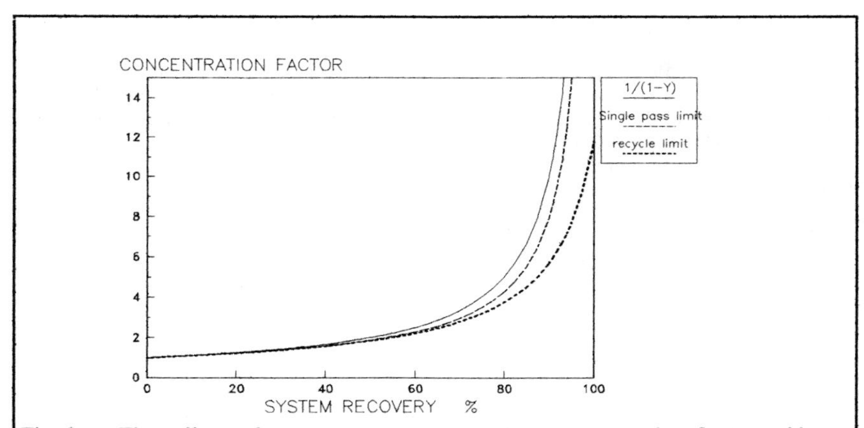

Fig 6. The effect of system recovery on the concentration factor with an element recovery of 25 % and an element rejection of 90 %.

As expected the system concentration factor only increases substantially at large system recoveries, and under these conditions there are large deviations from the simple limits established above. It is clear though that to achieve the highest concentration factor then the element should be operated at the highest possible element recovery compatible with fouling, and polarisation. One notable feature of the above is that as the system recovery approaches unity the concentration factor enhancement approaches a finite value. While in practice other things such as the osmotic pressure effects will reduce this limit further it does provide a more useful upper bound on the concentration factor that could be achieved. Indeed by taking the limit of total system recovery an upper bound can be established in terms of the element rejection properties and the element recovery.

$$\lim_{Y \to 1} CF = \frac{1}{(1 - Y_e)(1 - R_e)}(1 - Y_e(1 - R_e))$$

which for high rejecting elements approximates to

$$\lim_{Y \to 1} CF \cong \left(\frac{1}{1 - Y_e}\right)\left(\frac{1}{1 - R_e}\right)$$

This equation shows, as might be expected, that the higher the element rejection the higher the concentration factor that can be achieved.

3.4 Standing Concentration

One important internal factor in a recycle system is the actual solution concentration being fed to the element. Increasing the recycle ratio increases the amount of high concentrated reject that is returned and hence raises the concentration of the feed being delivered to the element. This has a number of consequences. In particular, it might significantly increase the osmotic pressure of the feed in comparison to the applied pressure, which would result in lower rejection and lower flow.

In terms of an operating system the actual concentration being fed to an element can be calculated using the relation

$$\frac{c_F^{st}}{c_F} = \frac{CF}{CF_e} = 1 + R\left(\frac{Y - Y_e}{1 - Y}\right)$$

which can be derived from the mass balance on the system. Even if the element recovery is not known a simple upper bound can be obtained.

$$c_F < c_F^{st} < c_F\left(1 + R\left(\frac{Y}{1 - Y}\right)\right)$$

In terms of design the first equation given in section 3.2 for the system concentration factor can be used to give a simple expression which can be used to estimate the system performance.

3.5 ILLUSTRATIONS

3.5.1 Example 1

The following example illustrates the use of some of the above equations in considering the feasibility and requirements of a particular application. The problem is essentially the treatment of a dilute soft water with a total dissolved solids (TDS) of 500 ppm at low pressures of about 100 psi. For water conservation reasons they would like to operate at 75 % recovery and achieve a 95 % system rejection, while operating the element at around 30 % recovery.

From section 3.4 it follows that the feed concentration will in this situation rise to about 1500 ppm. From section 3.3 the concentration factor would be 3.85, which would give a reject concentration of about 1900 ppm. The implications of this is that the osmotic pressure has increased from about 5 psi to about 17 psi, which would represent at least a 10 % decrease in flow. Turning to the rejection requirements the formula in section 3.3 can be used to calculate that a membrane with at least a rejection of greater than 98.2 % will be required. Further refinements [3] would indicate that a membrane with a rejection of at least 98.4 % is required

3.5.2 Example 2

This second example involves the interpretation of results obtained from commercial recycle system treating municipal water. The rejection (reject based) is reported at 85 % when operating at 80 % system recovery, while the element recovery is estimated to lie between 20 and 30 %.

From the formulae in section 3.2 the system rejection can be shown to be 95 %. With regard the element rejection it must have a value between 95 and 99 %. The exact value depending on the element recovery. From the formula given in section 3.2 figure 7 has been constructed showing the dependency of the element rejection on the element recovery

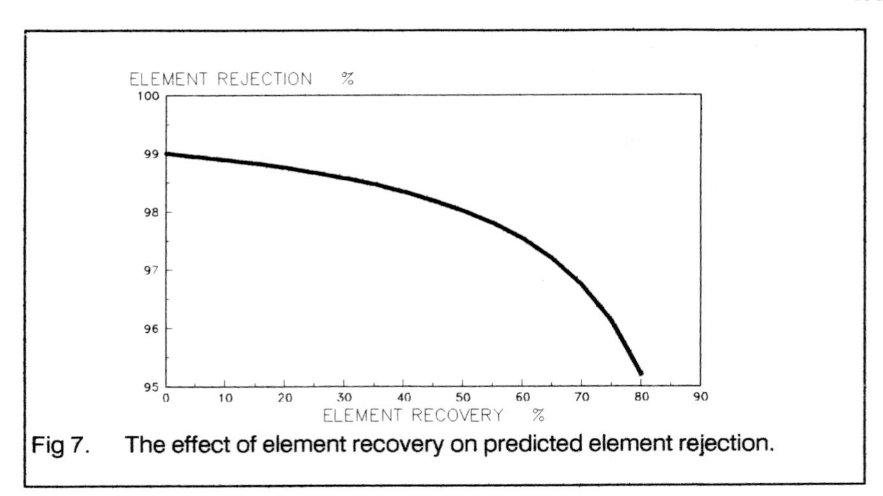

Fig 7. The effect of element recovery on predicted element rejection.

As can be seen the dependency of the element rejection on element recovery is weak in the range of interest, and would indicate that the element rejection is about 98.5 %.

4 SUMMARY

Reverse osmosis systems with recycle of the reject are commonly used to conserve water on small systems. This system design incurs both an energy and a performance penalty, when compared to a staged system. Simple relationships are given between the system performance and the element performance. In particular it is shown that if maximising concentration factor is the goal then a simple formula can be derived to estimate the maximum possible concentration factor that could be achieved. Not surprisingly this formula indicates that the higher the element rejection the higher the limiting concentration factor. Finally, it has been shown how these equations can be used for design and interpretation of data.

Definitions

c_F Feed concentration to system

c_F^{st} Feed concentration to element

c_P Permeate concentration

c_R Reject concentration

CF System Concentration Factor, ratio of concentration in reject to that fed to system

CF_e Element Concentration Factor, ratio of concentration in reject to that fed to element

Q_F Feed flow rate to system
Q_F^e Feed flow rate to element
Q_P Permeate flow rate
Q_R Reject flow rate

R System Rejection, defined with reference to concentration being fed to the system
R_e Element Rejection, defined with reference to concentration being fed to the element
R_R Reject Rejection, defined with reference to concentration of the reject stream

Y System Recovery, ratio of water recovered in permeate to that fed to system
Y_e Element Recovery, ratio of water recovered in permeate to that fed to element

Θ Recycle ratio, flow ratio of water being recycled to that being rejected from the system

References

1. R Rautenbach and R Albrecht "Membrane Processes" Publ: J Wiley & Sons, 1989, ISBN 0-471-91110-0

2. M Mulder "Basic Principles of Membrane Technology" Publ: Kluwer Academic, 1991, ISBN 0-7923-0979-0

3. C W Saltonstall and R W Lawrence, Desalination, 48 (1983) 25

ATTAINING AN EFFECTIVE MEMBRANE PROCESS

Alan Merry

PCI Membrane Systems Ltd

INTRODUCTION

An effective membrane system could be defined as one that solves problems, not create them. In doing this it should perform the duty that it was specified to do. In order to achieve this obvious target it is necessary to pay careful attention to the test work upon which the design will be based, the design itself, and the installation of the plant. There follows a discussion on the design of test programmes, and the common mistakes that can be made during them. Also discussed are some important points in the areas of design, installation, and plant operation.

DESIGN REQUIREMENTS

Before any trial programme can be designed, it is important to have a clear objective in mind. To assist with this is useful to consider the criteria by which the performance of a plant is judged, what can cause a plant to under perform, and what options there are for correction of mistakes. These criteria are:

Capacity
Composition of the retentate and permeate
Operating costs

Capacity

The capacity of a plant may be stated in terms of feed, permeate or retentate flows, depending on which is considered the most important. As all three are functions of each other, they can be affected by fouling.

Reversible fouling can be removed by cleaning, and as such is not normally a problem, unless it occurs very rapidly.

Irreversible fouling is an obvious problem. Therefore in order to determine the correct membrane area for a given capacity it is necessary to obtain data on the flux as a function of the concentration, to confirm the irreversibility of the fouling, and determine the effectiveness of the cleaning cycle.

In the event of an installation that is under capacity it is usually possible to correct the problem provided the fouling is reversible. Solutions include: to fit additional membrane area, to increase the frequency of the cleaning, or to change the cleaning regime used. In some cases, the problem can be corrected by changes to the treatment of the fluid upstream of the membranes plant.

Stream Composition

The normal criteria here is the concentration of certain species in either the permeate or the retentate. This too can be influenced by fouling, which may, due to loss of flux, result in the desired retentate concentration not being reached.

If a installation fails to match the specification, it is difficult to correct the situation. Sometimes it is possible to correct the situation by changing the operating conditions such as pressure or pH, but often a change of membrane is required.

Operating Costs

The two largest components of this are the power consumption, and the cost of replacing membranes. The estimation of power consumption is simple given the appropriate rheological data. If, however, significant additional membrane area is added to the plant, the power consumption will increase.

The cost of membrane replacement will be a function of the life of the membrane, which will be influenced by the cleaning regime employed, and the chemical and physical characteristics of the process fluid. Thus moving to an aggressive clean to maintain capacity may result in shorter membrane life.

TRIALS

Although some membrane applications are now considered to be mature and therefore do not require test work, many are new and it is necessary to conduct tests. An initial step might be to carry out a simple short test to determine the probable suitability of a membrane system for the duty. Assuming that it does look suitable, now is the time to determine what the final needs are:

Is a continuous process required, or would a batch process be more appropriate?

How long can the plant be taken off stream for cleaning, and how often?

What are the key criteria to judging success?

The simplest situation that can occur is when the pilot plant is the same configuration as that anticipated for the full scale plant, in which case the trials should be arranged to operate over similar time cycles as the final plant. Providing the operating conditions, and the concentration ranges are the same, then it is only necessary to scale up the membrane area in order to produce the process design.

Often, however, the pilot plant is a batch, or a two stage continuous plant, when the anticipated design is a three or four stage plant. In this case careful consideration has to be given to designing the trials, where it is important to consider the information sought, and the variables.

Information

The information required is generally:

- Flux as a function of concentration.
- Retention as a function of concentration.
- Degree of fouling.
- Effectiveness of cleaning.
- Probable membrane life.
- Quality of product.
- Pressure drops as a function of concentration.

Variables

These include:

- Membrane type.
- Module geometry
- Pressure.
- Temperature.
- Cross flow velocity.
- pH.
- Upstream treatment of process fluid.

A matrix of the above variables with information required yields a daunting number of tests. It is therefore normal to fix a number of the variable. For example, knowledge of other applications may suggest that the cross flow velocity should be fixed.

Other variables such as the module geometry may be dictated by the characteristics of the feed. For example, a feed carrying suspended material would normally exclude the use of a spiral wound element.

The programme of work can now be split into three basic steps:

1) Identify the conditions that are likely to be close to the optimum. This can usually be done in a batch trial under total recycle, with a scan of the variables such as pressure, temperature, and possibly pH. With temperature and pressure it is normal to start at the lower values, working up to the higher values, and then decreasing them again. A example curve, taken from PCI's data files, is shown in Figure 1.

2) During this phase, the bulk of the data that will be used for design will be collected. The objective is to collect flux and retention data over a range of concentrations. The trials are either batch runs, or continuous plant run at different concentrations. Each run should be at least five hours in duration, with some runs lasting longer than this if possible. Runs of less than three hours duration do not normally yield data suitable for design purposes.

   The usual problem here is in obtaining the retention data, as this involves analysis which is often expensive and time consuming. However, if the specification for the final installation includes stream compositions, it is vital that adequate data is collected.

It is important to clean the membranes effectively, which means that the water flux of the membranes must be measured after each clean. The water flux of the new membranes should also be measured. It is not necessary to clean the membranes back to the virgin state water flux, but it is important that the water flux stabilises after five or six runs, possibly at 70% to 80% of the initial water flux.

3) The objective of this phase is to confirm the data obtained in the previous stages, and to give confidence in the design. A number of runs should be carried out at conditions that are representative of the final design, including the duration of the runs.

Common Pitfalls

There are a number of common pitfalls and misunderstandings that can occur during trials. Some of them may be summarised as:

1) Insufficient foulant present during trials: Fouling of membranes is a complex issue, but there some basic forms

   a) Fouling caused by exceeding the solubility limit of a species: This usually shows up as a loss of flux in the last stages of a multi-stage plant, or towards the end of a batch cycle. Batch systems sometimes show a greater tolerance to this situation as the material can become stabilized during the high retention time, and the slow increase in concentration in the system. This can be a source of problems when scaling up from batch trials to a multi-stage plant.

      This type of fouling can often be controlled by adjustment of pH or temperature, or by limiting the target concentration factor.

   b) Fouling caused by a strong adsorbent: This is usually obvious as flux falls rapidly throughout the plant within a short space of time of introducing the process fluid into the plant.

      Although changing the pH may control this type of fouling, it is more probable that changing to a different membrane type will be necessary.

c) Fouling caused by weak adsorption or sedimentation: This type of fouling is indicated by loss of flux in the first stage of a multi-stage plant. Small batch tests often mask this type of fouling as it appears that fresh supplies of the foulant need to be introduced into the plant in order for the phenomenon to be observed.

The important concept is: Ensure that the ratio of volume of feed to membrane area used, taken between cleaning cycles, in the trials is similar to that expected on the full scale.

2) Failure to record water fluxes, or cleaning details.

3) Failing to record one key piece of data, normally a concentration, which renders the rest of the data useless.

4) Insufficient retention data: There are two common occurrences here: one is to just measure the permeate concentrations, and to ignore the retentate, which may mean that it is impossible to predict the composition of the permeate in the final installation if the plant configuration differs from the pilot. The other is to analyse only the final retentate and permeate, and assume that the same ratio of concentrations will be found in the final plant. Again, if there is a difference in plant configuration, the assumption is invalid.

5) Non representative feed: This may be due to aging of samples, poor agitation, shear damage, inappropriate sampling point, or contamination from a vessel or pipe.

6) Overlooking viscosity effects: This can occur if all the pilot work is done on a plant with only one channel through the system, when a full scale installation would normally be multichannel.

The pressure drop along a membrane is influenced by the permeation rate, so it is possible to have two parallel channels each with different fluxes, and thus different pressure drops. However, the channels would be connected together by a manifold, so the pressure drop would be the same for each channel. Normally the system would react by increasing the flow in the channel with the higher flux, and decreasing the flow in the other, providing the viscosity did not change significantly.

In the case of a significant change in viscosity, particularly in the case of shear thinning fluids, the flow along the higher flux channel will decrease. This leads to a further increase in the concentration, and hence viscosity. This cycle carries on until the channel becomes completely blocked.

PROCESS DESIGN

The designer now has his raw material: the data from the trials. They now have to calculate the membrane area required, how it is to be deployed in the plant,(i.e how many stages), and what the performance will be.

To do this the raw data must be processed into some form of scale up correlation, assuming that the pilot plant configuration is different from the projected plant. The two most useful are a correlation for the flux as a function of a characteristic concentration, and a correlation of the retention with concentration. The designer must then adjust these correlations by adding a safety factor. This is a judgement based on the number of trials, and the severity of the operating conditions. The design flux may therefore be between 50% and 90% of the correlation. As an example of the need for this, Figure 2 shows an index of the correlation of flux with concentration for reverse osmosis of tomato juice. The correlation for the test work was obtained on two sites in trials totaling over seventy hours. A two stage continuous plant was used. The other correlation was drawn from data obtained from full scale plant operating on three sites for forty to sixty days.

To produce the design is then a simple case of selecting the membrane area to give the required mass balance.

MECHANICAL DESIGN

The mechanical design is normally straight forward unless there is a special requirement, such as to avoid excessive shear on the product (whole milk, and egg are two examples), or particular engineering standards apply as may be the case in the chemical industry.

INSTALLATION

It is necessary to pay attention to the relative heights of tanks and pipework, and to ensure that the permeate pipework presents a minimal pressure drop at all times. This particularly applies to tubular systems where implosion of the membrane will occur if the permeate pressure exceeds the retentate pressure. This can occur if gas becomes compressed in the permeate pipework, if critical pumps are started, or stopped suddenly, or retentate pipes discharge below the level of the plant.

These problems can be overcome by ensuring a free exit for gas, starting and stopping centrifugal pumps against closed valves, and placing suitable antisyphon devices in the feed and retentate piping.

OPERATION

The keys to successful operation are attention to the cleaning cycle, ensuring that the feed does not change beyond the limits examined in the trials, and avoiding exceeding the specified performance.

CONCLUSIONS

In order to attain an effective membrane process it is of prime importance to conduct effective trials, and to have a clear objective. The application of realistic safety factors to data, and attention to detail during installation, and operation are also important.

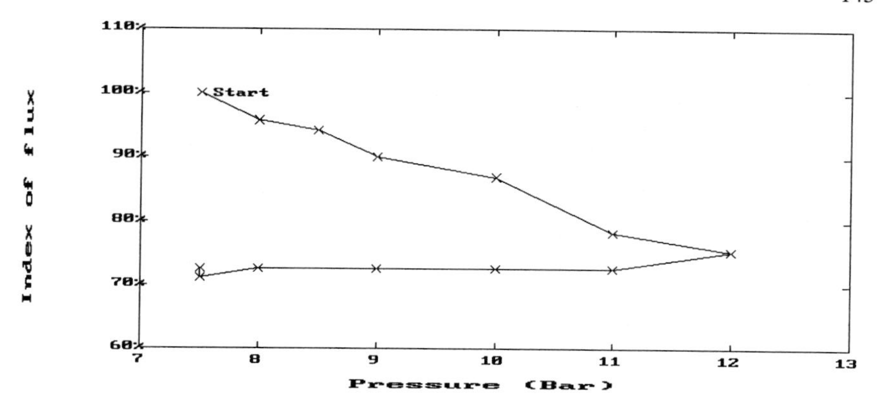

Figure 1. Example of the effect of pressure on flux, illustrating the need to run pressure scans from low pressure to high pressure. Data taken from UF trials on overflow from an anerobic digester operating on starch waste.

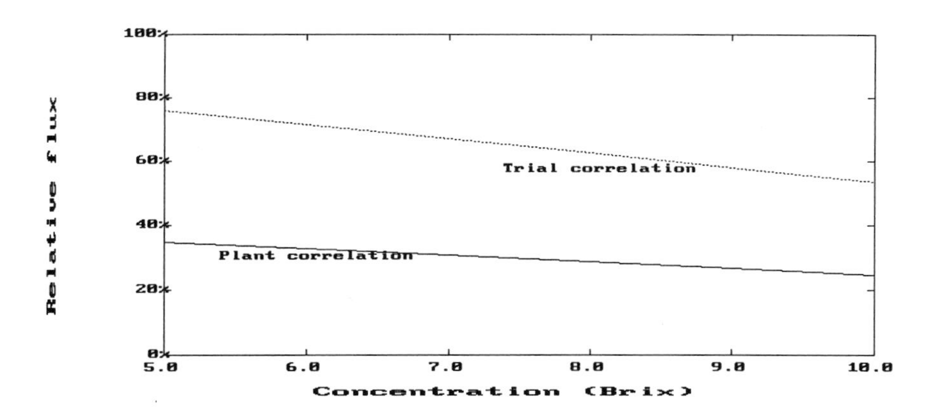

Figure 2. Comparison of flux/concentration correlations for trials and operating plant, illustrating the need to apply safety factors to trial data. Data taken from PCI files on tomato juice concentration by Reverse Osmosis.

WATER AND EFFLUENT

APPLICATIONS OF ION EXCHANGE MEMBRANES, CURRENT STATE OF THE TECHNOLOGY AFTER 45 YEARS

Wayne A. McRae, Consultant
Ackerstrasse 10
CH-8708 Männedorf

ABSTRACT

Ion Exchange Membranes ("IEM") were discovered by the author in 1948 at Harvard University. In July '48 he built (still at Harvard) a laboratory membrane chloralkali cell based on such membranes. The cell established proof of principle but the phenol sulfonic acid-formaldehyde membrane available was not stable. In 1949 the first laboratory electrodialysis ("ED") cell was built which used IEM.

In 1948 Ionics, Inc. was formed, inter alia to continue development of IEM and their applications. The first commercial ED plant was built by Ionics in 1954 and installed by that company in Saudi Arabia. The first commercial membrane chloralkali plants were installed by that company in 1972 and '73.

At present the principal applications of IEM are:

- desalting of brackish or potable water by ED. It is estimated that there are more than 5000 such plants worldwide, having a combined capacity of more than 1.5 million cu. meters per day of desalted water and more than 3 million sq. meters of membrane area. Included in this application is the partial demineralization of potable water before mixed bed ion exchanger ("IX") deionizers for pure and ultrapure water applications, e.g. electronic applications, feed to high pressure boilers. ED is indicated for the removal and concentration of nitrate and nitrite from potable water;

- concentration of seawater by ED for salt production. It is estimated that about 1.6 million metric tonnes of salt per year are produced by this method, representing about 0.6 million sq. meters IEM. A related application is the simultaneous concentration of industrial effluents and recycle of recovered water, e.g. concentration of blow-down from evaporative cooling towers in power plants and processing of metal treating wastes;

- electrolysis of sodium chloride brine to chlorine, caustic and hydrogen. It is estimated that more than 6 million metric tonnes of caustic (and a similar amount of chlorine) are produced annually by this method, more than 15 percent of world chlorine and caustic capacity and about 200,000 sq. meters of perfluoro IEM valued at more than $ 100 million. Within a decade or so the annual replacement market for chloralkali membranes should be approximately $ 250 million;

- separators in some alkaline batteries. It is estimated that about 0.75 million sq. meters of IEM are produced annually for this purpose.

Important but much smaller applications of IEM include the demineralization of cheese whey (e.g. for infant food), about 150,000 metric tonnes of 90% demineralized dry whey solids per year and the electrohydrodimerization of acrylonitrile to adiponitrile, about 67,000 metric tonnes of adiponitrile annually.

The author reviews important aspects of the above processes and apparatus and IEM for them.

1. A Brief and Selective History of Ion-Selective and Ion Exchange Membranes:

1.1 Ion-Selective Membranes Before 1948: Workers in this period included L. Michaelis (principally 1925 to 1932), T. Teorell (beginning in 1935), K.H. Meyer (principally 1936 to 1946), K. Sollner (beginning in 1930). Cation selective membranes were generally based on cellulose nitrate or cellulose films oxidized for example with sodium hypochlorite. Anion selective membranes were prepared by absorbing the natural, basic (pI 12), fish protein protamine onto cellulose nitrate or oxidized cellulose nitrate films. As early as 1937 Meyer had prepared anion selective membranes by condensing phthalic anhydride and triethanol amine and cation selective membranes by condensing phthalic anhydride and glycerin, in both cases on a supporting matrix. In 1940 Meyer prepared cation selective membranes by soaking cellulose film in acid dyes and anion selective membranes by treating with methyl iodide artificial sausage casing made commercially from animal hides.

All of the above membranes suffered from poor ion-selectivity, poor ionic conductivity and/or instability, sometimes all three such factors at once.

1.2 Ion-Exchange Membranes After 1947: The first membranes which were simultaneously highly ion-selective, highly ion conductive and stable were fabricated from phenol sulfonic acid and formaldehyde by the author at Harvard University in 1948. The unexpected and remarkable electrochemical properties immediately suggested the principal applications of IEM today: electrodialysis, chloralkali electrolysis and battery separators. At present more than 1 million sq. meters per year of IEM are made world-wide including about 0.75 million sq. meters for alkaline secondary batteries. More than 3.0 million sq. meters of hydrocarbon based membranes are installed in ED plants and about 200,000 sq.

meters of fluorocarbon based membranes in electrolytic chloralkali cells. Today, in contrast to 1948, it is a matter of common sense that low molecular weight organic acids and bases which are soluble and strongly ionized in water will also be strongly ionized when incorporated in water swollen but water insoluble polymers and that such strong ionization will of necessity give the polymers high intrinsic ion conductivity and high ion-selectivity. Once the attractive commercial properties of IEM were recognized many alternative methods of fabrication were soon developed. For example Ionics, Inc. (which was founded in 1948 inter alia to continue to develop the discoveries made at Harvard) soon replaced membranes made from compounds condensed with formaldehyde with membranes based on vinyl and related monomers e.g. copolymers of styrene and divinyl benzene containing sulfonate or quaternary ammonium as active bound groups, in all cases reinforced with fabric. (Ionics has installed ED equipment probably originally furnished with more than 1.2 million sq. meters of original membrane). Asahi Chemical Industry Co., Ltd. makes similar membranes. (ED equipment has been installed probably originally furnished with more than 0.4 million sq. meters of Asahi Chemical ED membranes of the above type). Based on a survey of recent patents and of published literature Ionics seems to be replacing some of its membranes based on yinyl aromatic monomers with membranes largely based on aliphatic vinyl monomers.

Asahi Glass Co., Ltd., of Japan, (not related to Asahi Chemical Industry) makes fabric reinforced IX membranes based on styrene-butadiene copolymers crosslinked for example by sulfuric acid or titanium tetrachloride (the latter dissolved in ether) and subsequently sulfonated to CX membranes or chloromethylated and aminated to AX membranes. (ED equipment has been installed probably originally furnished with more than 0.2 million sq. meters of Asahi Glass membranes of the above type).

Tokuyama Soda Co., Ltd., of Japan, fabricates IX membranes by coating a paste of e.g. polyvinyl chloride, dimethyl phthalate, styrene (or vinyl pyridine) and divinyl benzene, polymerizing and subsequently adding IX groups to the styrene (or vinyl pyridine) residues. (ED equipment has been installed probably originally furnished with more than 0.2 million sq. meters of Tokuyama Soda membranes of the above type).

In the People's Republic of China and in the Commonwealth of Independent States (formerly U.S.S.R.) IX membranes are prepared by blending IX resins with e.g. polyethylene, rubber, vinyl chloride-acetate copolymers and the like, for example on a roller mill and calendering the mixture into thin sheets which are generally reinforced with suitable fabric. The ratio of binder to IX resin is quite critical; more than about 35% binder leads to membranes having high electrical resistance whereas much less than about 35% leads to porous membranes with reduced ion selectivity. Such heterogeneous membranes are satisfactory for the demineralization of brackish and potable water. In the PRC ED equipment has been installed probably originally furnished with more than 1.2 million sq. meters of such membranes. Although it is extremely difficult to obtain reliable information on the use of ED in the CIS it appears that ED equipment has been installed there using more than 0.6 million sq. meters of such membranes. In the 1950's similar heterogeneous, bonded membranes were made in the U.S. by Rohm and Haas Co. and in the U.K. by Permutit Ltd. IONAC Chemical Co. of the U.S. (not related to Ionics, Inc.) makes limited quantities of such membranes.

Pall-RAI Research in the U.S. swells polyethylene or similar films with a mixture of styrene and divinyl benzene and copolymerizes the mixture with gamma (or similar radiation) subsequently adding IX groups to the styrene residues. As an order of magnitude guess, Pall-RAI produces perhaps 0.375 million sq. meters of CX membranes annually for separators in alkaline primary and secondary batteries. Similar membranes were made by the U.S. company AMF in the 1950's.

The above summary does not include producers of minor amounts of hydrocarbon based IX membranes and touches only briefly on the historical development of IX membranes. IX membranes are available from the Japanese producers which membranes have thin skins selective for univalent ions and are used primarily for the production of brine from seawater at current densities much less than the limiting current density (that current controlled by diffusion from the solution to the membrane). Similar skins are also useful to produce anti-fouling membranes and bipolar IEM (which "split" water into dilute acids and bases). Recent patent literature indicates an interest by the IEM industry in membranes based on non-crosslinked products. For example sulfonate or quaternary ammonium groups may be added to polysulfones, polyphenylene oxide and the like, the resulting poly ionomer dissolved in a suitable volatile solvent and cast into films with or without suitable reinforcing fabrics. The IX capacities are deliberately restricted to values which permit swelling but not solution in water. Frequently the method of manufacture results in essentially block copolymers.

1.3 IX Membranes for Chloralkali Electrolysis: In July 1948 the author built at Harvard University a laboratory chloralkali cell using a phenol sulfonic acid-formaldehyde CX membrane. The cell established proof of principle but the membrane was not stable. Patents were awarded in several countries granting inter alia broad claims to membrane chloralkali cells.

In the period 1950 tO 1957 Hooker Electrochemical Co. (now Occidental Chemical Co.) and Rohm and Haas Co. attempted to develop membrane chloralkali cells using heterogeneous membranes formed by milling impalpable CX resin powder into fluorinated polymer binders. The technology was not commercialized, probably for the most part because suitable dimensionally stable anodes were not available.

In the period 1958 to 1961 Ionics and Diamond Alkali Co. (now Diamond Shamrock Corp.) attempted to develop membrane chloralkali cells using reinforced CX membranes based on copolymers of divinyl benzene and acrylic and/or methacrylic acids. A three compartment cell was used comprising such a CX membrane which defined the cathode compartment and a porous diaphragm defining the anode compartment. Brine passed into the center compartment and through the porous diaphragm sweeping back chlorine and hypochlorite. The diaphragm was porous polytetrafluoroethylene supported on an expanded titanium anode platinized on the side away from the diaphragm. Membrane life of 6 months was demonstrated, degradation apparently being due to decarboxylation of the membrane in the strong caustic at cell operating temperature. The development was not commercialized.

DuPont claims (p 125, Modern Chlor-Alkali Technology, Vol. 3, Ellis Horwood Ltd., 1985) that XR/NAFION (TM) perfluorinated CX ionomer membrane chemistry was first discovered at duPont in 1962 and first demonstrated as a separator in a chloralkali cell in 1964. (It is not said which entity made this demonstration. No patent on such use seems to have been filed by duPont at the time and there is no question that when duPont had a feature article in duPont's corporate magazine on the development by Ionics of membrane chloralkali cells using duPont's XR membranes no mention was made of such a demonstration. D.J. Vaughan, DuPont Innovation 1973, 4(3), 10-13). In 1966 XR was used in General Electric Co.'s SPE fuel cells. In 1970 fabric reinforced XR was developed for Ionics. In 1971 DuPont developed asymmetric XR perfluorosulfonic acid membranes at the suggestion of Ionics. (These were the ancestors of duPont's present 300 series NAFION membranes). These were followed rapidly by membranes having cathode side skins of sulfonamide groups. Asahi Chemical Industry and Tokuyama Soda developed perfluorosulfonic acid membranes having cathode side skins of perfluorocarboxylic acids, the basis for most of the commercial membranes now used in membrane chloralkali cells. Asahi Glass developed homogeneous perfluorocarboxylic acid membranes.

As suggested above, at present most chloralkali membranes are based on laminates (with or without fabric reinforcement) of perfluorocarboxylic acids (on the cathode side, producing high current efficiency) and perfluorosulfonic acids (on the anode side, producing desired mechanical properties and low electrical resistance). Such membranes can be very sophisticated involving three or more perfluoro ionomer lamellae, gas release embossments or hydrophilic, inorganic coatings, all designed to save a few kilowatt-hours per tonne of chlorine. Although development of perfluoro CX membranes for chloralkali use has been carried out also by Dow Chemical Co., Hoechst A.G. and Tokuyama Soda among others, commercial production at present is limited to duPont, Asahi Chemical Industry and Asahi Glass. Chemistries are generally based on copolymers of derivatives of hexafluoropropylene (bearing the bound ionic groups) with tetrafluoroethylene. Such membranes typically operate at 90°C at current densities of 3500 amperes per sq. meter producing 30 to 35% caustic. Prices are roughly $ 600 per sq. meter. Under good operating conditions economic lifetimes are "three, four or more years" (E.I. Baucom, p 131 in"Modern Chlor-alkali Technology Vol. 5", ed. T.C. Wellington, Elsevier Applied Science, 1992). (A sq. meter of actual membrane area produces roughly 31 metric tonnes per year of chlorine and 35 m.t. of caustic 100% basis). At present more than 15% of world chlorine and caustic is supplied by membrane cells, that is more than 6 million metric tonnes yearly of chlorine and and more than 7 million m.t. of caustic 100% basis utilizing roughly 200,000 sq meters of membranes in more than 150 plants. The installed value of such membranes is therefore roughly $ 120 million. It is likely that almost no new large (e.g. in excess of 30 m.t. per day) chloralkali plants will be built using mercury or diaphragm cell technology since membrane cells are substantially more economic and environmentally friendly. (It is expected however that as mercury and diaphragm cells are replaced by membrane cells in merchant scale plants many of the cells so replaced will be installed in small plants in developing countries. In some cases the "replacement" of diaphragm cells with membrane cells consists of replacing the diaphragms per se with membranes and making the necessary plumbing changes). It may be predicted that early in the next century roughly 85% of world chlorine and caustic will be produced in membrane cells requiring roughly 1.3 million sq. meters of membranes valued in 1993 dollars at about $ 800 million and representing an annual replacement market of more than $ 250 million.

As noted above most membrane chloralkali plants produce 30 to 35% caustic which in large integrated plants may be economically evaporated as needed to meet the market for 50% caustic. Membranes have recently become available which make about 50% caustic directly (Y. Sajima et al., p 170 in "Modern Chlor-Alkali Technology Vol. 5", ed.T.C. Wellington, Elsevier Applied Science, 1992 and J. Powers, U.S. Pat. 4,900,408 (1990)) and should find application in some small chloralkali plants and also in larger plants where low cost electricity is available.

In order to make 30 to 35% caustic with membrane cells it is necessary to use ultrapure, essentially saturated, sodium chloride brine having for example not more than about 0.02 ppm of dissolved and suspended calcium. (Such purities can be easily obtained with modern chelating CX resins). Recent developments (see Powers referred to above) suggest that such purity requirement can be relaxed by a factor of ten or more by optimizing the chemistry of the membrane polymer thereby greatly improving the "forgivingness" of the membranes against plant upsets.

Compared to the first cation exchange membranes 45 years ago, the capabilities of present chloralkali membranes are a miracle, not possibly foreseen in 1948.

1.4 Bipolar IX Membranes: Such membranes have one surface consisting of CX resin and the opposite surface of AX resin. When direct electric current is passed through such a membrane in a direction to pull anions out of the interface between the anion and cation exchange resins and through the anion exchange resin, such interface rapidly becomes depleted of all ions other than those resulting from the dissociation of water. Dilute alkali can therefore be produced at the outer surface of the AX region and dilute acid at the outer surface of the CX layer. The concept seems first to have been disclosed by J. Frilette (J.Phys.Chem. 60 (1956) 435). An enormous effort has been applied by the AquaTech (TM) division of Allied-Signal Corp. to refine and commercialize such technology, so far at least without signal commercial success. The bipolar membranes are used in a more or less conventional ED stack together with conventional unipolar membranes. Such a stack will have many acid/alkali producing membranes between a single pair of end electrodes. The advantages of the process compared to the classic direct electrolysis seem to be that because only "end electrodes" are required:

- the cost of the electrodes used in direct electrolysis is avoided;
- the energy consumption (to produce oxidized and reduced products e.g. and hydrogen) at such electrodes is avoided.

The disadvantages appear to be:

- at present bipolar membranes are comparatively expensive;
- at present economic life of bipolar membranes is limited to about 1 year. (This appears to be due to the very high voltage gradients (order of magnitude 1 million volts per cm) pertaining at the interface between the AX and CX regions);
- practical current densities are limited to about 1000 amperes per sq. meter available area.

2. Applications of IEM:

2.1 ED: The use of direct current electricity to increase the rate of dialysis of electrolytes or to produce demineralized water from potable water has been known for about 100 years. Early cells used three compartments between a single pair of electrodes, the compartments being separated from each other by porous diaphragms. The latter were essentially not ion-selective and not intrinsically electrically conducting. This variety of ED is perhaps better regarded as double electrometathesis, anions in the central compartment being replaced by hydroxide from the cathode, cations by hydrogen ions from the anode. Such three compartment "ED" cells were sold before World War II for water demineralization.

In Zürich in 1940, K.H.Meyer and W. Strauss (Helv.Chim.Acta 23 (1940) 795-800) suggested a multiple compartment ED apparatus using many pairs of alternating, non-porous, ion-selective membranes between a single pair of electrodes. (The suggestion was based on studies by Prof. Meyer's group of the electricity generating organ in the electric eel). A laboratory scale proof-of-principle apparatus was built but the membranes were unstable and poorly selective.

The discovery of IX resins in sheet form by the author assured a commercial future for the electrodialysis process (pages xi, 4, 215 in "Demineralization by Electrodialysis", ed. J. Wilson, Butterworths Scientific Publications, London (1960)). The first commercial ED apparatus was sold by Ionics in 1954 to Arabian-American Oil Co. ("ARAMCO") and installed in Saudi Arabia for desalting brackish water. Since then more than 5000 ED plants have been installed world-wide for demineralization of brackish and potable water.

Ionics, Inc. has sold more than 2000 ED plants with a combined capacity of probably more than 600,000 cu.meters per day, originally furnished with probably more than 1.2 million sq. meters of membrane.

In addition to the above, more than 2000 ED plants have been built and installed in P.R. China by Chinese entities. These plants have a combined capacity of more than 600,000 cu. meters per day, originally furnished with probably more than 1.2 million sq. meters of membrane.

Although as mentioned above, it is difficult to obtain reliable information from the C.I.S., it is estimated that more than 1000 ED plants have been built and installed in the C.I.S. by C.I.S. entities. These plants have a combined capacity of probably more than 300,000 cu. meters per day, probably originally furnished with more than 600,000 sq. meters of membrane.

Further to the above, roughly 100 plants have been installed by others (e.g. Corning France (formerly S.R.T.I.), Portals Water Treatment Ltd. (formerly Permutit-Boby)).

The success of ED in demineralization of water has apparently been due to its greater tolerance of particulate and fouling matter compared to reverse osmosis, greater "forgivingness" of process upsets and of unskilled operators, simplicity in design and construction of ED stacks compared to RO modules, the ability to inspect, clean or replace one membrane at a time, and, at least in the case of Ionics, the existence of a comprehensive, global sales and service network and a vertically integrated manufacture. The introduction of periodic, symmetric current reversal ("EDR") by all manufacturers (U.S. Pat. 2,863,813) has greatly increased tolerance to fouling and scaling matter and reduced, often eliminated, the need for continuous dosing of chemicals and for chemical cleaning.

The economic life of IX membranes in the demineralization of water is generally in excess of 10 years in well operated plants.

Trends in the field appear to be:

- reduction in pumping and direct ED energy and increase in electric current density;

- the use of ED as pretreatment before IX deionizers to reduce the chemical regeneration of the latter;

- the use of hybrid processes for producing ultrapure water e.g. trains comprising cross-flow UF or microfiltration ("MF") to reduce particulates and fouling; EDR to reduce scaling constitutes and to produce a concentrated reject; RO to reduce silica and non-ionized organics; and/

or electrodeionization("EDI") to remove silica and 90% or so of remaining electrolytes; and mixed bed IX deionizers to produce 18 megohm product. Such trains have very low chemical and operator requirements. EDR reduces or eliminates the scaling tendency of subsequent RO; RO and/or EDI reduce or eliminate the need for on-site chemical regeneration of mixed bed IX:

- the use of EDR and the above mentioned hybrid processes in Build, Own, Operate and Transfer plants in which the manufacturer of the demineralization plant owns and operates the plant, selling water to the user or water distributor.

2.2 Concentration of Seawater by ED: In terms of membrane area, this application is by far the second largest user of IEM. The application is dominated by the Japanese manufacturers:

- Asahi Chemical Industry Co., Ltd., about 5 plants, about 800,000 metric tonnes of salt per year, about 400,000 sq. meters of installed membrane area;
- Asahi Glass Co., Ltd., about 2 plants, about 370,000 m.t. salt per year, about 185,000 sq. meters installed membrane area;
- Tokuyama Soda Co., Ltd., about 3 plants, about 400,000 m.t. salt per year, about 200,000 sq. meters installed membrane area.

Seawater is concentrated to 18 to 20% dissolved solids (using monovalent ion selective, skinned, membranes). Reversal is not used. Salt is produced from the brine by evaporation and crystallization. Seven plants are in Japan, one each in S. Korea, Taiwan and Kuwait. None are justified on economic grounds compared to imported solar or mined salt. (It is the policy of the Japanese government that all "domestic" salt must be made in Japan and sold to the Japan Salt Monopoly Corporation).

Note that the osmotic pressure difference between about 19% sodium chloride solution and partially depleted seawater is about 200 atmospheres at 25°C, well beyond the range of RO.

Closely related to the above is the simultaneous concentration of industrial effluents and recycle of recovered water. Such applications will probably increase as environmental restrictions increase. Examples are the concentration of blow-down from cooling towers in power plants and the processing of metal treatment wastes.

2.3 Production of Chlorine and Caustic from Salt: The first commercial, electrolytic, membrane chloralkali plants were built by Ionics in 1972 and '73. The capacities were small, resp. 60 and 200 metric tonnes of chlorine per year. In 1975 Asahi Chemical Industry Co., Ltd. started a membrane chloralkali plant of about 35,000 m.t. per year as part of one of its own chloralkali plants in Japan. Since then more than 150 membrane chloralkali plants have been installed around the world having a combined capacity of more than 6 million m.t. of chlorine annuallly, furnished with about 200,000 sq. meters of membrane. The principal suppliers of such plants have been (in alphabetical order): Asahi Chemical Industry Co., Asahi Glass Co., Chlorine Engineers Corp., Oronzio DeNora Technologies S.p.A., ICI Chemicals and Polymers Ltd., Ionics, Inc., Lurgi GmbH, OxyTech Systems Inc., Uhde GmbH, often with or through other engineering and construction firms.

There is no question that the development of modern, sophisticated, membrane chloralkali technology was due directly to the unfortunate Minamata disease (not in fact caused by effluents from mercury chloralkali plants) which led the Japanese government to request the phased replacement of mercury cells in Japan

In addition to the trends noted above in the section "IX Membranes for Chloralkali Electrolyzers" the following may be added:

- The 3rd Ministerial Conference on the North Sea resolved that all mercury cells should be phased out in Europe by 2010. These and other regulations are expected to bring about major conversion from mercury cells to membrane cells long before 2010 (R. Curry in "Modern Chlor-Alkali Technology Vol. 5", Ed., T. Wellington, Elsevier Applied Science 1992, page 186). As a result of restrictions on chlorinated solvents and CFC's and reduced use of chlorine in the pulp and paper industry and in water treatment, there is no doubt that some mercury cell plants will be shut down and not replaced;

- pressures against transport of chlorine (particularly export of chlorine) and storage of chlorine should result in substitution of large merchant plants by smaller on-site membrane chloralkali plants;

- many on-site chloralkali plants will probably be Built, Owned, Operated and Maintained by the plant supplier, selling chlorine and caustic on a toll basis to the local user.

2.4 Electro-De-Ionization ("EDI")/Filled Cell ED/Electrically Regenerated Ion Exchange: This concept was apparently first annunciated by W. Walters et al. (J.Industr.Engng.Chem. (1955), 47, page 61 et seq.). The process was developed by Ionics in the 1960's (U.S. 3,149,061) and more recently by Millipore Corp. (U.S. 4,632,745). At least the demineralization compartments of ED apparatus are filled with anion and/or cation exchange beads or fibers or the IEM are embossed and in contact with each other. (In reversing type EDI both compartments are so filled). At low current densities the filling acts to augment the surface area of the membranes resulting in a very large increase in allowable current density. At high current densities water splitting occurs as in the case of normal ED and the IX filling is converted largely to the hydroxide and hydrogen forms resp. The apparatus is then essentially IX regenerated in real time electrochemically. The process has begun to find commercial applications during the last few years almost always as part of a hybrid process for producing demineralized water. When essentially 18 megohm water is required EDI reduces the load on mixed bed IX by an order of magnitude or or more and the need for regeneration and regeneration chemicals by about the same factor. Applications are for feed water for high pressure boilers and for ultrapure water for the electronics industry. When less than 18 megohm water is satisfactory EDI can entirely replace mixed bed IX. Typically EDI will be preceded in a hybrid plant by some combination of EDR, RO, UF or MF, IX water softening and/or absorbent for organics. The tendency of demineralized water producers to reduce chemical requirements therefor seems to assure a future for EDI.

2.5 IX Membranes for Alkaline Primary and Secondary Batteries: Such membranes are used in fuel cells,Ag/Zn, AgO/Zn, Ag_2O/Zn, Hg/Zn, HgO/Cd, HgO/Zn, MnO_2, Ni/Cd (vented), Ni/Zn and Zn/air cells. They reduce diffusion of silver, mercury and hydroxide ions and growth of zinc dendrites and improve cell shelf-life. Thicknesses are as low as 25 micrometers and electrical resistances as low as 0.045 ohms per sq. cm. in 40% KOH at 24°C (literature Pall-RAI Research). Costs are as low as about $ 6.00 per sq. meter in quantity (but can be much higher for special varieties). World-wide volume appears to be about 0.75 million sq. meters per year.

2.6 Miscellaneous Applications: The largest of these applications is the demineralization of whey and non-fat milk for food and feed applications. During the manufacture of cheese or of milk casein a serum is often produced containing much of the albumin, globulin, lactose and minerals present in the milk plus acid and/or salt added during manufacture. Such albumin and globulin is a valuable food or feed source but use is generally limited by the electrolytes present. For the past 30 years ED and EDR have been used to remove excessive electrolytes, generally from concentrated serum. There are approximately 70 such ED or EDR plants world-wide having about 35,000 sq. meters total installed membrane area and producing about 150,000 tonnes per year of demineralized whey solids. Some of such plants remove in excess of 90% of the ash content of whey. Others remove roughly 50% of such ash, most of the remainder being removed by strong acid cation exchange in the hydrogen cycle followed by weak base anion exchange in the free base cycle. Much of the highly demineralized product is used as a component of mothers' milk replacement.

Other applications include:

- electrohydrodimerization of acrylonitrile to about 67,000 tonnes per year of adiponitrile for manufacture of Nylon 66 (Asahi Chemical Industry);

- recovery of valuable components from metal plating or treating effluents (including recovery of hydrofluoric and nitric acids by bipolar ED from stainless steel spent pickle liquor);

- deashing of beet, cane or other sugar juices and molasses. AX membranes are generally subject to fouling by medium molecular weight organic carboxylic acids in such solutions. Although fouling resistant AX membranes have been proposed, the generally short processing season for such sugar solutions leads to high capital charges. Further the additional sugar which could be recovered through desalting is not justified in view of the agricultural policies of the sugar producing countries. There are few, if any, ED plants desalting sugar solutions;

- deacidification/acidification of fruit juices. There are few, if any, plants on a commercial scale although there have been pilot applications;

- desalting of soy sauce, amino acid solutions, fermentation products etc. There are probably a few small ED plants in such applications;

- demineralization of blood plasma. There are few such plants.

TREATMENT OF ACID EFFLUENTS BY ELECTROMEMBRANE PROCESSES

Claude GAVACH

University of Montpellier II - CNRS URA 330
B.P. 5051 - 34033 Montpellier Cedex - France

SUMMARY

Some industrial waste are composed of mixtures of acids and metal salts which are traditionaly treated by precipitation with chalk. The electromembrane processes (electrodialysis (ED) and electro-electrodialysis (EED)) are offfering an alternative way of treatment leading to the recovery and the reconcentration of the acid.

The application of these membrane processes to the treatment of this kind of effluent results from the development of new types of ion permeable membranes: anion permeable membranes with a reduced proton leakage and cation permeable membranes specific to proton.

In this paper we examine the transport properties of these special membranes and also the factors influencing the performances of ED and EED when the membranes are applied to the treatment of acidic waste waters.

INTRODUCTION

Ionic membranes are also called ion-exchange or ion-permeable membranes. They are composed of a swollen polymeric sheet bearing fixed ion exchanging sites: sulfonic or carboxylic groups for the cation permeable membrane (CPM) and amine or quaternized ammonium groups for the anion permeable membrane (APM).

Besides their applications as electrolysor diaphragms (for instance in the chlor-alkali process or in water electrolysis) and as fuel cell separators, this kind of membrane is used in separation techniques. In this field, the main industrial application is being covered by electrodialysis (ED) which is used at a large scale for desalting brackish waters, brine production from sea water and demineralization of whey. Moreover, this technique has also a lot of small scale applications in the domain of industrial effluent treatment. Only salts can be removed from aqueous media but not the bases because the APM are destroyed in high pH medium, nor the acids because the APM used in ED is overly permeable to protons.

Recently special commercial APM's have been developed to make ED suitable for the treatment of acidic effluents. By this technique the acids can be eliminated from the waste waters and recycled into the industrial process. The majority of acidic waste waters are produced by metallurgy industries (ore hydrometallurgy, pickling, surface treatment) and contain both acids and metallic ions. In this paper we examine how ED as well electro-electrodialysis (EED) can be applied for the treatment of these effluents.

THE MEMBRANES

In contact with acidic solutions the conventional APM shows a loss in permselectivity because these membranes are hightly permeable to protons.

This proton leakage through polymeric anion exchange membranes is due to the swelling of the membrane material; the adsorbed water being a proton conductor. This is the reason why for a long time, ED could not be applied to the treatment of acidic solutions.

Three commercial membranes have been recently developped for this special purpose: the AAV-Selemion membrane produced by Asahi Chemical, the ACM-Neosepta membrane produced by Tokuyama Soda, and the ARA membrane produced by Morgane. Despite the fact that the compositions of the membranes have not been revealed, it can be established that their conductivity becomes hight enough for ED applications only when the membranes are equilibrated with acidic media [1]. In contact with aqueous solutions having pH values greater than 4, the AAV and ARA membranes are insulators, and the ACM gets higher electrical resistance. These results reveal that the ion exchanging sites of these membranes are weak base functions. The physico chemical factors at the basis of this reduced proton leakage are under investigation. Today, the origin of the low permeability of protons through theses membranes is still not completely elucidated, but we can presume that it is the same for the three membranes because they exhibit several properties which are strangely similar.

When these membranes are equilibrated with an acidic aqueous solution, they contain a certain amount of acid which is present in the interstitial phase. The number of equivalents of adsorbed acid in the membrane can be two times higher than the number of ion exchanging sites. The shape of variation of the number of adsorbed acid versus the external acid concentration (fig 1 and 2) is the same for the three membranes and the two studied acids (HCl and H_2SO_4). Moreover the shape of the obtained curves is completely different from the theoretical variation which can be expected from the classical Donnan co-ion exclusion theory [2]. In the later case, the concavity of the curves is oriented to the ordinate and not to the abcissa as in the experimental curves.

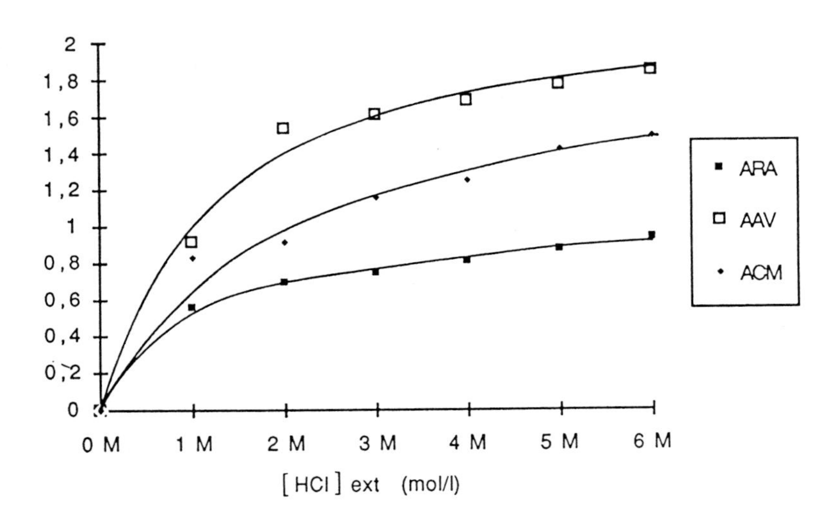

Figure 1: Sorption equilibrium of HCl in the membranes. Variation, versus the external acid concentration, of the amount of HCl in the interstitial phase of the membranes.

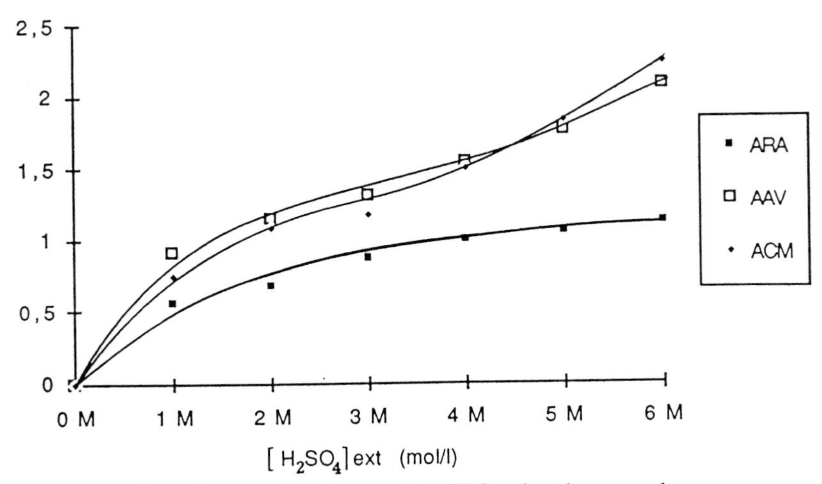

Figure 2: Sorption equilibrium of H_2SO_4 in the membranes. Variation, versus the external acid concentration, of the amount of H_2SO_4 in the interstitial phase of the membrane.

Both fig 3 and 4 show the variation of the conductivity of the membranes equilibrated with aqueous solutions of HCl or H_2SO_4 as a function of the acid concentration in the external solution. The membrane resistance has been measured at 1000 Hz [3, 4]. In all cases the values of the conductivity is at least 10 times lower than those of the equilibration solutions and the curves show a more or less pronounced maximum. A maximum also exists with aqueous solutions of acid. For H_2SO_4 it corresponds to about 4 moles/l.

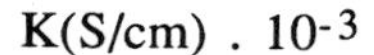

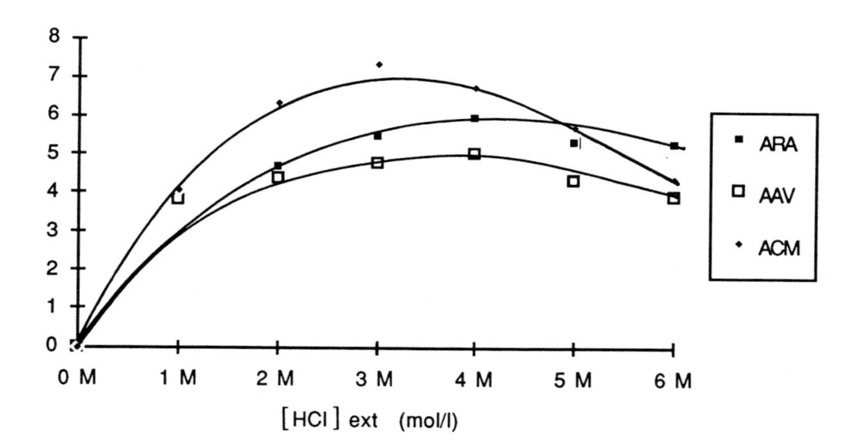

Figure 3: Conductivity of the membranes equilibrated with HCl aqueous solutions.

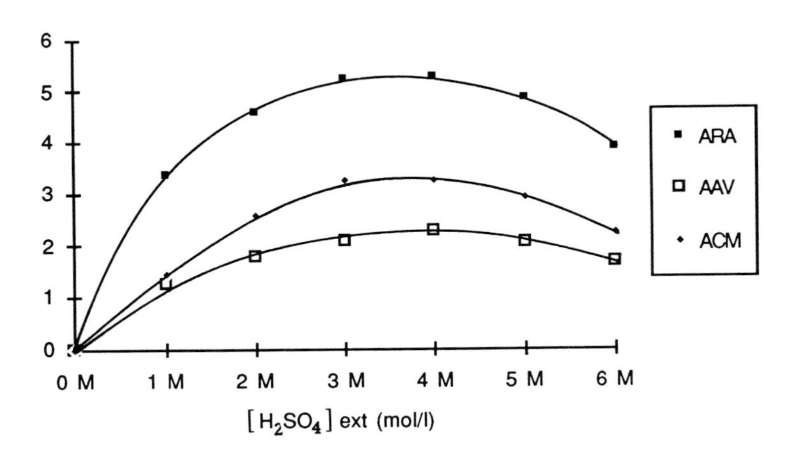

Figure 4: Conductivity of the membranes equilibrated with H_2SO_4 solutions

The figures 5 and 6 show the variation of the transport number of the anion when the membrane is placed between two identical acid solutions. The current density is 0,1 A/cm^2. These measurements have been accomplished using radioactive tracers [3, 4, 5, 6]. With sulphuric acid the variations of the transport number of sulfate ions is calculated according to the assumption that SO_4^{2-} is the permeant anion, and is similar for the three membranes. While in contact with HCl solutions, the AAV and the ARA membranes show very similar variations which are different from those obtained with H_2SO_4. With HCl, despite the fact that at the equilibrum with the ACM membrane, the curve of the amount of the adsorbed acid is placed between the curves obtained with the two other membranes, the measured values of the chloride ion transport number are notably higher with the AAV and the ARA membranes.

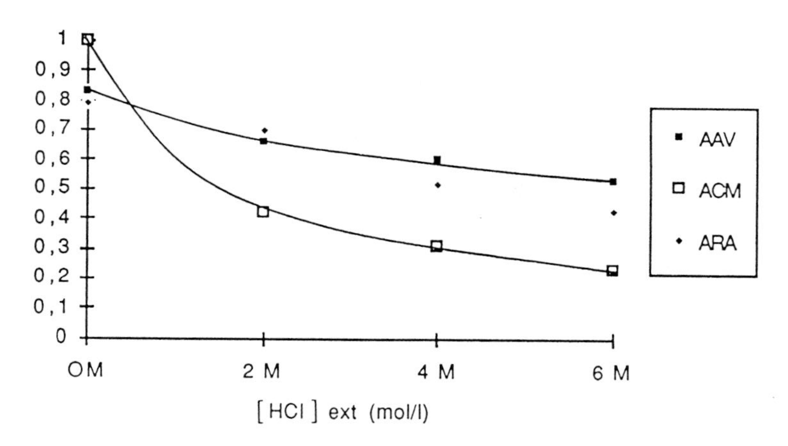

Figure 5: Transport number of the Cl^- ion in the membrane in contact with HCl aqueous solutions. The HCl concentrations are equal in the two solutions. Current density 100 $mA.cm^{-2}$.

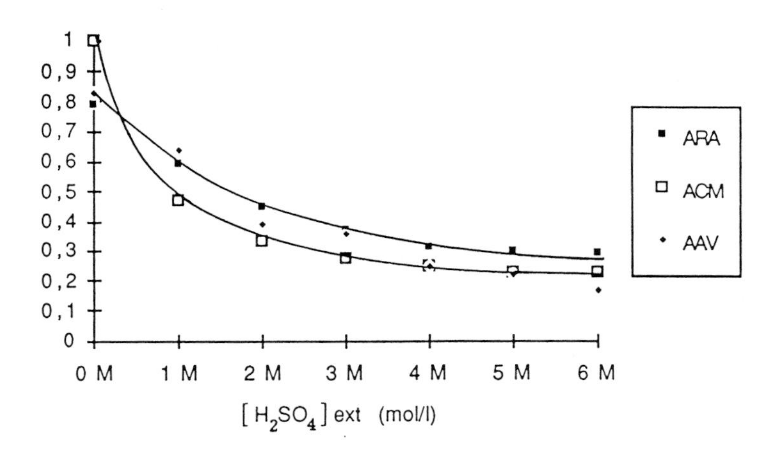

Figure 6: Transport number of sulphate anion in the membranes in contact with sulphuric acid solutions.
The acid concentration are equal in the two solutions.
Current density: 100 $mA.cm^{-2}$.

ED can also be applied to the separation of acid from metallic salt if one of these special APM is associated with a CPM which is much more permeable to proton than to bivalent metallic ions. For the application of ED to the reconcentration of sodium chloride from sea water, new CPM have been developed in order to reduce the flux of Ca^{2+} with respect to those of Na^+ ion. The transport properties of this kind of CPM have been intensively studied by Sata and coworker [7, 8, 9] who also showed that the adsorption of a polycation layer onto the membrane surface reduces the flux of calcium with respect to those of sodium.

The relative transport number between a monovalent cation M^+ and a bivalent B^{2+}, $P_{M+}(B^{2+})$ is defined as follows:

$$P_{M+}(B^{2+}) = \frac{t_{M^+} \cdot C_{B2+}}{t_{B2+} \cdot C_{M+}} \qquad [2]$$

Where t is the transport number and C the ionic concentration in the aqueous solution.

The transport competition between Cu^{2+} and H^+ has been studied with the CMS membrane, a monovalent cation specific membrane, produced by Tokuyama Soda. We could show [10] that under current densities ranging from 50 to 200 $mA.cm^{-2}$, $P_{H^+}(Cu^{2+})$ increases from 0,33 to 0,66. The concentration polarisation effect has to be taken into consideration in order to account for this result. At high current density, the concentration ratio between H^+ and Cu^{2+} in the aqueous layer in contact with the membrane are lower than in the bulk. By absorbing a layer of cationic polyelectrolyte at the surface of a perfluorinated CPM (Morgane CRA), we could show that the values of the transport numbers of proton and nickel ions were respectively 0,9 and 0,1. While in the absence of polycation layer they were respectively equal to 0,7 and 0,3. The aqueous solution in contact contains $NiSO_4$ (0,5 M) + H_2SO_4 (0,5 M).

ELECTRODIALYSIS

On fig 7 we recall the basic principle of ED. Under the effect of the electrical field, the electrolyte M^+A^- is transfered from the diluate to the concentrate with 100% current efficiency (CE) only if both the APM and the CPM are idealy permselective.

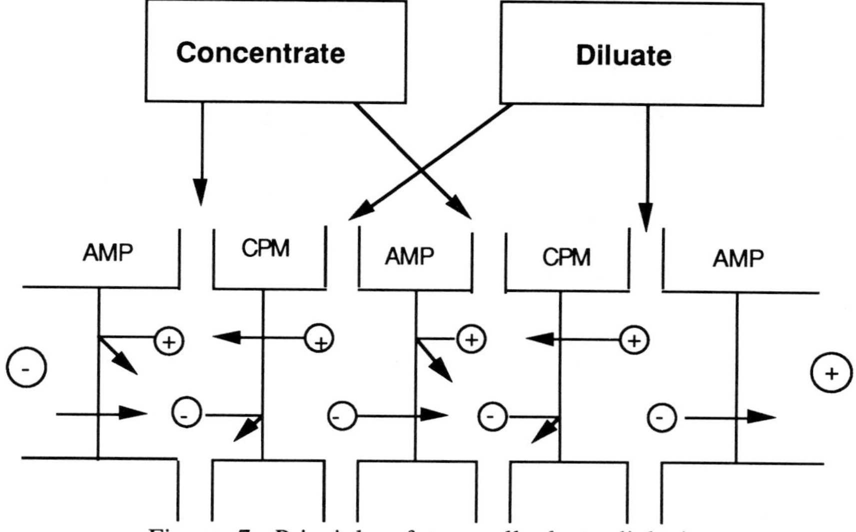

Figure 7: Principle of two-cell electrodialysis

With symetrical systems i.e when the electrolyte concentration is the same in the diluate and in the concentrate, the expression of current efficiency is the following:

$$CE = 1 - t_{CPM}(A^-) - t_{APM}(M^+) \quad [1]$$

$t_{CPM}(A^-)$ and $t_{APM}(M^+)$ are the transport numbers of the co-ions. When the treated electrolyte solution is a hightly dissociated acid solution, the proton leakage through the APM will reduced CE because $C_{APM}(H^+)$ is no longer zero and will increase with the acid solution concentration as shown on fig 5 and fig 6.

The ED of the HCl solution has been studied using a laboratory cell with a 40 cm^2 membrane area. In order to maintain a quasi-constant HCl concentration in the diluate, its initial volume was 5 l while those of the concentrate was 0,2 l. The APM was an ARA Morgane membrane and the CPM a conventional ED membrane (CMV Selemion). In fig 8 the variation with time of the HCl concentration in the concentrate for four different current densities is plotted. The initial HCl concentration in the two circuits is 1 M. During this ED, water is transfered from the diluate to the concentrate whose volume variation is plotted in fig 9.

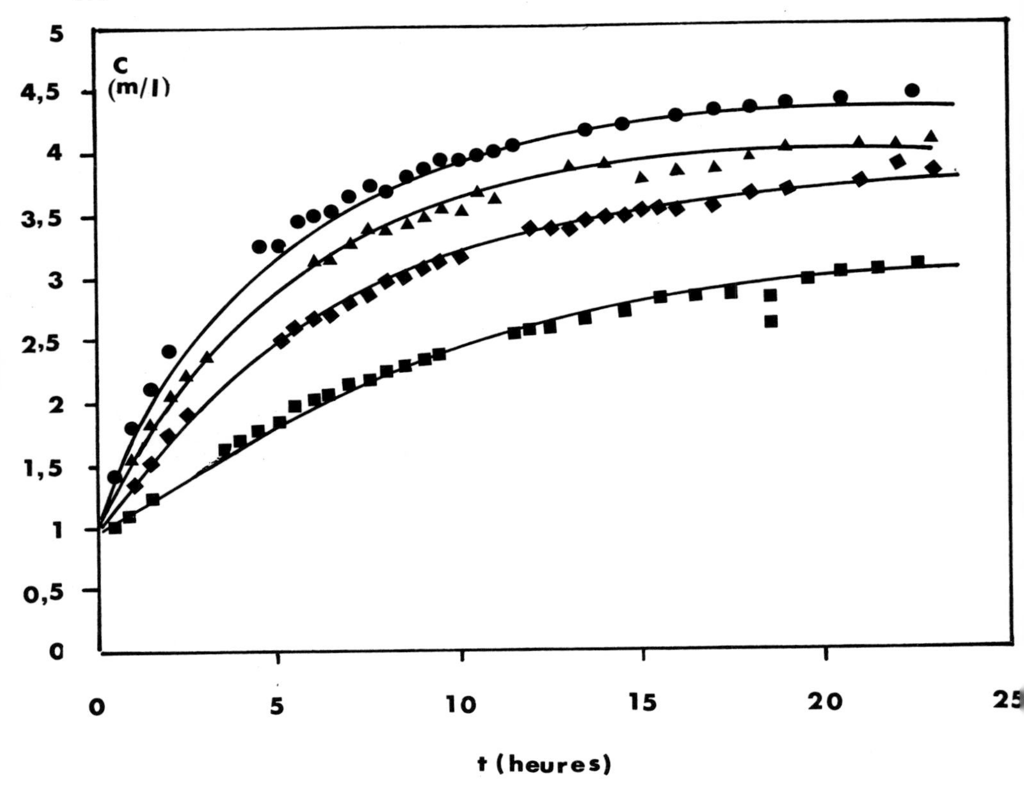

Figure 8: Electrodialysis of HCl solutions:
Variation of the HCl concentration in the concentrate
Initial conditions: Diluate: V = 5 l, C HCl 1 mole/l
Concentrate V = 0,2 l, C HCl 1 mole/l

Current density ($mA.cm^{-2}$): ■ 50; ◆ 100; ▲ 150; ● 200
Membrane area: 40 cm^{-2} APM: ARA Morgane
CPM: CMV Selemion

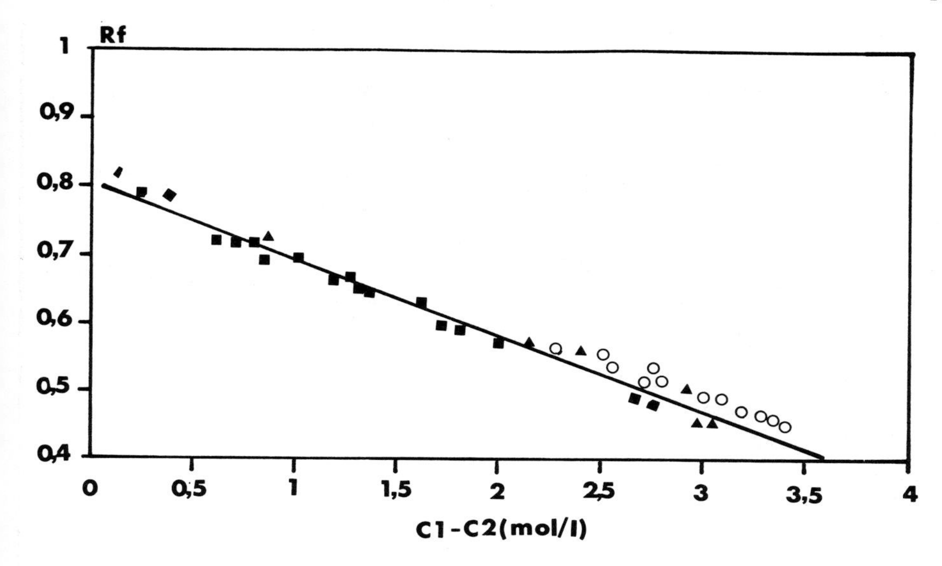

Figure 9: Electrodialysis of HCl: Variation of the concentrate volume. Symbols and experimental conditions: see legend of fig 8.

In fig 10 are schematically represented all the matter transport phenomena which occur during this ED operation assuming that the CPM is ideally not permeable to the Cl^-.

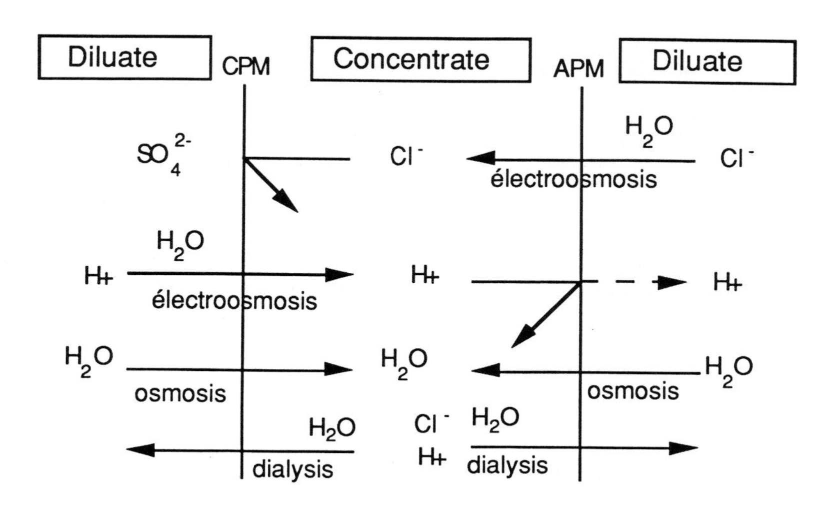

Figure 10: Schematical representation of the different membrane phenomena accuring during electrodialysis of HCl

Besides the ionic electro-transports (Cl^- through the APM, and proton through both the CPM and the APM) and the water electro-transport which is composed of a pure electroosmosis process through charged pores and of the transport of water molecules dragged by the movement of ions, we must also take into consideration the electrically silent matter transport which could result from the creation of a HCl concentration difference between the diluate and the concentrate. This electrolyte concentration difference will induce a diffusion of acid (dialysis) through the proton permeable APM and also osmotic water fluxes through the two membranes. We can distinguish the effect of the electrotransport of proton through the APM on the total current efficiency from the effect of HCl dialysis by plotting the values of the current efficiency versus the concentration difference of acid. The experimental data are given on fig 11.

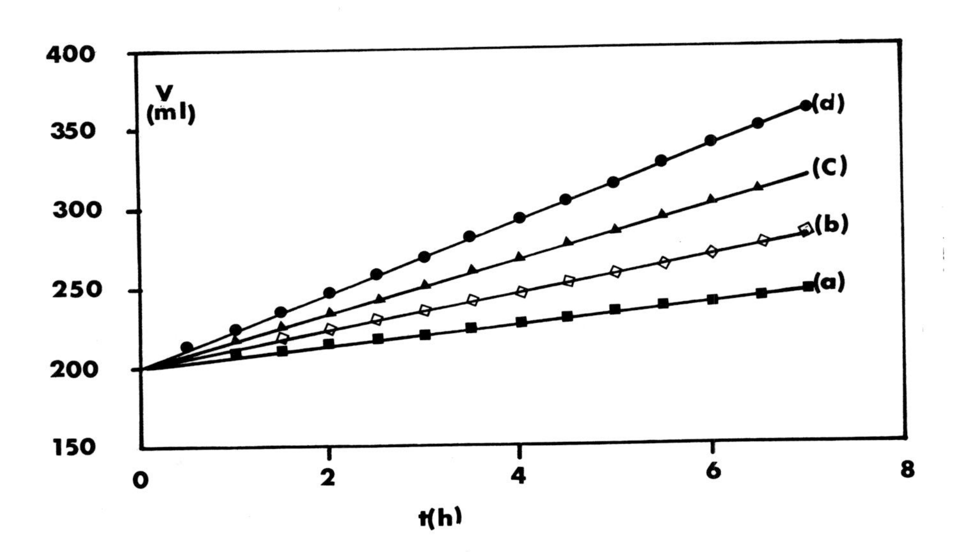

Figure 11: Electrodialysis of HCl
Variation of the current efficiency versus the HCl concentration difference between the concentrate and the diluate.
Symbols, and experimental conditions: See legend of fig 8.

We find that the current efficiency follows a linear variation when plotted versus the HCl concentration difference between the diluate and the concentrate. These results are very interesting because they confirm the proposed transport model. The extrapolated value of the current efficiency for C_c - C_d = 0, i.e when equation (1) is valid, give the values of the Cl^- transport number determined by radiotracers measurement (fig 5). In the considered case where HCl concentrations are 1 M in the diluate, $t_{APM(H^+)}$ = 0,16.

Considering linear variation with time of the diluate volume described by fig 11

$$V_D = V^{\circ}_D (a + bI)t \qquad [3]$$

also and the linear variation of the current efficiency as a function of (C_c - C_d). We can predict the variation of the HCl concentration in the concentrate when the HCl concentration is constant in the diluate.

Moreover, a CPM which is much more permeable to proton than to metallic cations can be associated to one of these special APM with a reduced proton leakage to form a couple of membranes suitable for the separation of acid and metal salts. The basic principal is schematized on the fig 12. With such a membrane stack, an effluent composed of a mixture of acid and of metal ion can be treated by ED. Purified and reconcentred acid can be obtained in the concentrate and, in its final composition, the diluate will contain almost the totality of the initial metal with a low amount of residual acid.

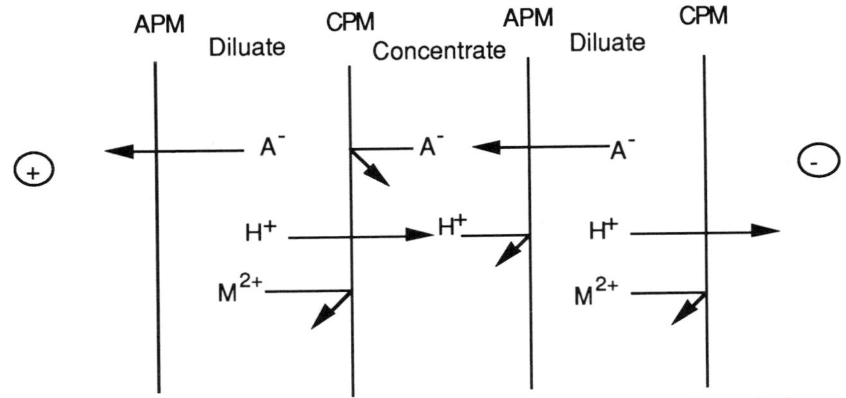

Figure 12: Principle of separation of acid from metallic salt by electrodialysis using a proton selective membrane

ELECTRO-ELECTRODIALYSIS (EED)

From a mixture of acid and metal salt, acid with higher reconcentration and purity levels can be obtained by using electro-electrodialysis instead of ED.

An EED cell is nothing but a electrolysis cell with a ion permeable membrane separating the catholyte from the anolyte and used for separation purpose. The basic principal of treatment by EED for this kind of effluent is schematically given on fig 13 in the special case where the anion cannot be oxydized at the anode.

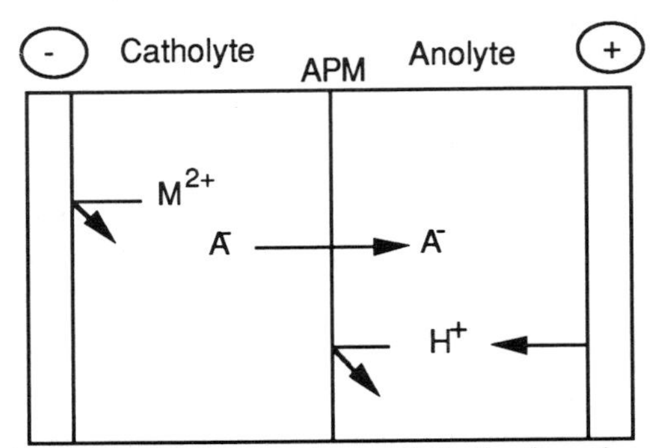

Figure 13: Principle of electro-electrodialysis for the separation and reconcentration of acid from effluent composed of a mixture of an acid and a metallic salt.

The effluent is introduced into the cathodic half-cell. When an electrical current flows through the EED cell, three elementary processes take place at the same time:

- At the cathode, the reduction of the proton and metal ion giving gazeous hydrogen and metal deposition.

- At the anode, the oxydation of water creating protons and oxygen in the anolyte.

- Through the APM, there is a transfer of anions from the catholyte and eventually following a flux of this protons in the opposite direction.

The final result from these elementary phenomena will be the transfer of the acid from the catholyte to the anolyte. The current efficiency of this operation is decreased by the proton leakage through the APM. The metallic salt is also removed from the catholyte and the metal is obtained under the form at a deposit with an high degree of purity.

Higher degree of acid reconcentration can be obtained by EED because the water flux to anolyte is negative due to the water consumption resulting from the electrochemical reaction.

Sulphuric acid with a normality higher than 12 M has been obtained by this electromembrane technique.

Moreover this reconcentrated acid is metal free because the movement of metal ions to the anolyte is hindered both by the electrical field and the permselectivity of the APM.

But EED shows certain disavantages with respect to ED due to the investment cost. In a EED cell, the surface area of the metallic electrodes is two times the surface of the membrane while in industrial ED stack, one pair of electrodes is associated to more than two hundred pairs of membranes. Both the investments and the energy cost are more important with EED.

For the industrial application of EED, chlor-alkali technology is not suitable because the technique needs extractable cathodes for removing the metal deposit. In addition gazeous oxygen and hydrogen form by products of the process.

CONCLUSION

With the development of new ion permeable membranes (APM with a reduced proton permselectivity and CPM more permeable to proton than metal ion), ED can be applied to the treatment of effluents composed of a mixture of an acid with a metal salt. But the reconcentration level of the acid is reduced by the water transport through the APM and the CPM and as the CPM is not completely specific to proton, the reconcentrated acid will contain a certain amount of metal. Pure acid with an higher reconcentration level can be obtained by EED which also leads to the total removal of the metal which is obtained under the form of a cathodic deposit. But EED has not yet got an industrial development because this technology has some technical commercial disavantages

REFERENCES

(1) I. TUGAS, G. POURCELLY and C. GAVACH
Submitted to J. Membrane Sciences.

(2) F. HELFFRERICH, Ion Exchange, Mc Grow-Hill Book Compagny, Inc. Chap 5 (1992)

(3) I. TUGAS, G. POURCELLY and C. GAVACH
Proceedings of the International Symposium of Functionalized Dense Membranes and Membrane Processes. Pont à Mousson 1991 - Club Membranes EDF.

(4) J. SANDEAUX, A. FARES, R. SANDEAUX and C. GAVACH
Submitted to J. Membrane Sciences.

(5) A. LINDHEIMER, J. MOLENAT and C. GAVACH
J. Electroanal. Chem. 216 (1987) 71.

(6) J. SANDEAUX, R. SANDEAUX and C. GAVACH
J. Membrane Sci. 59 (1991) 265.

(7) T. SATA, R. IZUO and R. TANAKA
J. Membrane Sci. 45 (1989) 197.

(8) T. SATA and R. IZUO
J. Applied Polymer Sci. 41 (1990) 2349.

(9) T. SATA and R. IZUO
J. Membrane Sci. 45 (1989) 45.

(10) M. SIALI, G. POURCELLY, J.M. JANOT and C. GAVACH
Submitted to J. Membrane Sci.

Salt Transport Phenomena Across Charged Membrane Driven by Pressure Difference

Akira YAMAUCHI and Yasuko TANAKA
Department of Chemistry, Faculty of Science, Kyushu University 33
Hakozaki, Higashi-ku, Fukuoka 812(Japan)

Abstract

Present study aims to seek a way for potential application to salt enrichment through fundamental analysis of salt transport phenomena on a basis of nonequilibrium thermodynamics. In this work, 3 different charged membranes were investigated. Under the appropriate osmotic pressure generated by sucrose, the volume flux, Jv and salt flux, Js were observed in the membrane-KCl salt solution systems and the filtration coefficient, Lp, reflection coefficient, σ and solute permeability, ω which are characteristic of the membrane, were estimated. Furthermore, the membrane conductance, κ, and electroosmotic permeability, β were obtained.These parameters were discussed in relation to the salt and solvent transport from phenomenological and electrochemical consideration.

Keywords: negative reflection coefficient; pressure dialysis; charged mosaic membrane; salt transport

Introduction

Pressure dialysis is a promising technique for producing pure water or concentrating sea water instead of conventional membrane separations such as the electrodialysis or reverse osmosis methods. In spite of several advantages predicted by the foreseeing works, the practical application has not yet been succeeded to the present. Sollner in 1932 pointed out that a membrane having two cation and anion exchange sites exhibits the unique membrane phenomena[1] and later Kedem & Katchalsky theoretically predicted the presence of a circulating current in such a system as the result of analysis owing to linear phenomenological equation in nonequilibrium thermodynamics[2]. As mentioned above, the membrane suitable to pressure dialysis would be required to involve two opposite charged sites in its body. Among several approaches to development of such a functional membrane[3-7], recently Miyaki et al. have obtained a novel membrane, so-called the charged mosaic membrane, in which the opposite charges were arranged with parallel array each other across the insulating layers[6-7]. Though the membrane characteristics were already discussed from the practical standpoints, the detailed behavior of salts within the membrane is remained unknown and should be clarified for understanding of salt transport under the pressure difference.

On the other hand, we have investigated the membrane phenomena in the amphoteric ion exchange membrane or the ion exchange membrane-electrolyte systems[8-9]. The amphoteric membrane also has two opposite charges which are homogeneously distributed within membrane, and may be applicable to the pressure dialysis. Therefore, it is worthwhile to examine whether the circulating currents or the excess salt flux will occur under the pressure difference between two electrolyte solutions across the membrane.

This study aims to compare the effect of the different charged states in the membranes on the salt transports, to see how the salts behave within the membrane during transport process and to get the fundamental informations for potential application to membrane separations.

Experimental

Materials: The amphoteric ion exchange membrane coded as CVS(abbreviated as AM hereafter) and the cation exchange membrane coded as CL-25T(abbreviated as CM) were supplied from Tokuyama Soda Co.Ltd. The charged mosaic membrane(abbreviated as MM) was purchased from Tosoh Co.Ltd. All the membranes were kept in the distilled water for long time until they were used after usual conditioning. The membrane characteristics are given in Table 1. The chemicals used in this study were all the reagent grades and the distilled water were used in the experiments.

TABLE 1

Characteristics of membranes

Membranes	Amphoteric ion exchange	Charged mosaic	Cation exchange
Ion exchange Capacities(meq./g)	0.192(cation) 0.183(anion)	1.06(cation) 1.04(anion)	1.63
Water content(gH_2O)	0.21	—	0.30
Thickness(mm)	0.12	0.14	0.16

Glass cells for measurement: A glass cell takes the shape of double structure in order to circulate a temperature-regulated water and its volume of chamber is 25 cm^3. In experiment, the membrane was tightly clamped between two glass cells so as to avoid the leak of solution. For the measurement of volume flow, the horizontal capillary with a graduated scale was attached to the cells. In case of salt flow measurement, the conductance electrode was inserted into one side of cells.

Flux measurement: The volume changes in the capillary were measured against time and the volume flux was determined from the slope of the straight line. For the salt flux, the concentration changes vs. time were determined by using conductance electrode and the flux was evaluated from the slope in the same manner as the volume flux. The osmotic pressure gradient was produced by adding the sucrose of appropriate concentration into one side of the chambers. KCl were used as solute in this study. During experiment, the systems were kept at constant temperature, 25 ^oC and were stirred by the Teflon coated magnetic stirrers.

Results and Discussion

The amphoteric ion exchange and charged mosaic membranes may be possibly advantageous for pressure dialysis or diffusion dialysis of salt solution owing to the presence of opposite charges in its bodies if they do not form the ion pairs. In our study, the above two membranes and the cation exchange membrane were used to observe the volume flux and salt flux in membrane-electrolyte solution system.

According to Kedem & Katchalsky[2], the practical phenomenological equations at nonequilibrium state are at constant temperature and total current, I = 0, described as

$$J_v = L_p(\Delta P - \Delta\Pi_i) - \sigma L_p \Delta\Pi_s \quad , \qquad (1)$$

$$J_s = C_s(1 - \sigma)J_v + \omega\Delta\Pi_s \quad , \qquad (2)$$

where Jv and Js are volume and salt fluxes, respectively: ΔP, mechanical pressure, $\Delta\Pi_i$ and $\Delta\Pi_s$, osmotic pressure due to impermeable and permeable solutes; Cs, mean concentration, respectively. Furthermore, Lp, σ and ω denote filtration coefficient, reflection coefficient and salt permeability which characterize the membrane.

In eqns.(1) and (2), if one of the driving forces is canceled, each characteristic parameter, Lp, σ and ω can be described as

$$L_p = (J_v/ (\Delta P - \Delta\Pi_i))_{\Delta\Pi_s, I} \quad , \qquad (3)$$

$$\sigma = (1/L_p)(J_v/ \Delta\Pi_s)_{\Delta P - \Delta\Pi_i, I} \quad , \qquad (4)$$

$$\omega = (J_s/ \Delta\Pi_s)_{J_v, I} \quad . \qquad (5)$$

The experimental works were performed under such conditions that the above parameters can be obtainable. On the present system, only the osmotic pressure gradients were imposed and the mechanical pressure was not applied.

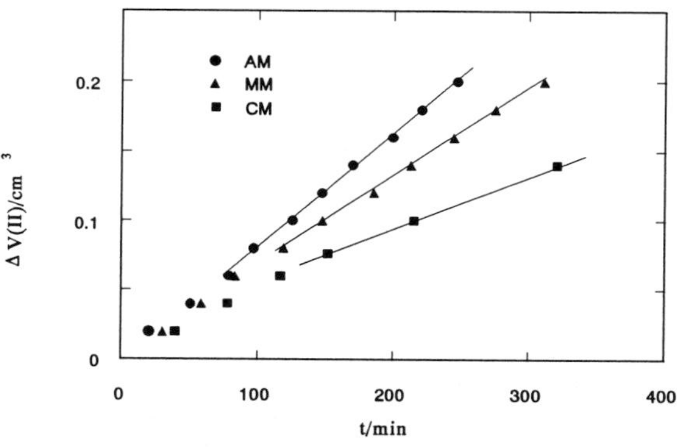

Fig. 1. Volume change vs. time in W(I) : 0.5M SA(II) system.

The filtration coefficients, Lp's were obtained with the systems in which the pure water (Phase I) and 0.5 moldm^{-3} sucrose solution(Phase II) were poured into glass chambers across the membranes. The volume changes showed the flows from Phase I to Phase II. Figure 1 indicates the relation between the volume changes and the elapsed times about three different membranes. Provided that the linear portions reflect the steady state in the system, the slopes divided by the effective membrane area become to be the volume flux, Jv. Then, Lp can be estimated by dividing Jv with the force imposed to the system, $\Delta\Pi_i$. The results were given in Table 2.

TABLE 2

Filtration coefficient, reflection coefficient, and solute permeability

Membrane	Lp $10^{-13}cm^3dyn^{-1}sec^{-1}$	σ	ω $10^{-15}moldyn^{-1}sec^{-1}$
AM	3.04	0.73	0.21
MM	2.78	-4.17	4.91
CM	1.62	1.0	0

As for the reflection coefficient, σ, the experiments were carried out in the concentration cell system, i.e., 0.01M KCl-0.1M KCl system without the added sucrose solution. The dependences of the volume changes on time were shown in Fig.2. In the similar manner to Lp, σ was estimated from the slope with taking into account the imposed force, $\Delta\Pi_s$. It should be noticed in Table 2 that the σ in MM system indicates the negative value owing to the volume flux against the osmotic flow.

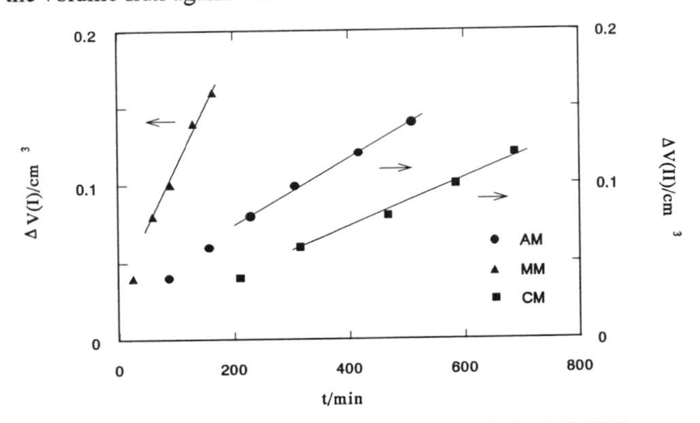

Fig. 2. Volume change vs. time in 0.01M KCl : 0.1M KCl system.

Figure 3 shows the experimental results concerning the solute permeabilities, ω's. The cell system is essentially the same as the case of σ but the 0.5M sucrose solution was added to the Phase II in the system. The estimated values, ω's were also given in Table 2. In this case, to satisfy the condition, Jv = 0, the chamber of the Phase II was closed with a glass plug and the concentration change of KCl in Phase I was observed by inserting the conductance electrode. The result about CM was omitted in Fig.3 because the concentration change was too small.

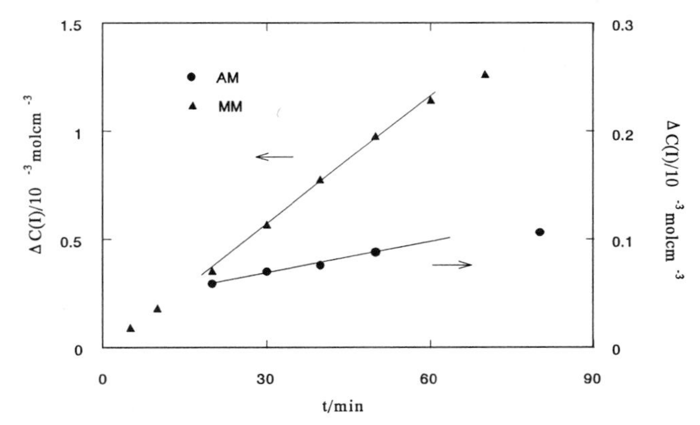

Fig.3. Salt concentration change vs. time in 0.01M KCl(I) : 0.1M KCl + 0.5M SA(II) system.

The obtained Lp in Table 2 is regarded as a measure to know what degree a membrane filtrates the solvent, i.e., the water per unit pressure in this case. The values indicated the same magnitude as those in literature, and were the orders of 10^{-13} $cm^3 dyn^{-1} sec^{-1}$[3]. However, comparing the three membranes, the sequence of the values became AM>MM >CM. In general, the filtration coefficients depend on the pore size in the membrane. The difference in this system is likely caused due to the other than the pore size because the skeletons of the ion exchange membrane have the dense networks and the sizes are considered to be almost the same. In the present situation, the water molecules within the membrane are under the influence of both cationic and anionic exchange groups. The transport of water molecule through MM and AM may be accelerated by two opposite charged sites. However, at present, the exact reason for the difference can not be explained.

σ reflects the immigration rates of the solute and solvent through the membrane. σ = 1 means that the solute is perfectly interrupted by membrane barrier and only the solvent molecules can permeate the membrane. σ = 0 means that the solute and solvent do pass freely through the membrane and there are no selectivity between the two species. The values of σ in Table 2 became to be depended on the membrane species. σ in CM gave unity within experimental errors and this suggested that there is no transport of the solute molecules. In the case of AM, the value turned out to be between 0 and 1, and the result suggested that the membrane can transport a more or less the solutes. Interestingly, σ in MM became the negative reflection coefficient. This tells us that the solute can be transported in preference to the solvent against the osmotic pressure.

The solute permeability, ω about three different membranes indicated the results with the same tendency as expected from the value of σ. The magnitudes were the orders of 10^{-15} mol dyn^{-1} sec^{-1} which consisted with those in literature[3]. As shown in Table 2, ω in the charged mosaic membrane was larger value than that in the amphoteric ion exchange membrane. In constrast with that, the cation exchange membrane did not transport the salt as given in Table 2. This means that the former two membranes would be useful to concentrate the salt solution by means of the pressure dialysis. Thus, for two membranes having the possibility of salt enrichment, more advanced study was developed below.

When two opposite ion exchange groups are arranged in parallel, Kedem & Katchalskey presented the relation between fluxes and forces as follows[2],

$$J_V = L_{p\,m}(\Delta P - \Delta\Pi_i) - \kappa(\beta_a - \beta_c)\Delta\Pi_s / 4C_s F \quad , \qquad (6)$$

$$J_s = -\kappa(\beta_a - \beta_c)(\Delta P - \Delta\Pi_i)/4F + \kappa\Delta\Pi_s / 4C_s F^2 \quad , \qquad (7)$$

where κ, β and F denote the membrane conductance, electroosmotic coefficient and Faraday constant, respectively. Further Lpm and ω_m indicate the filtration coefficient and solute permeability in the membrane with charges in parallel array, respectively and are given as the following equations,

$$L_{p\,m} = (L_{p\,c} + L_{p\,a})/2 + \kappa(\beta_a - \beta_c)^2 / 4 \quad , \qquad (8)$$

$$\omega_m = (\omega_c + \omega_a)/2 + \kappa / 4C_s F^2 \quad . \qquad (9)$$

Provided that two opposite charges in the amphoteric ion exchange membrane also assume to be the parallel array microscopically, κ and β in eqns.(8) and (9)can be estimated from the results in Table 2. The results were given in Table 3. In calculation, Lpc and Lpa assumed to be equal to Lp of the cation exchange membrane and ω_c and ω_a were neglected as zero because each part of ion exchange sites can not individually contribute to salt transports.

TABLE 3

Membrane conductance and electroosmotic coefficient

	κ	β
	$10^{-2}\Omega^{-1}cm^{-2}$	$10^{-2}cm^3coul^{-1}$
AM	0.446	2.01
MM	10.9	0.326

The obtained membrane conductance, κ is closely related to ω, and reflects a circulating current which occurs within the membrane [2-3]. Also from the comparison with AM and MM in Tables 1 and 3, it can be speculated that there exist the correlation with the ion exchange capacity. This exhibits that the introduction of ion exchange groups into membrane is the condition to get the effective ion enrichment. On the other hand, the electroosmotic coefficient, β represents water transport per unit coulomb and the value indicates the quantity which the salt flow is accompanied with. Accordingly small value of β in MM compared with that of AM supports that the efficiency of ion enrichment will be superior over AM.

Acknowledgement

The present work was partly supported by Grant in Aid for Scientific Research No.92009 from The Salt Science Research Foundation.

References

1 K.Sollner, Biochem. Z., 244(1932)370.
2 O.Kedem and A.Katchalsky, Trans. Faraday Soc., 59(1963)1918, 1931.
3 J.N.Weinstein, B.J.Bunow and S.R.Caplan, Desalination, 11(1972)341.
4 C.R.Gardner, J.N.Weinstein and S.R.Caplan, Desalination, 12(1973)19.
5 J.Shorr and F.B.Leitz, Desalination, 14(1974)11.
6 Y.Miyaki and T.Fujimoto, Membrane, 8(1983)212.
7 Y.Matsushita, H.Choshi, T.Fujimoto and M.Nagasawa, Macromolecules, 13(1980) 1053.
8 A.Yamauchi, Y.Okazaki, R.Kurosaki, Y.Hirata and H.Kimizuka, J. Membrane Sci., 32(1987)281.
9 Y.Nagata, K.Kohara, W.Yang, A.Yamauchi and H.Kimizuka, Bull. Chem. Soc. Jpn., 59(1986)2689.

The Use of Membranes in the Activated Sludge Process

Ian M. Reed, Denise L. Oakley and Linda Y. Dudley
AEA Environment and Energy, B353, Harwell Laboratory,
Didcot OX11 ORA

INTRODUCTION

Conventionally, "activated sludge" is treated by gravity sedimentation to produce a clarified overflow and thickened sludge bottoms, a portion of which may be recycled to the aeration vessel. Membrane processes have been tested as a polishing step for final effluent to remove particulates and micro-organisms, thus producing a very high quality effluent. An alternative is to use membrane processes as the primary solid/liquid separation stage in order to replace sedimentation and improve recycle of the activated sludge. These applications have been studied by a number of workers (Fane (1), Kanayama et al (2), Sammon and Stringer (3), Suwa et al (4)). A compact membrane-based activated sludge process could find application in cases where a high discharge quality must be guaranteed or where space limitations prohibit the use of gravity sedimentation.

MOD (5) has investigated the use of a hollow fibre microfiltration system to treat overflow from settling to ensure effluent quality. Under normal conditions this system successfully produced water of very high quality. However, when the carry over from the sedimentation tank contained even moderate levels of suspended solids the membrane module rapidly blocked and fluxes fell to near zero. The aim in this paper is to demonstrate that membranes can be used reliably to separate activated sludge and to study the factors affecting membrane performance.

This paper describes the results of trials on a range of ultrafiltration and microfiltration membrane systems operating on an activated sludge from an extended aeration process receiving domestic sewage. The performance of a number of different membrane geometries, both flat sheet and tubular, is described and the behaviour of each is assessed. A comparison is also made between the performance of several membrane types. These membranes were made from 4 different materials and ranged in pore size from a molecular weight cut-off of 25,000 to a nominal pore size of 0.2μm.

MATERIALS AND METHODS

Activated sludge

Activated sludge, derived almost exclusively from domestic sources, was collected from a treatment works in Stanford-in-the-Vale. The mixed liquor suspended solids concentration (MLSS) was approximately 3800 mg/l. No signs of chocolate mousse formation were evident on sludge collection days and it was reported that it had not been a problem at this site.

Samples of activated sludge were withdrawn directly from the aeration tank at the treatment plant and stored in carboys prior to use for a maximum of 2 weeks.

A validation run was also carried out using activated sludge from a pilot scale activated sludge plant at HMS Sultan, Gosport. The sewage was derived exclusively from a domestic source and was collected by a vacuum system similar to that used on ships. Unsettled activated sludge was drawn directly from the aeration stage.

Membrane test rigs

Tubular test rig The membrane test rig is shown schematically on Figure 1. This rig was specially constructed so that a range of different ultrafiltration and microfiltration modules could be used. Activated sludge is circulated around the system by a fixed speed lobe pump, which is capable of delivering 30 l/min at pressures of up to 6 bar. Control of the operating pressure and the crossflow rate are achieved by adjusting the flow control valve and the pump by-pass. The recirculating liquid is passed through a heat-exchanger in order to prevent the sludge from becoming too hot. Both the retentate and permeate streams are recycled to the feed tank so as to maintain a constant solids concentration.

The following variables were measured throughout each run; inlet and outlet pressure to the module, feed temperature at the heat-exchanger outlet and in the feed tank, recirculation flowrate, and permeate flowrate. Pressures were measured using Transamerica type BHL-3020-01 transducers linked to a Maywood Instruments D 2000 digital pressure indicator. Temperatures were monitored using chrome/alumel thermocouples in stainless steel sheaths. These were connected to a Comark 2501 electric thermometer via a Comark 2037 selector switch. Flowrates were measured manually.

The retentate flowrate was determined using a volumetric jug and the permeate flow was determined gravimetrically.

Flat sheet test rig Flat sheet membranes were tested using a DDS Minilab 10 system. The Minilab 10 was a small flat plate membrane system designed to accommodate four flat sheets of membrane in series with a total effective membrane area of 336 cm^2. The feed channels in the module had rectangular cross-section, 1mm in height and 1cm in width. There were two separate permeate channels each serving a pair of membranes, thus allowing two types of membrane to be tested simultaneously. Feed was recirculated using a fixed speed gear pump. Pressure and recirculation rate were regulated by diaphragm valves.

Membranes and modules

Table 1 gives the material, nominal pore size and configuration of the membranes used in this study. Four membrane materials were tested; two in the ultrafiltration range and two microfilters with 0.2 μm nominal pore sizes.

Membrane Material	Nominal Pore Size	Module Configuration
polyethersulphone	25,000 cut-off	tubular 12.5 mm diameter
PVDF	200,000 cut-off	tubular 12.5 mm diameter
polypropylene	0.2 μm	tubular 5.5 mm diameter
prototype MF	0.2 μm	tubular 4 mm diameter

Table 1 List of membranes used in the study

Membrane equipment

The following membrane systems were studied using the membrane test rig:
(i) PCI Membrane Systems MIC-RO 240. This consists of two tubular membranes, each 12.5 mm in diameter and 61 cm in length, retained by a porous stainless steel support and connected in series. This gives a total effective membrane area of 240 cm^2. The membranes themselves were cast onto a paper tube, which fits inside the porous support,

and could be replaced by removing the module end caps. This system was used to test the polyethersulphone membrane.
(ii) Enka Microdyn MD 020 TP 2N. This module was made up of three tubular membranes, in parallel, each with an internal diameter of 5.5 mm and 0.75 m in length. This gives a total membrane area of 360 cm^2. The polypropylene membrane elements were fusion welded to the module housing.
(iii) A prototype membrane module 1.0 m in length containing seven 4.0 mm diameter tubular membranes (effective membrane area 850 cm^2). Individual membrane elements were not replaceable.
(iv) DDS Minilab 10. For description see above.

Analytical methods

Bacterial analysis Counts of colony forming units in feed and permeate samples were taken at intervals during selected runs. Samples were plated onto; i) tryptone soya agar (TSA) plates to detect total aerobic colony forming units and ii) MacConkey agar (code CM7b) plates to determine the coliform concentration.

Chemical oxygen demand (COD) Chemical oxygen demand of selected feed and permeate samples was measured using a Hach Model 45600 COD reactor. Samples were analysed using the digestion method described in the manufacturers instructions.

Mixed liquor suspended solids (MLSS) The solids content of feed and permeate samples was determined by measuring the dry weight content. A known mass of sample (approximately 100 g) was evaporated to dryness in an oven set at 105°C. The weight of the dried sample was measured and the dry weight content calculated based on the assumption that the density of the liquid sample was 1 g/ml.

Membrane trials

Each membrane was tested in a series of runs each of 2 hours duration. As far as possible, operating conditions were kept constant throughout the run. However, due to the large cooling capacity of the chiller unit, the temperature of the circulating sludge usually fell slightly during operation.

At the end of each trial the membranes were cleaned in place using the alkaline cleaning agent Ultrasil 11. The standard cleaning procedure was as follows:
(i) dissolve 25 g of Ultrasil 11 in 10 l of warm (50°C) water,

(ii) drain the rig and circulate the cleaning solution for 30 minutes at low transmembrane pressure,
(iii) drain the rig and rinse with cold tap water,
(iv) repeat step (iii) until the circulating water remains clear.

RESULTS AND DISCUSSION

Tests on the Minilab 10 system

The Minilab 10 system was tested with a sample of activated sludge from Stanford in the Vale. After only 10 minutes operation, pressure within the system began to rise rapidly and the membrane ruptured. On inspection, it was discovered that the channels connecting each of the membrane plates had become blocked with flocs of activated sludge, even though the feed channels were much more open than in the hollow fibre module tested by MOD. The sludge flocculates so quickly that any constrictions in the flow path or flow dead-spots can very rapidly become plugged. As a result the remaining tests concentrated on the use of wide channel membrane geometries.

Performance of the polyethersulphone membrane

Figure 2 shows the flux history of the polyethersulphone ultrafiltration membrane. Three two hour runs were performed. The breaks in the curve indicate where the system was shut down for cleaning. The membrane was operated at a mean pressure of 2.1 bar and a Reynold`s number of 29,000 (crossflow velocity - 3.1 m/s). Runs 1 and 3 were carried out with well mixed activated sludge (total solids concentration 2.8 g/l and 2.3 g/l respectively). Run 2 was performed with partially settled activated sludge at a solids concentration of 8.6 g/l.

Flux behaviour for each run was typical of crossflow membrane processes, i.e. there was a rapid initial flux decline over a period of 30 minutes followed by a slow progressive decrease. For all three runs the final flux was in the range 36 to 41 l/m^2h. As can be seen from the initial fluxes for each run, cleaning with Ultrasil 11 gave almost 100% flux recovery.

Performance of the PVDF membrane

Figure 3 shows the flux history of the PVDF membrane, which had a nominal molecular weight cut-off of 200,000. Five runs were carried out using activated sludge from Stanford in the Vale at MLSS levels of 2.1 g/l, 0.9 g/l, 1.7 g/l, 1.0 g/l, and 2.1 g/l. Operating conditions were the same as for the polyethersulphone membrane. The final flux declined progressively from 52 l/m^2h for the first run to 37 l/m^2h in run five. Nevertheless performance was satisfactory throughout and cleaning gave good flux recovery.

Performance of the polypropylene membrane

In all, five runs were performed with the polypropylene membrane. Runs 1 to 4 were carried out with sludge from Stanford in the Vale. The sludge was well mixed but measured MLSS for the four runs was, in sequence, 7.5 g/l, 5.6 g/l, 4.9 g/l, and 4.2 g/l. Run 5 was performed using activated sludge from HMS Sultan. The MLSS was 5.2 g/l. The system was run at a Reynold`s number of 22,000 (crossflow velocity - 5.2 m/s). Runs 1 and 2 were carried out at a transmembrane pressure of 2.4 bar. This was reduced to 1.4 bar for subsequent runs.

The complete flux history for this membrane is shown in Figure 4. Again breaks in the curve indicate where the system was shut down for cleaning. In the first run the final 'steady' flux was only 33 l/m^2h, which was considerably less than in the subsequent runs. This suggests that the membrane, which is naturally very hydrophobic, might not have been completely water wetted. Run 4, which was carried out at 1.4 bar transmembrane pressure, produced a final flux in excess of 50 l/m^2h compared to 43 l/m^2h for Run 2 operating at 2.4 bar. This slightly surprising result was probably due to a faster rate of membrane fouling at the higher pressure. In addition, water wetting of the membrane might have been somewhat better in Run 4, due to the action of the cleaning agent, leading to greater permeability. Although the pore size of the polypropylene membrane was very much greater than that of the polyethersulphone membrane (microfiltration versus ultrafiltration) there was little difference in the permeate fluxes that each produced. This is not uncommon in crossflow membrane processes and occurs because, given sufficiently permeable membranes, flux is controlled by material from the process stream depositing on the membrane.

As for the polyethersulphone membrane, cleaning polypropylene membranes with Ultrasil 11 gave excellent flux restoration.

Figure 5 shows the performance of the polypropylene membrane with the activated sludge from HMS Sultan over a continuous period of 12 hours. Performance was similar to that obtained with the standard activated sludge. Flux was stable over the entire period of operation. Table 2 shows the bacterial content and chemical oxygen demand (COD) for the raw activated sludge and selected permeate samples from this run. No micro-organisms were detectable in any of the permeate samples tested and membrane processing gave 20-30 fold COD reduction. The final effluent COD was 150 mg/ml.

	Chemical Oxygen Demand (mg/l)	Total Aerobic Units (CFUs/ml)	Total Coliforms (MacConkey) (CFUs/ml)
Activated Sludge	4800	$1.5x10^6$	$2.6x10^6$
Sludge Supernatant	2400		
Permeate t=10 mins	260	NDG	NDG
Permeate t=465 mins	185	NDG	NDG
Permeate t=685 mins	150	NDG	NDG

Table 2 Permeate Quality for Polypropylene Membrane

CFU - colony forming units
NDG - no detectable growth

Tests on the prototype microfiltration membrane

A prototype tubular MF membrane with a nominal pore size of 0.2 μm was also tested on samples of activated sludge from Stanford in the Vale. Runs were carried out at a Reynolds number of 13,000 (crossflow velocity - 4 m/s) and a mean transmembrane pressure of 1.8 bar. The flux fell steadily to around 7 l/m^2h after 2 hours operation (see Figure 6). This was less than 20% of the fluxes achieved with the other two membrane types. Permeate quality achieved using this membrane was also poor. Statistically significant levels of micro-organisms were detected. This suggests that the true pore size of the membrane was somewhat larger than stated and may account for the low flux. If the pores are large enough they may be blinded by particulate material from the activated sludge. It appears therefore that choosing a membrane with too large a pore size can be counter productive.

Effect of membrane type on performance

Filtration rate did not appear to be sensitve to the type of membrane used. The PVDF, polyethersulphone and polypropylene membranes all gave fluxes in the range 40-50 l/m^2h even though they were made of different materials and the pore sizes varied by a factor of about 50. This is common in crossflow membrane processes since, provided there is no specific interaction between membrane and particulate matter in the feed, fluxes are determined primarily by fouling layers that build up on the membrane surface. In the case of the prototype membrane just such an interaction appears to have taken place. In this case the suspended solids appear to be of approximately the same size as the membrane pores and, as a result, caused pore blinding.

Effect of solids concentration on membrane performance

The performance of the membrane did not appear to be sensitive to the solids concentration as is illustrated in Figure 7. Flux for the polyethersulphone membrane at 8.6 g/l solids was only slightly less than at 2.8 g/l. It is likely that similar trends would be observed with other membrane types.

CONCLUSIONS

1. The work conducted here has clearly demonstrated that membrane processes can be used successfully to clarify activated sludge with MLSS of up to 8.6 g/l.
2. Similar performance was obtained on both the activated sludge from HMS Sultan and that taken from a largely domestic source in Stanford-in-the-Vale, upon which most of the tests were carried out.
3. Membrane processing can produce a clarified product which is essentially free of viable micro-organisms and has a low COD.
4. Performance was found to be relatively insensitve to the type of membrane used, provided membrane pore size was not too large. However, the prototype MF membrane produced poor quality permeates and unreliable fluxes, probably because the membrane pore size was too large.
5. Narrow channel flat plate membranes have been found to be susceptible to channel blockage by flocculated activated sludge. Wide channel systems are the most suitable for processing activated sludge.
6. Testing at various dilutions of the activated sludge suggests that fluxes in wide channel systems are affected only slightly by changes in the solids concentration of the feed. As a result membrane processes could be used in place of sedimentation and sludge recycle.
7. In terms of ease of cleaning, stability of flux and overall flux levels the best performance of the membranes used in this study was given by the Enka membrane. This membrane is a polypropylene microfilter with a rated pore size of 0.2 µm. The

membrane gave stable operation over a period of 12 hours when tested with activated sludge from HMS Sultan. Membranes could also find a role in the clarification of final effluent.

8. From the trials on the Enka membrane it appears that increasing the applied pressure can increase the rate of membrane fouling. Operation should therefore be carried out at the lowest pressure that gives the desired permeate flowrate. Applied pressure can be increased progressively during a run to maintain this flowrate.

ACKNOWLEDGEMENTS

This work was financed by the Ministry of Defence (Navy). The authors would like to thank Thames Water for supplying activated sludge from their Stanford-in-the-Vale plant and Wayne Grocutt for operation of the plant at HMS Sultan.

REFERENCES

1. Fane A G (1986). "Ultrafiltration: Factors Influencing Flux and Rejection." from "Progress in Filtration and Separation" Ed Wakeman R, p101. Elsevier Science Publishers.

2. Kanayama H, Tomoyasu T, Katayama S (1987). "Water Treatment by Use of a Membrane Bioreactor System." Proc. 1987 International Congress on Membrane and Membrane Processes, Tokyo 1987.

3. Sammon D C, Stringer B (1975). "The Application of Membrane Processes in the Treatment of Sewage." Process Biochemistry, March 1975, p4.

4. Suwa Y, Suzuki T, Toyohara H, Yamagishi T, Urugushigawa Y (1992). "Single-stage, Single-sludge Nitrogen Removal by an Activated Sludge Process with Cross-flow Filtration". Water Research, vol. 26, no. 9, p1149.

5. MOD. Private communication.

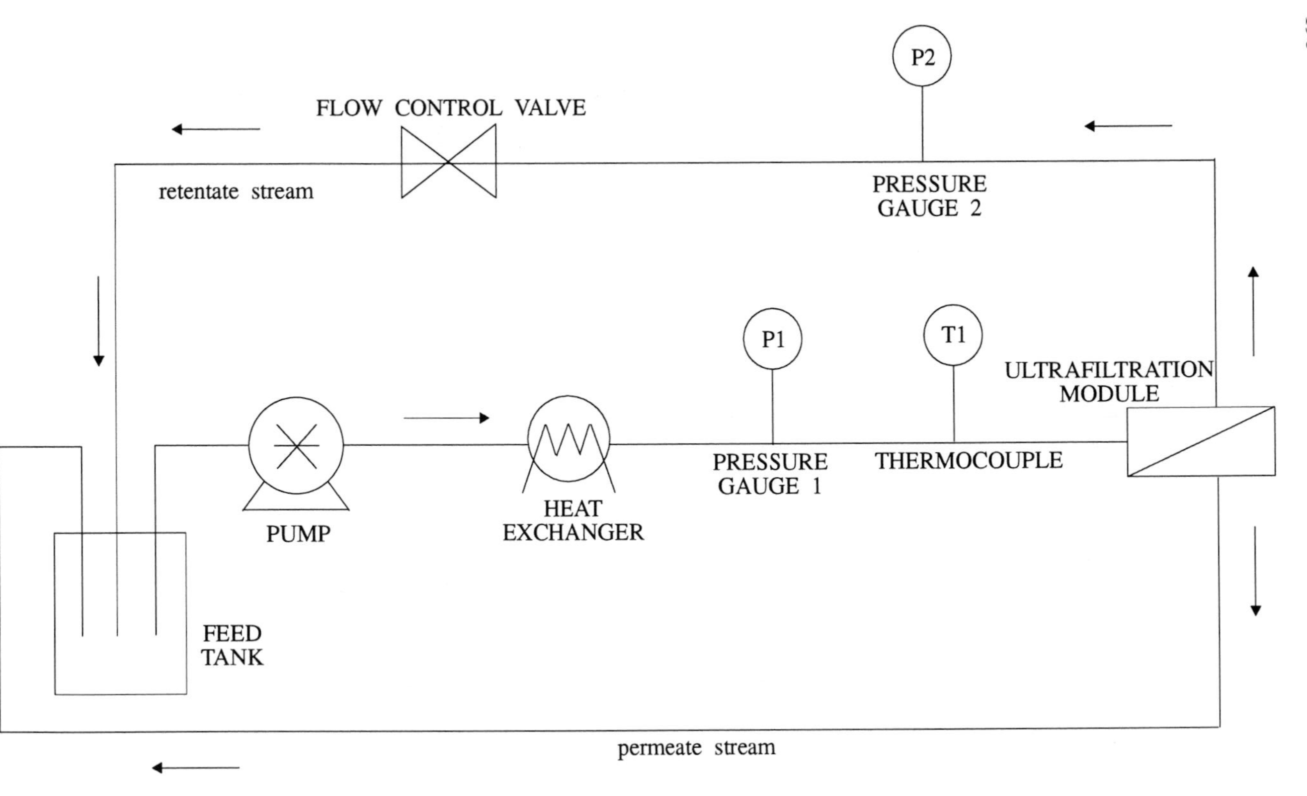

FIGURE 1 SCHEMATIC DIAGRAM OF PCI MEMBRANE UNIT

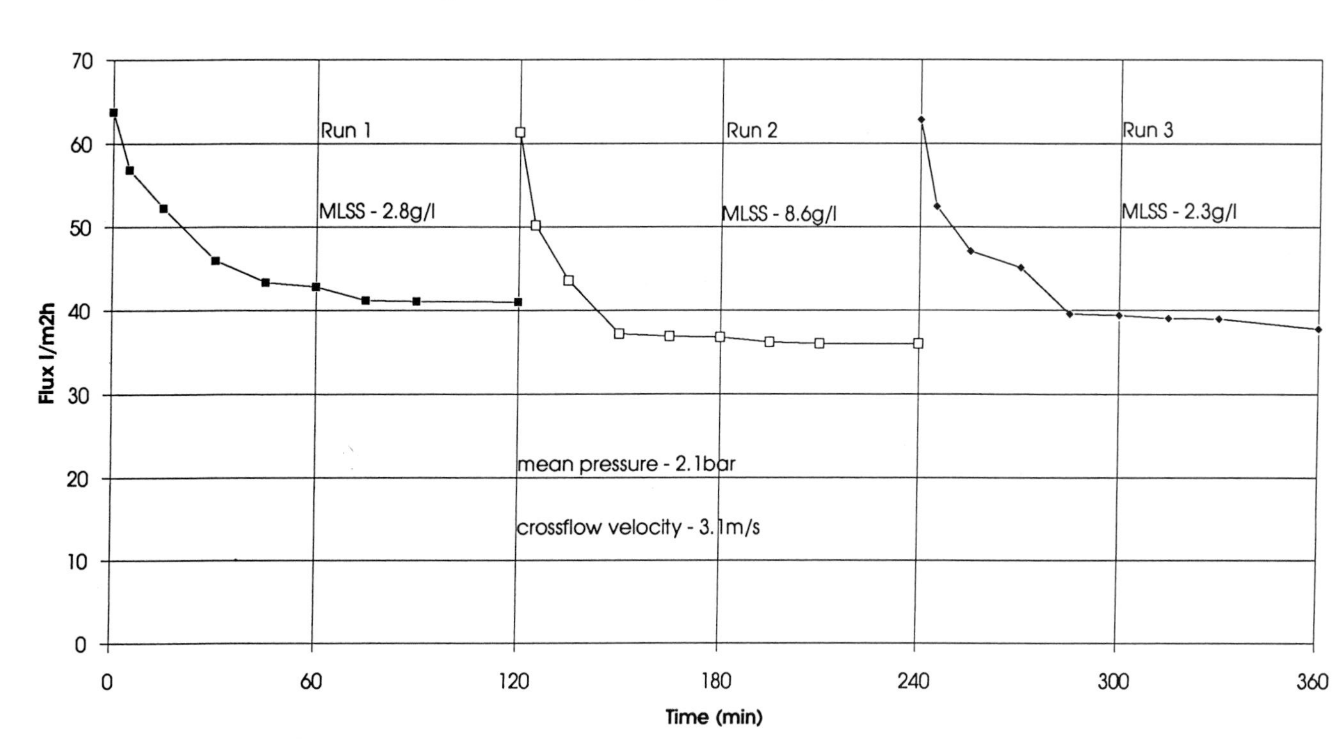

Figure 2 Flux History of Polyethersulphone Membrane

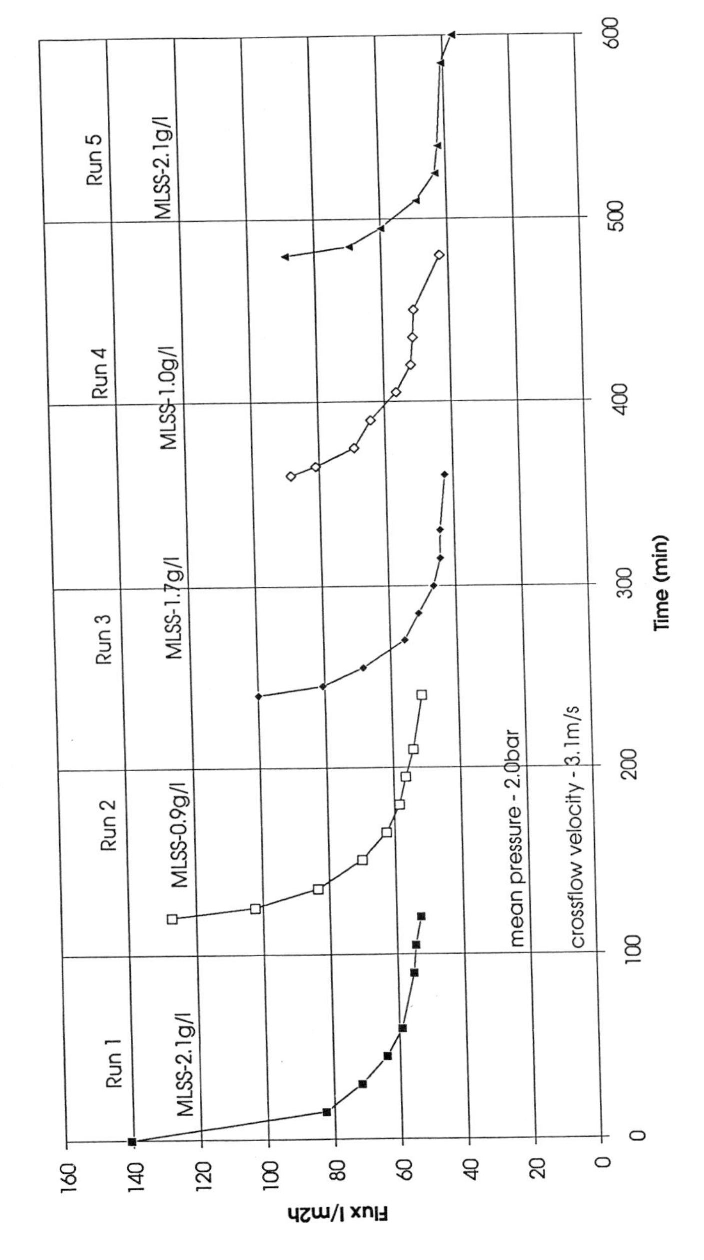

Figure 3 Flux History of PVDF Membrane

Figure 4 Flux History of Polypropylene MF Membrane

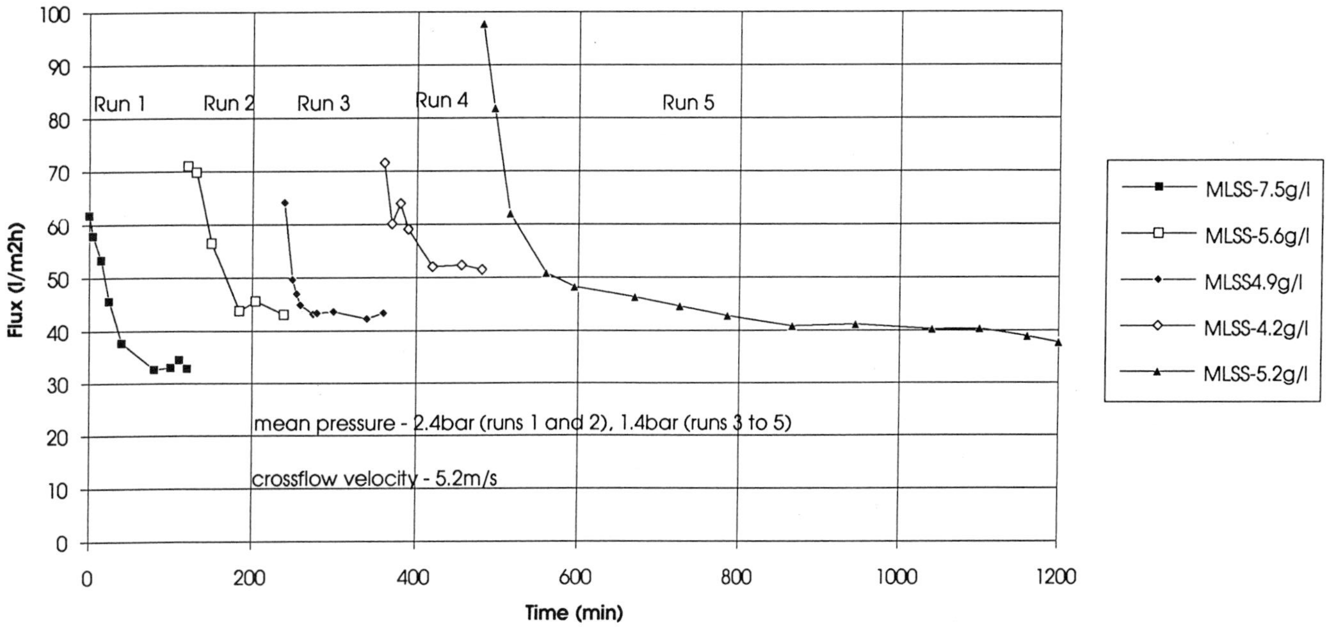

Figure 5 Performance of Polypropylene MF Membrane Activated Sludge from HMS Sultan

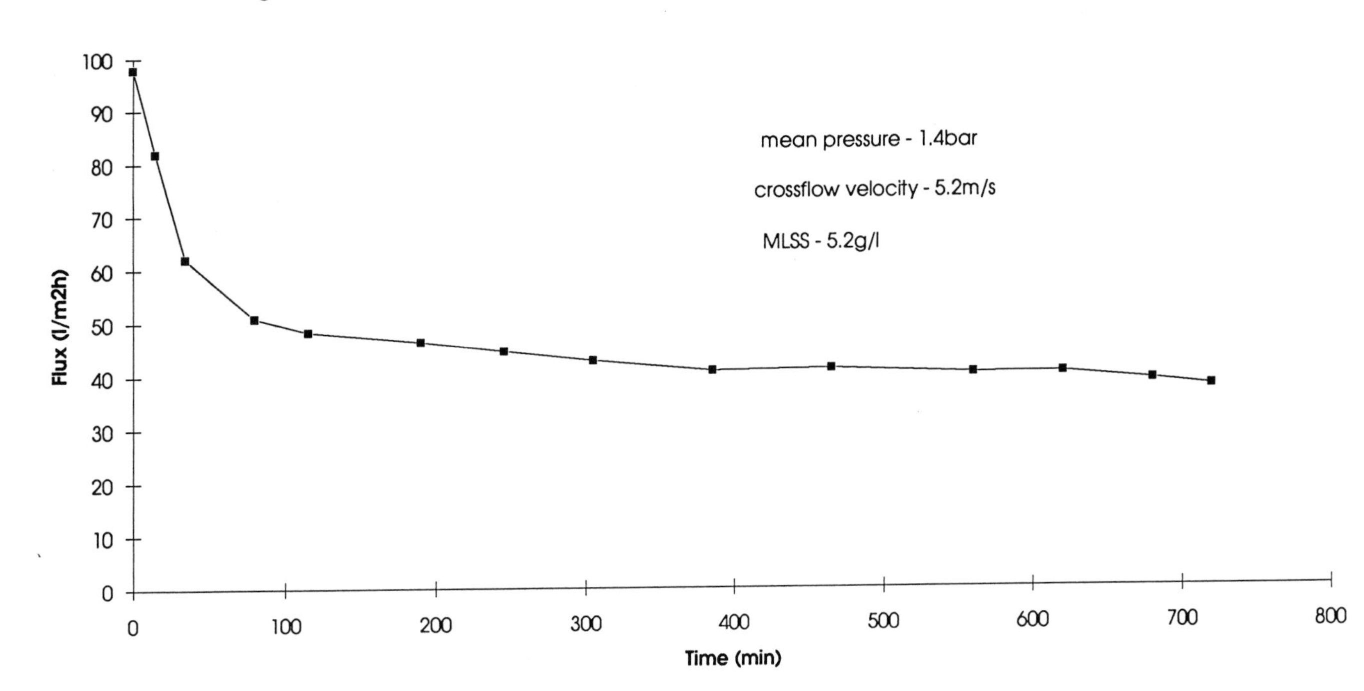

Figure 6 Flux of Prototype MF Membrane

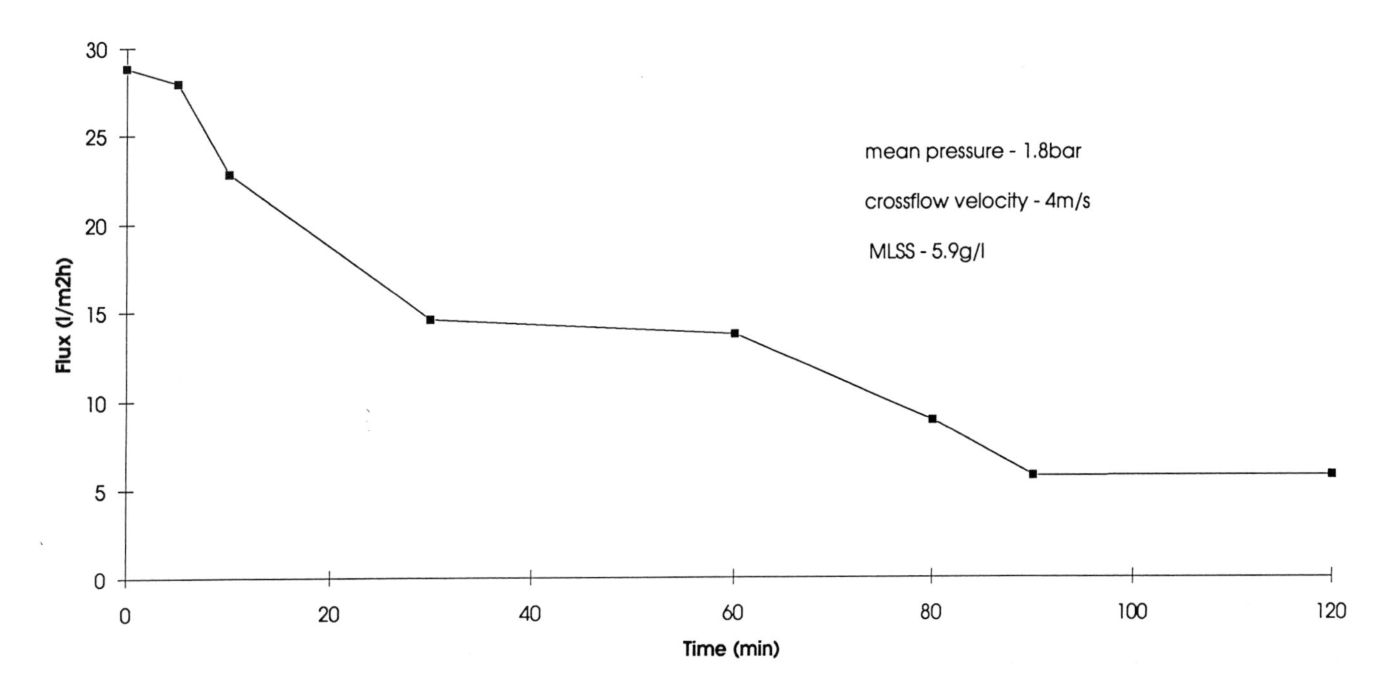

Figure 7 Effect of Solids Concentration on Performance of Polyethersulphone Membrane

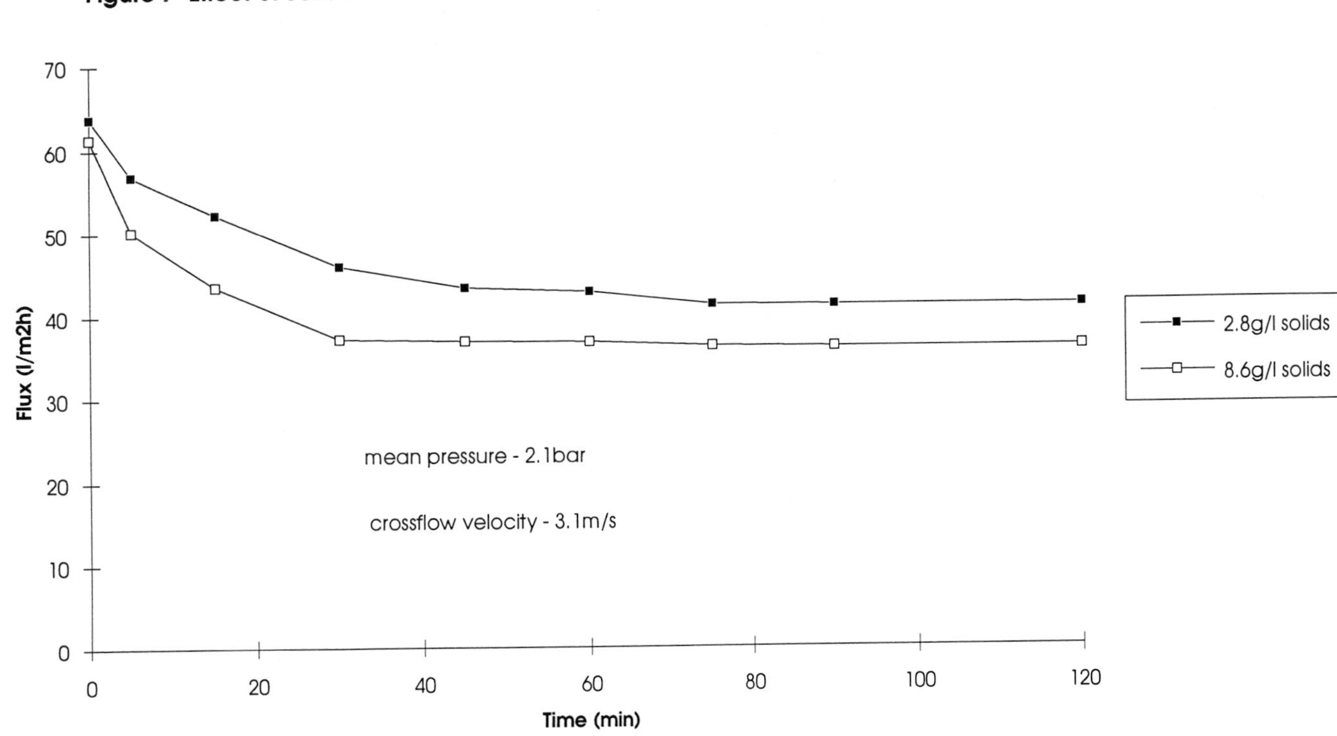

THE EFFECT OF ULTRA- AND NANOFILTRATION ON THE REMOVAL OF DISTURBING SUBSTANCES IN THE PAPER MACHINE WATER CIRCULATION SYSTEM

Jutta Nuortila-Jokinen, Susana Luque*, Lena Kaipia and Marianne Nyström

Laboratory of Technical Polymer Chemistry, Department of Chemical Technology
Lappeenranta University of Technology, P.O. Box 20, SF-53851 Lappeenranta, Finland
* Department of Chemical Engineering, University of Oviedo, 33071 Oviedo, Spain

ABSTRACT

Today the trend in the Pulp and Paper industry is towards totally closed water circulation systems because of stringent environmental legislation and economical aspects. Therefore, in the paper machine water circulation system studied here, the interest has been focused on the clear filtrate, which is formed after a fibre recovery unit. Today most of the clear filtrate is discharged as effluent and it constitutes more than half of the total effluent load of a paper mill. If this clear filtrate could be reused in the paper machine water system it would radically reduce the total effluent load and the consumption of fresh water of the paper mill.

In this study the removal of disturbing substances from the clear filtrate by ultra- and nanofiltration was studied with both modified and unmodified membranes. Also a combined ultrafiltration/nanofiltration method was tested. The results were judged by regained pulp brightness.

The results from the ultrafiltration experiments showed some improvement in brightness. Generally the long chained lignins and extractive substances were retained well in the retentates. The reduction of short chained lignin residuals and carbohydrates was modest and neither was the reduction of COD very impressive. The modification usually improved the reduction of long-chained negatively charged components and increased flux, when good electrostatic repulsive conditions were achieved.

In the nanofiltration experiments the brightness was totally regained and the reductions of all different substances as well as of some ions were very good. The flux achieved was rather small for a commercial application, either because it was a low flux membrane or because the flux was reduced due to fouling of the membrane.

In the combined ultrafiltration/nanofiltration experiments the ultrafiltration permeate was used as feed solution. The results showed less improvement in brightness than expected, although the reductions of COD, anionic trash and lignin were very good. The sugars were not equally well retained.

This study showed that by nanofiltration the clear filtrate can be cleaned from brightness decreasing and disturbing substances and it could therefore be recycled in the paper machine water circulation system without hesitation.

INTRODUCTION

Today the trend in the Pulp and Paper industry is to close up their water circulation systems as far as possible because of stricter legal requirements of effluent treatment and waste water loadings as well as because of economical aspects.

In the white water system of a paper mill there are many benefits in closing the system /1/. First of all the consumption of fresh water can be minimized. Also the need of added chemicals is reduced and there are also less losses in fibre, filler and fines. In a closed white water system the temperature is elevated compared to an open system thus reducing the cost of heating. On the other hand the closure causes increased concentrations of suspended and dissolved solids in the water circulation system. Therefore, the closure can lead to increased amounts of deposits in the system as well as to increased corrosive effects.

Today most of the clear filtrate is discharged as effluent from the paper machine water circulation system and it constitutes more than half of the total effluent load of a paper mill. The reuse of the clear filtrate would radically reduce the total effluent load and the consumption of fresh water in the paper mill water system.

In a previous study the use of ultrafiltration (UF) for the purification of different streams in the Pulp and Paper industry was studied /2/. In this study the focus is on the possibilities to remove brightness decreasing and disturbing substances, such as pitch, lignin residuals and hemicellulose, from the clear filtrate of a paper machine by ultra (UF)- and nanofiltration (NF) with modified and unmodified membranes.

MATERIALS

The clear filtrate

The clear filtrate studied here is formed after a fibre recovery unit (save all disc) in a paper machine water circulation system as shown in Fig. 1. Today the secondary use of the clear filtrate is minimal compared to the stream, which is led to the channel. Therefore this study is focused on the possibilities to reutilise the clear filtrate instead of discharging it as effluent. The paper machine in question uses peroxide bleached mechanical pulp. The clear filtrate is rather free from fibres and long chained molecules, but it contains fines, salts, residuals of lignin, hemicellulose and cellulose, pitch components as well as fillers and retention agents. Its pH value is about 5 and its conductivity 0.94 mS/cm. The polysaccharide content of the filtrate is usually less than 100 ppm and it can be expected that the sugar molecules are rather short chained. The content of lignin residuals and extractive substances is relatively low, but anyway the water decreases the brightness of pulp about 1-3 units. The chemical oxygen demand, COD, of the filtrate is 700 - 2000 mg/l.

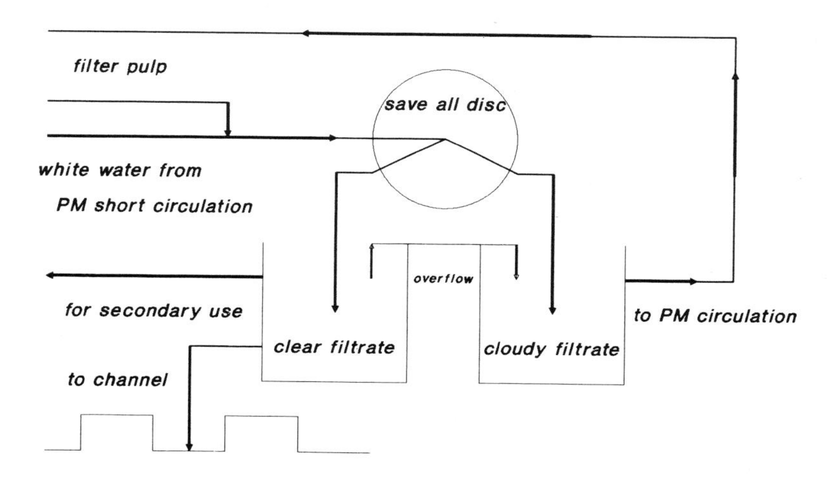

Figure 1. The clear filtrate formed after the fibre recovery unit in the paper machine water circulation system.

The modification agents

The modification agents used were all long chained negatively charged polyelectrolytes. Dextran sulphate (DEXSU) provided by Pharmacia is a polydisperse anionic dextran, M_w = 500 000 g/mol. Its ability to make a polysulphone membrane more negatively charged has been verified by streaming potential measurements /3/. The anionic polyelectrolytes Fennopol A 305 and A 321, produced by Kemira, are copolymers of acrylamide and acrylate. The Fennopol A 321 has a lower viscosity and is more anionic than Fennopol A 305.

METHODS

The **ultrafiltration** experiments were carried out in a tubular PCI module described in detail elsewhere /4/. The PCI membrane used was a PU 608 made of polysulphone with a cut-off = 8 kD. The effective membrane area is 452 cm^2.

The **nanofiltration** experiments were carried out in a laboratory flat sheet module made of polycarbonate. The effective membrane area is 53 cm^2. The centrifugal pump used was a Hydra-Cell D-10XBSVHHHH. The filtration unit can be used up to pressures of 25 bar, but in this study only pressures up to 12 bar were used. The membranes used were NF40 (FilmTec), NTR-7410 and NTR-7450 (Nitto Denko). The NF40 membrane is a thin film composite membrane made of aromatic polyamide. The NTR-7410 and NTR-7450 membranes consist of a sulphonated polyether sulphone skin layer on a polysulphone support layer. All of the membranes are negatively charged. The pure water fluxes at 25 °C for the NF40, NTR-7410 and NTR-7450 membranes were about 15, 120 and 60 $l/(m^2h)$ at 6 bar respectively.

The water used throughout all experiments was ion exchanged and filtered through a Millipore RO unit giving conductivity < 1 μS/cm.

The original clear filtrate as well as the resulting concentrates and permeates were analyzed for ionic content (conductivity, pH, ion chromatography (IC) and atomic absorption spectroscopy (AAS)), lignin residuals (UV/VIS-absorption at 280 and 400 nm), sugar content (anthron-sulphuric acid colour method), chemical oxygen demand, COD (Hach reactor, standard SFS 5504) and anionic polyelectrolytes (cationic demand (CAD) titrated with 0.001 N poly-DADMAC using Mütek particle charge detector). The brightness decreasing effect was tested either by preparing brightness cakes from a 3 % fibre suspension of bleached birch and the water to be tested (incubation for 20 hours, 40 °C) or by filtering the sample through a cake made of eucalyptus pulp and pure RO water (100 ml 3 times through, 40 °C). The brightness was measured by an Elrepho 2000 spectrophotometer. The results have been judged by the regained brightness calculated as follows:

$$B\% = \frac{B_p - B_f}{B_w - B_f} * 100\% \qquad (1)$$

where
B% the regained brightness in per cent
B_p the brightness given by the permeate
B_f the brightness given by the feed solution
B_w the brightness given by pure RO-water.

The reductions of different substances have been calculated by comparing the concentrations of the substance in permeate and in feed:

$$R=(1-\frac{c_p}{c_f})*100\% \qquad (2)$$

where R reduction in per cent
c_p the concentration in permeate
c_f the concentration in feed.

RESULTS AND DISCUSSION

The **ultrafiltration experiments** with unmodified and modified membranes were made alternating pressure, flow rate and temperature. The brightness tests were carried out with brightness cakes made of bleached birch pulp and the water to be tested. The means of reductions and regained brightness of all experiments are shown in Table I and Fig. 2. It can be seen that membrane modification with Dextrane sulphate (DEXSU) gives improved reductions for lignin residuals and COD. The brightness was regained totally or even better than for the RO-water. With the Fennopol modifications it can be seen that the more anionic nature of Fennopol A 321 causes better reductions of sugar, lignin and anionic trash content compared to Fennopol A 305. The reduction in COD with Fennopol modifications is smaller than for the unmodified membrane or with DEXSU modifications. Neither is the brightness regained with Fennopol modifications very impressive. Anyway, it can be seen, that by ultrafiltration long chained lignins, which are at least partly responsible for the brightness decrease in the pulp, can be removed from the clear filtrate and that DEXSU modifications of the PU 608 membrane enhances the removal of short chained lignins, polysaccharides and COD. The DEXSU modification also increased the flux, because of the good electrostatic repulsive conditions achieved.

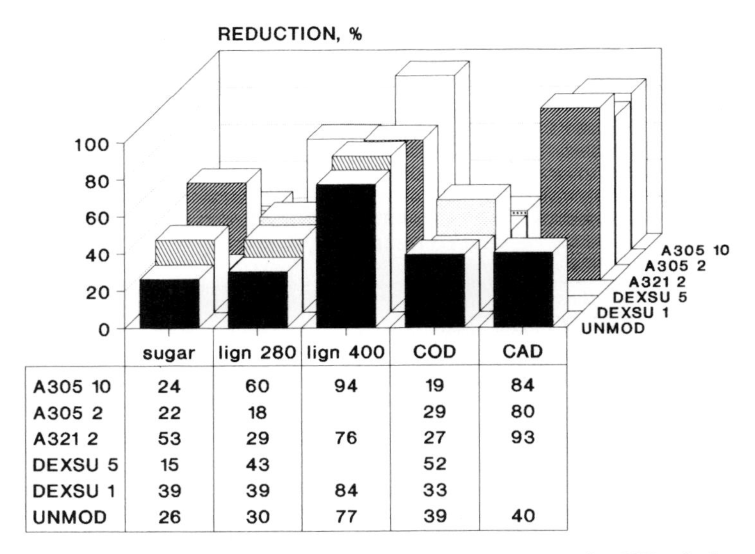

	sugar	lign 280	lign 400	COD	CAD
A305 10	24	60	94	19	84
A305 2	22	18		29	80
A321 2	53	29	76	27	93
DEXSU 5	15	43		52	
DEXSU 1	39	39	84	33	
UNMOD	26	30	77	39	40

Figure 2. The means of reductions of different components in UF of clear filtrate, using unmodified and modified PU 608 membranes. Modifications used: Dextrane sulphate (DEXSU) 1 and 5 ppm, Fennopol A 321 2 ppm, Fennopol A 305 2 and 10 ppm.

Table I Operating conditions and achieved brightness values in UF of clear filtrate. Unmodified and modified PU 608 membranes. Modifications used: Dextrane sulphate (DEXSU) 1 and 5 ppm, Fennopol A 321 2 ppm, Fennopol A 305 2 and 10 ppm. The brightness values have been scaled so that the brightness for RO-water is 87.0 %.

	UNMODI-FIED	DEXSU 1 ppm	DEXSU 5 ppm	FENNOPOL A321, 2 ppm	FENNOPOL A305, 2 ppm	FENNOPOL A305, 10 ppm
PRESSURE, bar	1.5-4.0	2.5-4.0	2.5	2.5	1.5-2.5	1.5-2.5
FLOW RATE, m/s	0.41-0.69	0.41	0.45-0.96	0.45-0.69	0.53-0.66	0.53-0.66
TEMPERATURE, °C	25	25	25-35	25-35	30-40	40
BRIGHTNESS:						
RO-water	87.0	87.0	87.0	87.0	87.0	87.0
Original	84.0	84.2	85.0	85.8	85.2	80.4
Permeate	85.0	87.2	87.1	86.2	85.9	84.2
REGAINED %	33	107	105	33	39	58

In the **nanofiltration experiments** the FilmTec NF40, Nitto Denko NTR-7410 and NTR-7450 membranes were used. The last mentioned membrane was also modified with 0.1 w-% Dextrane sulphate solution, called here NTR-7450M. The operation conditions were always kept the same: pressure 6 bar, flow rate 2 m/s and temperature 40 °C. The fluxes and flux reductions of the clear filtrates after filtration of 1/3 of the volume at 40 °C were:

Membrane	Flux $l/(m^2h)$	Flux reduction %
NF40	15	7
NTR-7410	48	57
NTR-7450	33	61
NTR-7450M	30	52

As it can be seen from Table II the brightness of the test pulp (samples filtered through a test pulp cake made of eucalyptus,) was totally regained. Also the reductions for long chained lignin residuals and anionic trash measured as CAD were > 90 %. The polysaccharides were retained almost as well: reductions > 87 %. The COD as well as short chained lignin reductions varied, the COD reduction being from 47-78 % and lignin reduction 59-93 %. (Fig.3)

Table II The gained brightness values in the nanofiltration of clear filtrate. Membranes used NF40, NTR-7410, NTR-7450 and NTR-7450M (modified with 0.1 w-% Dextrane sulphate). Eucalyptus pulp.

Brightness	NF40	NTR-7410	NTR-7450	NTR-7450M
RO-water	89	89	89	89
Original	88	88	89	89
Permeate	89	89	90	90

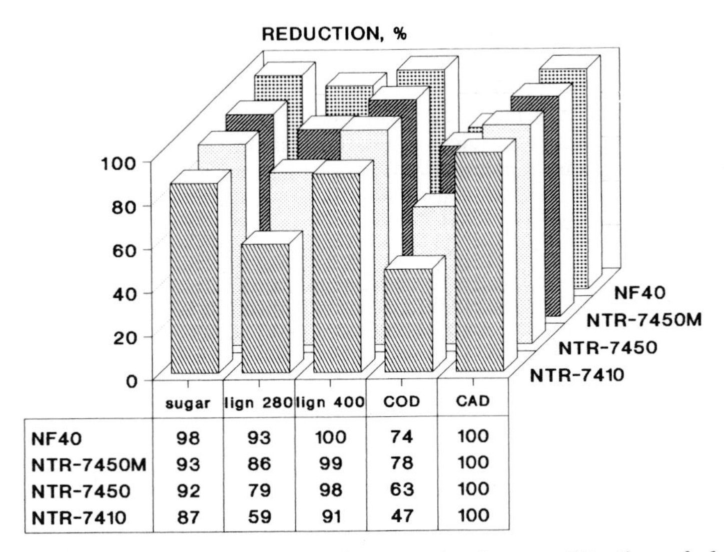

	sugar	lign 280	lign 400	COD	CAD
NF40	98	93	100	74	100
NTR-7450M	93	86	99	78	100
NTR-7450	92	79	98	63	100
NTR-7410	87	59	91	47	100

Figure 3. The reductions of different substances in the nanofiltration of clear filtrate. Membranes used NF40, NTR-7410, NTR-7450 and NTR-7450M (modified with 0.1 w-% Dextrane sulphate).

The reductions of different ions were studied as well (Fig. 4). Multivalent anions and cations are retained well by all nanofiltration membranes tested, but the reduction of Cl^- is poor. The poor reduction is probably caused by the presence of sulphate ions, which according to literature decrease the retention of Cl^-. It is possible that the positively charged ions are retained in the fibrous sublayers of the Nitto Denko membranes. The reduction of Fe^{3+}-ions was very good and also the reduction of Ca^{2+} was from satisfactory to very good (75-98 %). The removal of Fe^{3+} is important due to the fact that these ions cause increased corrosive effect when enriched in the water circulation system. As important is the removal of Ca^{2+}-ions because of their ability to form insoluble calcium soaps, which form large colloidal particles when enriched in the system. Other important ions which most probably exist in the clear filtrate, such as Al^{3+}, were not analysed.

The PU 608 polysulphone membrane with 5 ppm DEXSU modification and the NF40 membrane were selected to the **combined ultrafiltration/nanofiltration** experiments. In these experiments the clear filtrate was first ultrafiltrated and then the UF permeate was used as feed solution in the nanofiltration experiments. In

UF a pressure of 2.5 bar and a flow rate 0.78 m/s was used. In NF the operation conditions used were: pressure 6 bar and flow rate 2 m/s. The temperature was 40 °C in all experiments. The average flux (40 °C) in UF was 122 l/(m^2h) and in NF 20 l/(m^2h) and the flux reductions were 9 and 3 % respectively. Although the overall separation was fairly good, the brightness did not improve as much as expected, the total regained brightness being only 11 %. This is probably because the UF did not perform as well as in our previous experiments and therefore the UF permeate used as feed solution in NF contained more of the brightness decreasing agents than it was supposed to. This can be seen from the poor reductions of lignins, most of all the long chained lignins, carbohydrates and extractive substances (Fig. 5). The reason for this behaviour in UF might be due to the rather high flow rate used suggesting that the reversible adsorption or concentration polarization layer acts as a prefiltering bed.

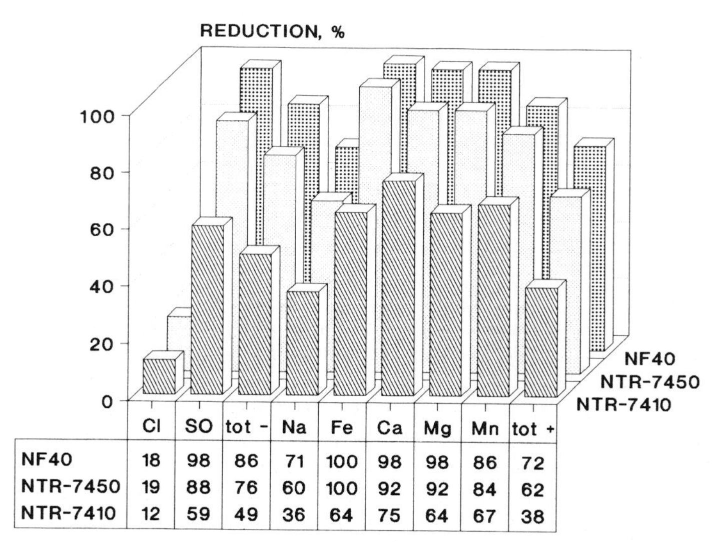

	Cl	SO	tot -	Na	Fe	Ca	Mg	Mn	tot +
NF40	18	98	86	71	100	98	98	86	72
NTR-7450	19	88	76	60	100	92	92	84	62
NTR-7410	12	59	49	36	64	75	64	67	38

Figure 4. The reductions of important different ions in NF of clear filtrate. The tested ions were: Cl^-, SO_4^{2+}, Na^+, Fe^{3+}, Ca^{2+}, Mg^{2+} and Mn^{2+}. The reduction of total -/+ charge means the reduction of all negative/positive ions tested for in the clear filtrate. Membranes used NF40, NTR-7410 and NTR-7450.

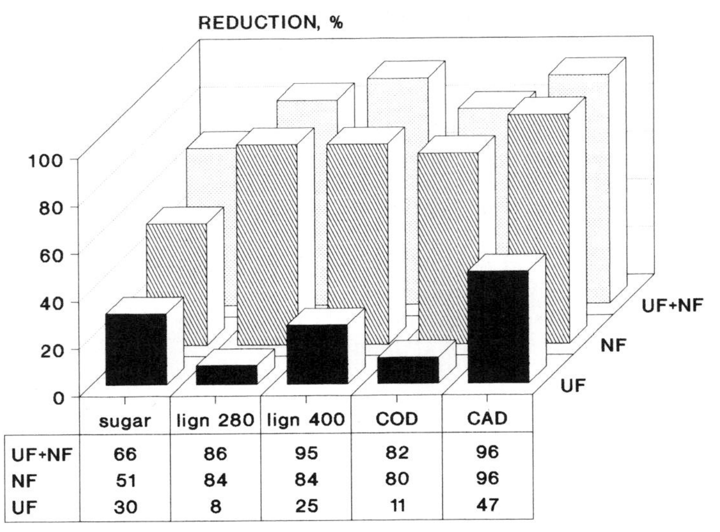

	sugar	lign 280	lign 400	COD	CAD
UF+NF	66	86	95	82	96
NF	51	84	84	80	96
UF	30	8	25	11	47

Figure 5. The reductions of different substances in combined UF/NF of clear filtrate. Membranes used PU 608 with Dextrane sulphate (DEXSU) 5 ppm modification (UF) and NF40 (NF).

CONCLUSIONS

The results from the ultrafiltration experiments of the clear filtrate show some improvement in brightness. In most of the experiments the long-chained lignins and the extractive substances were retained well in the retentates. The reduction of short-chained lignin residuals and carbohydrates was modest, from 15 to 35 %, and neither was the reduction of COD very impressive. The modification usually improved the reduction of long-chained negatively charged components and improved the flux.

Nanofiltration with the NF40 membranes gave the best results. The brightness was totally regained and all reductions of different substances were > 93%, except for the COD reduction, which was 74 %. Also the reduction of important ions like Fe^{3+} and Ca^{2+} was very good, 100 % and 98 % respectively. This shows that NF of clear filtrate produces a very clean permeate, which does not contain brightness decreasing agents. Hence this permeate can be used in the white water circulation system of a paper machine without hesitation. If the flux could be improved to the same level as usually obtained with UF, the economics would also become feasible.

In the combined UF/NF experiments the results showed far less improvement in brightness than expected, although the total reductions of different substances were good or very good. Further studies of combined UF/NF are to be done in order to achieve better results in UF. The importance of UF would mainly be in cleaning the clear filtrate from particulate matter like fines and clay.

Thus it has been shown that the permeate resulting from nanofiltration of the clear filtrate is very clean and practically free from most of the brightness decreasing substances as well as from corrosive ions (Fe^{3+}) and colloid forming ions (Ca^{2+}). Therefore this permeate can be reused in a paper machine water circulation system.

ACKNOWLEDGEMENTS

The authors wish to thank The Finnish Academy of Science for financial support. They would also like to thank all those students of Lappeenranta University of Technology, who have worked in this project, for their technical assistance.

REFERENCES

1 Panchapakesan, B. "Closure of mill whitewater systems reduces water use, conserves energy", Pulp & Paper, (1992) 57-60.

2 Nyström, M., Uusluoto, T., Nuortila-Jokinen, J. "Removal of disturbing substances by ultrafiltration from make-up waters in the pulp and paper industry; Effect of electrostatic repulsion." Submitted.

3 Wahlgren, M.C., Sivik, B., Nyström, M. "Dextran modifications of polysulfone UF-membranes: Streaming potential and BSA fouling characteristics", Acta Polytech. Scand., Ch-series, **194** (1990) 1-18.

4 Nyström, M. "Ultrafiltration of O/W emulsions stabilized by limiting amounts of tall oil", Colloids Surf., **57** (1991) 99-114.

The use of Microfiltration to Remove Colour and Turbidity from Surface Waters Without the Use of Chemical Coagulants.

R. A. Morris, I. Watson, S. Tsatsaronis,
Memtec Limited, Windsor, NSW, 2756, Australia

Abstract

Microfiltration units have been used consistently over the past four years as alternatives to conventional water purification plants for the removal of taste, odour and colour from various surface water sources in Australia.

Data has been collected at various Australian sites, namely:- Argyle, Western Australia; North Richmond, New South wales; Bantry Bay, New South Wales; Molendinar, Queensland, and Tooberac, Victoria. In addition, a long term pilot study was conducted at Fredericks Reservoir, Maryland, United States of America. Data from these studies is reviewed.

The water quality emanating from the Continuous Microfiltration (CMF) systems achieves or betters relevant current Australian and international guidelines. Physical parameters show the filtrate to have a true colour of less than 15 Platinum Cobalt Units (PCU) and typically 5 PCU or less in most cases. The turbidity of the filtrate is down to 0.2 NTU even when the feed varies from 5-500 NTU. In addition, iron levels are typically reduced to less than 0.2 mg/l. Taste and odour were removed in tests of a site affected by algal blooms.

1. Introduction

Conventional water plants for the treatment of surface waters use chemicals such as aluminium sulphate (alum) to coagulate particles. These particles are removed by clarification and sand filtration. In addition to the coagulant, caustic soda or lime addition may be required to correct pH. Chlorine is finally added to provide disinfection.

Conventional treatment may be affected by feed quality variations which in turn can affect treated water quality. The use of chlorine does not effectively kill chlorine resistant microorganisms such as giardia lamblia or cryptosporidium. An additional problem is the possible effect downstream of chlorinated by-products[1].

Memcor® microfiltration units have been used consistently over the past four years as alternatives to conventional water purification plants for the removal of taste, odour, and colour from various surface water sources in Australia[2]. Microfiltration is a single stage treatment and does not involve the complexities of chemical addition for particulate removal. The membrane acts as a barrier to suspended matter, bacteria, and algae. Hence, the microfiltration process is not affected by variation in feed quality (insoluble material) and filtrate quality is consistent. The process is also an effective means of providing primary disinfection by removal of suspended organics and microorganisms which in turn minimises chlorine addition downstream. More than 99.99% removal of giardia lamblia and cryptosporidium cysts were observed in laboratory studies conducted by Hibler[3] at Colorado State University and Godfree[4] at Altwell Ltd., UK, respectively.

The main reasons for the ability of the continuous microfiltration to produce a high quality filtered water are the pore size of the membrane and a novel gas backwash process. The system is engineered to remove particles down to 0.2μm in size and control the membrane fouling (gel) layer effectively. The feed is forced through the filter under pressure, hence further material is removed on the dynamic membrane (gel) layer. The gel layer is produced from bacteria, cells, debris and other material found in conventional water sources. The gel layer reduces the effective pore size to less than 0.2μm which enhances filtration rating. All the feed is filtered and no settling is required, hence, even low density material is removed.

The feed contaminants collected on the membrane surface are removed efficiently by a novel gas-backwashing technique. At regular intervals, air at 600 kPa (90 psi) is injected into the centre of the fibres. This air is then released through the walls of the membrane matrix and shakes loose any build up of contaminants. The air is forced through the pores of the membrane and dislodges surface contaminants. Feed is introduced to the module and a concentrated stream of contaminants is removed from the module. This backwash process takes approximately 90 seconds to complete. The backwash process is totally automatic and the backwash stream can be diverted for disposal. A diagram showing a cross section of a single membrane hollow fibre during normal operation and gas-backwashing is seen in Figure 1.

Figure 1. The membrane under normal operation and during gas backwash.

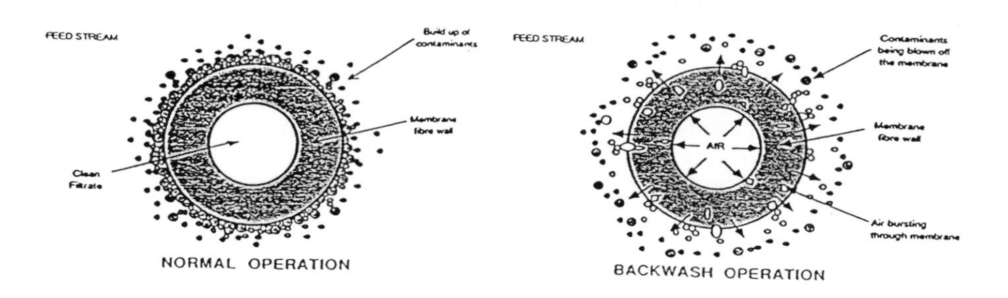

Many communities receive water which does not meet the Australian guidelines, yet even these guidelines are not in the forefront of standards around the world[5]. One key measure of water quality is turbidity - a measure of water clarity. The Australian guideline is currently a turbidity of 5 NTU. The World Health Organisation guideline is 1 NTU for disinfection, while the guide level under the EC drinking water directive is 0.4 NTU. The microfiltration process consistently achieves turbidities of less than 0.2 NTU and in most cases is less than 0.1 NTU[6].

Another key measure of water quality is colour - typical maximum permissible guideline levels around the world is less than 20 PCU. The term "true colour" is used to indicate water from which turbidity has been removed. The term "apparent colour" refers to soluble material and suspended matter. Consequently, the CMF unit directly removes the suspended matter fraction of the apparent colour value. Where soluble contaminants are present, some coagulant addition may be required. The level of chemical addition is typically less than conventional methods.

Data has been collected at various Australian surface water sites, namely:- Argyle, Western Australia; North Richmond, New South wales; Bantry Bay, New South Wales; Molendinar, Queensland; and Tooberac, Victoria. In addition, a long term pilot study was conducted at Fredericks Reservoir, Maryland, United States of America. In each case the microfiltration plants were installed as single-stage treatments with only mechanical screening prior to microfiltration. Data from these studies is reviewed. Some comparative tests using chemical coagulants prior to microfiltration were undertaken.

2. North Richmond/Bantry Bay, NSW, Australia

Memtec have carried out two extensive pilot trials in the Sydney area, namely at North Richmond Water Treatment Works and Bantry Bay Reservoir. The turbidity of the raw water at North Richmond is prone to large fluctuations in turbidity. During periods of heavy rainfall the Hawkesbury river can flood which causes the increase in turbidity. Turbidities range from 5-150 NTU. However, the majority of the time the turbidity of the microfiltered filtrate was less than 0.2 NTU. In addition, this trial showed a reduction in iron levels to less than 0.1 mg/l. Colour was not measured at North Richmond. At Bantry Bay the apparent colour of the water emanating from the reservoir was 7.0-15.0 platinum cobalt units (PCU). After microfiltration the true colour of the water was 2.0-5.0 PCU. In addition, iron levels were reduced from 0.4 mg/l to <0.01 mg/l. Refer Table 1.

Table 1. Bantry Bay Reservoir, Water Quality Tests

	Raw Water	Microfiltration Filtrate	EC Level Guide/Maximum
Turbidity NTU	1.2 - 4.6	0.2 Mean	0.4/4
Iron mg/l	0.4 - <0.1	0.01	0.05/0.2
Colour PCU	7.0 - 15.0 (Apparent)	2.0 - 5.0 (True)	1/20

3. Argyle, Western Australia

A continuous microfiltration plant was installed at Argyle Diamond Mine, (ADM) in December 1991. The ADM is situated at the top end of North West Australia in a very remote location. The ADM uses the microfiltration plant to service approximately 300 residents in the Argyle village and occasionally produces process water for the alluvial plant. The plant was initially fitted with 300-m^2 membrane surface area and is currently fitted with 600-m^2 to provide 90 m^3/h capacity.

Previously water provided to the residents straight from Lake Argyle was of very poor quality. The CMF unit produces water with low suspended solids (0.06ppm) even when wide fluctuations in the feedstream quality occur from the wet season to the dry season (300ppm-5ppm). The apparent colour of the water was reduced from 90-100PCU to a true colour of 10-15 PCU, in wet season tests. Refer Table 2.

Table 2. Argyle Water Quality Tests

	Raw Water	Microfiltration Filtrate
Turbidity (NTU)	3 - 500	0.2 Mean
Suspended Solids (ppm)	5 - 300	0 - 0.6
Colour (PCU)	100 - 90	<15 - 10
Total Iron ppm	0.05 - 0.4	<0.1
pH	7.5 - 8.8	7.5 - 8.8
Conductivity ($\mu S/cm$)	190 - 200	200 - 210
Alkalinity $CaCo_3$ppm	17 - 24	17 - 24

4. Tooborac, Victoria, Australia

The water quality of a reservoir at Tooborac in Victoria was extremely poor due to algal blooms and the related taste and odour problems. A continuous microfiltration unit (60-m^2) was installed in March 1991 in order to provide water for the local town (500 citizens). Plant capacity is 7 m^3/h. Microfiltration removed the taste and odour problems associated with the algal blooms. Refer Table 3.

Table 3. Tooborac Water Quality Tests

	Raw Water	Filtrate from Memtec CMF
Turbidity NTU	4.7 - 29	0.1
Conductivity μS/cm	545	540
Colour PCU	10 - 100 (apparent)	2.5 - 10 (true)
Iron (mg/l)	0.6	<0.1
pH	7.2 - 8.8	7.2 - 8.8
Alkalinity $CaCo_3$	80	80
Taste	ND	Excellent
Odour	Weedy/muddy	None
Total Aerobic Count/ml	3100	<1
E. Coli/50ml	<1	<1
Algae/ml Algae Species - Brateacoccus - Anabaena (blue green algae - only present Dec. 91) - Dinobryn - Peridinium	430,000	<50 limit of detection

5. Fishing Creek Reservoir, Frederick, Maryland, USA

The Surface Water Treatment Rule (SWTR) in the United States requires the use of a treatment process, usually filtration and disinfection, to produce potable water from a surface water source. A trial was conducted in conjunction with the Johns Hopkins University at the Fishing Creek Reservoir, Frederick, Maryland[7,8]. The trial assessed the performance of microfiltration in relation to the SWTR. The plant operated continuously for a period of over 600 days[9], at a nominal flux rate of 120 lph/m^2 and trans-membrane pressure of 35 kPa.

The microfiltration process consistently produced a turbidity of less than 0.2 NTU despite variations in feed turbidity. Total coliforms were reduced from 100 units/100ml to the level of detection. The ability of the process to remove Giardia Lamblia was tested via seeded challenges of Giardia Muris. Results indicated complete removal of Giardia cysts at a challenge level of 10^4 - 10^5 cysts.

Tests were conducted to assess the performance of microfiltration for the removal of trihalomethane formation potential[10]. Approximately 20% of the THM precursors were removed by microfiltration directly. Trials using chemical coagulants (alum at 14 mg/l and ferric chloride at 12 mg/l) reduced THMFP by approximately 60%. The coagulant addition did not significantly alter flux rates.

Table 4. Water quality Tests without coagulation pretreatment.
Fishing Creek Reservoir, Frederick, Maryland, USA

Parameters	Raw Water	CMF Filtrate
Turbidity, NTU	0.5 - 12	<0.2
pH	6.68 - 7.41	6.6 - 7.4
Alkalinity, mg/l	2.1 - 4.5	2.1 - 4.4
Particle count, particle/ml	400 - 1000	<50
Particle size distribution, microns	0.5 - 15	<0.5
Total coliform bacteria, /100ml	10 - 100	<1
Heterotrophic plate count cfu	10 - 1000	1 - 100
MS2 Bacteriophage (Seeded) Log PFU/100ml	5.69 - 6.82	4.02 - 6.08
Total organic carbon, mg/l	1.3 - 5.4	0.8 - 4.4
Dissolved organic carbon, mg/l	0.9 - 3.9	0.78 - 3.5
Trihalomethane formation potential, μg/l	58.6 - 634.7	42.3 - 400
Giardia Muris (Seeded), /100ml	$3.5x10^4$-$5x10^5$	0

6. Molendinar, Queensland, Australia

A study was conducted at the Gold Coast City Council's Molendinar Water Purification Plant to assess microfiltration as an alternative to provide future treatment capacity.

The existing plant treats up to 150 MLD by conventional means. The existing plant which has a capacity of 150 MLPD treats surface water from Hinze Dam (12 km away) by conventional means: alum coagulation, clarification, filtration and chlorine disinfection. In addition, chloride dioxide is added prior to filtration for soluble manganese oxidation and removal.

Pilot studies were conducted both with and without coagulant addition to measure removal of turbidity and colour. The duration of each test varied from several days for each coagulant test to ten days for the test with no coagulant. Flux rates were maintained at greater than 100 lph/m^2.

The tests are summarised as :

i) No coagulant.
ii) Alum dose at 25.6 ppm.
ii) Ferric dose at 12.0 ppm ferric chloride.
iii) Ferrous sulphate dose at 25 ppm.

The results are shown in Table 1.5. The water colour was reduced to less than 15 PCU by microfiltration alone. The addition of coagulant further reduced the colour level. Alum and ferric chloride coagulant produced a filtrate colour level of 2.5 PCU whilst ferrous sulphate reduced colour to 5 PCU.

Table 5. Molendinar, Queensland Water Quality Tests.

	Raw Water	Filtrate from Memtec CMF
Turbidity NTU	8.9 - 2.0	In all cases 0.1 - 0.2
Colour (PCU)	30 - 20 (apparent)	< 15 no chemical dosing 2.5 alum dosing 2.5 ferric dosing <5 ferrous dosing
pH	7.6 - 6.8	7.6 - 6.8
Alkalinity $CaCO_3$ppm	20 - 30	N/A

7. Conclusions

The continuous microfiltration process can provide effective single-stage treatment of surface waters for colour, turbidity, algae and microorganism removal. Colour is removed to the true colour level of the feed. Where this exceeds the requirements for treatment the addition of coagulant prior to microfiltration can be applied. The microfiltration process provides primary disinfection which includes the removal of chlorine resistant microorganisms such as giardia lamblia and cryptosporidium. Long-term studies of pilot and operational plants have demonstrated stable operation and filtrate quality despite variations in feed quality.

References

1. Hileman, Bette. "Cancer Risk found from Water Chlorination". Pg 7-8 July 13 Chemical and Engineering News 1992.

2. Butler, R. Membrane Treatment Plants for Packaged Potable Water Treatment and Re-use Applications - A Comparative Analysis. New Technologies in Water and Waste Water Management. The Asian Food Industries, Kuala Lumpur, August 1992.

3. Hibler, C., Personnel Communication, 1987.

4. Godfree, A., Personnel Communication, 1990.

5. 40 CFR Parts 141 and 142: Drinking Water; National Primary Drinking Water Regulations; Filtration, Disinfection, Turbidity, Giardia Lamblia, Viruses, Legionella, and Heterotrophic Bacteria; Final Rule. Federal Register. 54(124):27486-27541 (June 1989).

6. MacCormick, A.B. Developing a Global Demand for Unique Australian Water and Sewage Technology, Paper to the National Engineering Conference, Newcastle, March 1992.

7. Olivieri, V.P., Parker D.Y., Willinghan, G.A., Vickers J.C., "Continuous Microfiltration of Surface Water". AWWA Membrane Processes Conference. Orlando FL. March 10-13 1991.

8. Olivieri, V.P., Parker, D.Y., Schrott, D.F., Willingham, G.A, and Vickers, J.C. "Continuous Microfiltration for the Treatment of Surface Water". American Filtration Society Conference, Atlanta, GA, October, 1991.

9. Vickers, James C., Willinghan, G.A., Parker, D.Y., Schrott, D.F., "Treatment of Surface Water Using Continuous Microfiltration", Memtec America Corporation and Johns Hopkins University, 1992.

10. Parker, D.Y. "Removal of Trihomethane Precursors by Microfiltration". Johns Hopkins University, 1992.

Energy Efficient Crossflow Microfiltration

Roger S. White, Fairey Industrial Ceramics Ltd., Tecramics Division, Filleybrooks, Stone, Staffordshire, ST15 OPU.

The principal of crossflow filtration can be seen from figure 1. The product to be filtered is passed over a porous surface, which can be tubular or flat in form. The speed at which the product passes across the surface tends to maintain the direction of flow whilst the filtrate passes through the membrane by virtue of a pressure differential (the Trans-Membrane-Pressure - TMP). A multi-channel membrane element is shown in figure 2.

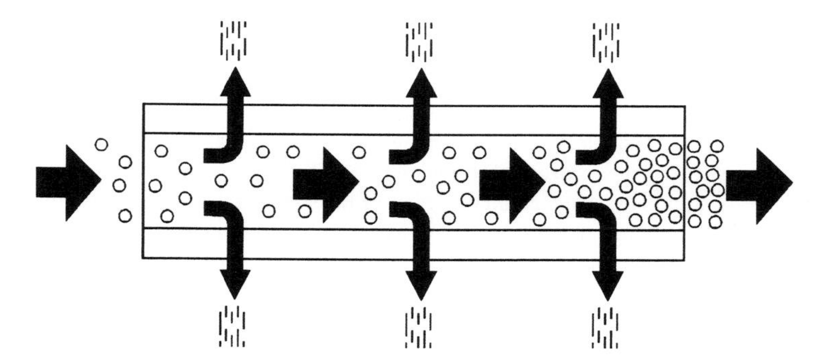

Figure 1. Principle of Crossflow Microfiltration.

If the speed is reduced or the TMP increased beyond optimum operating parameters then the suspended solids will tend to block, or blind, the pores of the membrane. The pore-sizes of membranes range from reverse osmosis (RO) through ultra-filtration (UF) to microfiltration (MF).

In the past, crossflow filtration systems have been restricted in their applications to medium- or high-value products because of the relatively high pumping energy requirement. In order to achieve the required crossflow velocity a high crossflow volume is required and hence high-powered pumping systems. A crossflow velocity in the region of 2 to 6 m/s is generally considered necessary to provide the surface shear on the membrane to break-up the fouling

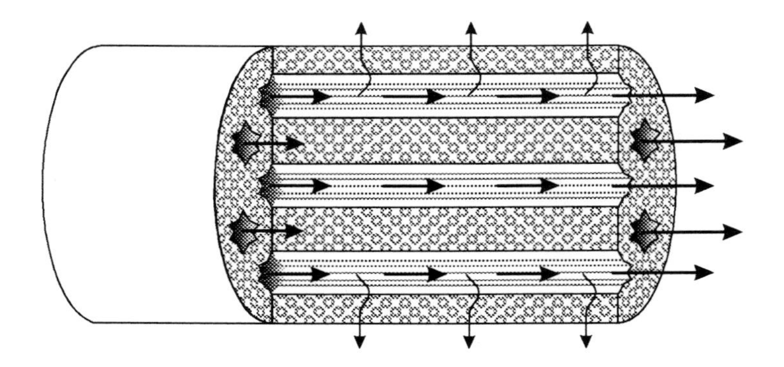

Figure 2. Multi-Channel Membrane.

layer. The optimum velocity depending on the product being processed: its viscosity, suspended solids, etc.

Diameter/Circumference Relationship - Tubular Membranes

With a constant crossflow velocity it is beneficial to design a system with the maximum circumference (membrane area) and the minimum cross-sectional area (retentate flow). In a circular design the circumference increases in direct proportion to the diameter:

$$\text{Circumference} = \pi D$$

whereas, the cross-sectional area increases as the square of the diameter:

$$\text{Cross-sectional area} = \pi D^2 /4$$

Therefore, it is desirable to have a greater number of the smallest diameter flowpaths (of the minimum acceptable diameter considering particle size and viscosity) rather than a large diameter single flowpath requiring high crossflow volumes at a similar crossflow velocity.

An Energy Efficient Alternative Configuration

The crossflow volume (m^3/h) equals the cross-sectional area of the flowpath (m^2) multiplied by the average velocity (m/hr).

Crossflow volume (m^3/h) = cross-sectional area (m^2) x average velocity (m/s) x 3600

To reduce the crossflow volume (and hence the pumping energy), whilst maintaining the velocity, the appropriate variable to decrease is the cross-sectional area of the flowpath. Typically, flowpaths have been circular in section, although square sections have also been produced. Initially thoughts suggested the fitting of a turbulator within the circular flowpath. This would, theoretically, introduce turbulence with subsequent reduction in fouling. The presence of the turbulator would also reduce the cross-sectional-area and hence the pumping energy input. The main problem with a turbulator is locating it within the membrane and the introduction of crevices. Rubbing movement of the turbulator against the membrane was also considered a problem. Having rejected turbulators alternatives were sought. This resulted in the star-shaped flowpath design. From this configuration, figure 3, the following enhanced features result:

Reduced cross-sectional area.
Induced turbulence/Reynolds number.
Increased perimeter (increased membrane area).
Crossflow pressure drop

Reduced Cross-Sectional Area

With a star-shape cross-section having an inner tip radius approximately half the outer tip radius, the cross-sectional area is reduced by nearly 50%. Maintaining the same velocity as for a circular flowpath, the crossflow volume, therefore, is reduced by nearly 50% and the pump energy input is reduced by a similar amount. This reduction may not appear significant for small systems but on larger systems considerable savings can be made. In the processing of low value products, such as waste materials, the reduction in pumping power may mean the difference in the viability of using crossflow filtration.

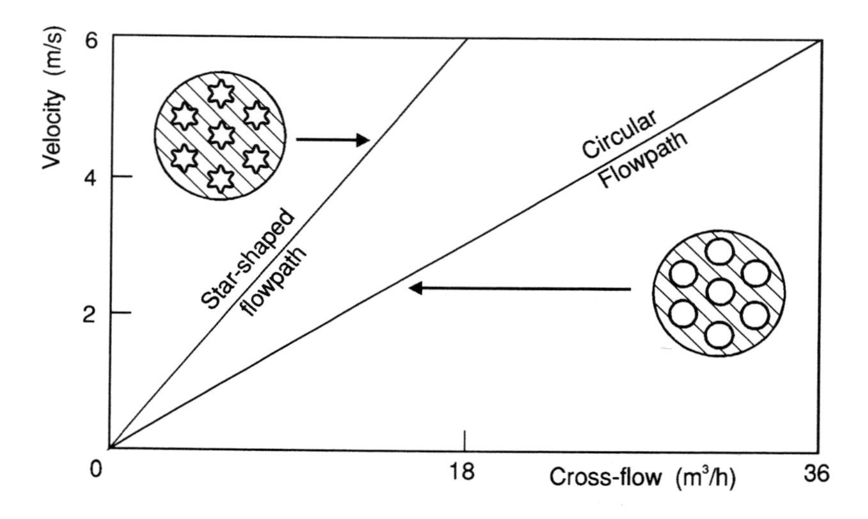

Figure 3. Effect of Crossflow Channel Section on Crossflow Volume.

Induced Turbulence/Reynolds Numbers

The design of membranes is such that a turbulent flow condition exists to break-up the fouling layer at the surface of the membrane. This requirement is similar to that demanded by heat exchangers, both tubular and plate type, to increase heat transfer coefficients. For a circular flowpath a minimum Reynolds number in the region of 2000-3000 is generally considered necessary to pass from laminar to turbulent flow at the interface.

$$\text{Reynolds Number} = \frac{\text{Density x Average Velocity x Diameter}}{\text{Viscosity}}$$

Returning to the heat exchanger comparision, various systems exist to lower the point where laminar flow changes to turbulent flow. On a tank cooling jacket where dimples are introduced into the jacket surface the change from laminar to turbulent flow can occur at a Reynolds number in the region of 400-800 with a corresponding drop in crossflow velocity. Similarly for a plate heat-exchanger with turbulence-inducing surfaces, the change from laminar to turbulent flow occurs at 150-200 resulting in a dramatic reduction in crossflow velocity.

The precise Reynolds number required to introduce turbulent conditions through the starshape flowpath is believed to be in the range 1000-1500, although the actual value has yet to be evaluated. Theoretically, this dramatically reduces the crossflow volume needed. Combining this with the reduction in crossflow volume, due to the reduced flowpath cross-sectional area, it is anticipated an overall reduction of crossflow volume of 70-80% is possible. This figure will obviously vary with the type of fluid being filtered.

Increased Perimeter

The perimeter of the chosen star-shape is approximately 15% greater than a circle of equivalent outer tip diameter. The increased perimeter, therefore, increases the membrane surface area by 15% per unit length. Yet another feature assisting in reducing energy input. Generally, the single elements are bundled together in parallel and installed in a housing, figure 4. The configuration shown contains 19 elements giving a total membrane area of about $1m^2$.

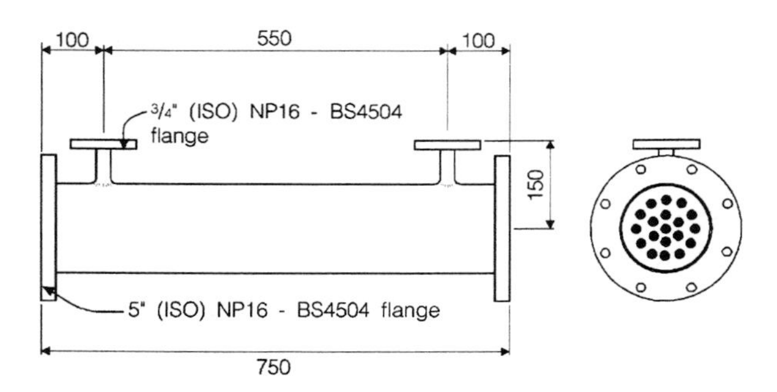

Figure 4. A Multi-Element Module

Crossflow Pressure Drop

Although there are reductions in energy by virtue of the reduced crossflow volumes, there is, of course, a slight increase in pressure drop per unit length because of the reduced cross-sectional area. Tests on water show a pressure drop increase at constant crossflow velocities of about 15%. However, as previously discussed the crossflow velocity can be reduced dramatically with the starshape while maintaining the necessary turbulence to break up the membrane fouling layer. The actual pressure drop is, therefore, considerably less per unit length, as a result of the reduced crossflow velocity, than that of the comparable circular flowpath.

Comparison of Energy Input for Star-Shaped and Circular Flowpaths on an Industrial Scale Plant Example.

Design Parameters

Process Volume	400,000 Litres/Day
Operation Hours Per Day	20 Hours
Assumed Flux	185 Litres/m^2/h
Membrane Area	108m^2
Crossflow Velocity	6m/s

This, for example, would require six baseplate systems each of 18m^2 arranged in the format shown in figure 5 giving a membrane area of 108m^2. The arrangement shown has eight parallel pairs of membranes.

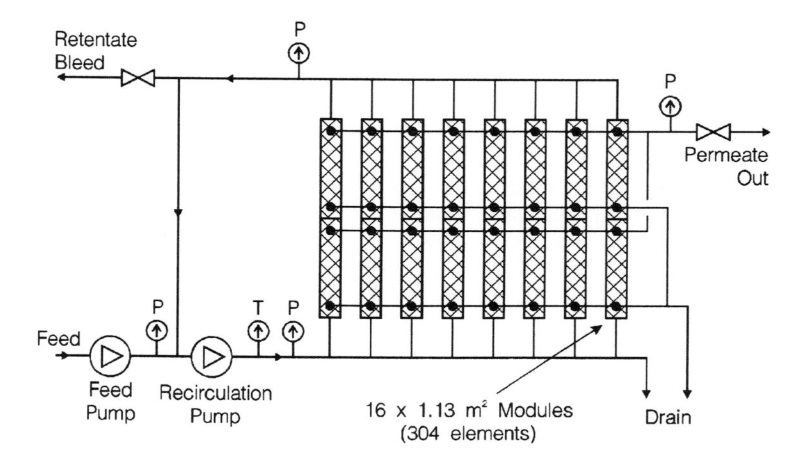

Figure 5. Typical Industrial Scale Plant - 18 m^2 Membrane Area

The pressure drop through the membranes is approximately 2 x 1bar = 2bar. The flow-rate per pair of membranes is $18m^3/h$ at 6m/s for the star-shaped membrane and $36m^3/h$ at 6m/s for the circular membrane as shown in Figure 3. For the eight flowpaths (Figure 5) this requires a pumping duty of:

$144m^3/h$ at 2 bar - star-shape flowpath
$288m^3/h$ at 2 bar - circular flowpath

Pump Selection Star-Shape Flowpath

Using the pump performance curve shown in figure 6 from a well-known centrifugal pump manufacturer it can be seen the required pump power is as follows:

Installed kW = 15 kW
Absorbed kW = 14 kW

Circular Flowpath

A similar exercise using the pump performance curves in figure 7 results in the following power requirements:

Installed kW = 37 kW
Absorbed kW = 29 kW

For the six baseplate system the pumping absorbed would be:

Circular = 6 x 29 kW = 174 kW
Star-shape = 6 x 14 kW = 84 kW
Difference = 90 kW.

Running Cost Savings

In cost terms, using £0.05 per kWh, the use of the star-shape membrane could save £4.50 per hour or up to say £22,500 per year (5000 hours): a considerable saving. Obviously this is a very simplistic comparison and the assumed flux may differ between circular and star-shape flowpaths. Of the tests to date only one specialised material has shown an inferior flux using the star-shape. In nearly all applications the cross-flow velocities required have been less for the star-shape due to the induced turbulence created by the irregular shape. The feed pump performance and energy input would be similar for both applications since this is related to the permeate flowrate and system feed pressure.

Other Energy Input Savings

Some separation process require the energy absorbed from the pump by the process liquid to be removed by refrigeration for example. In this case the savings in running costs may be doubled using the star-shape membrane.

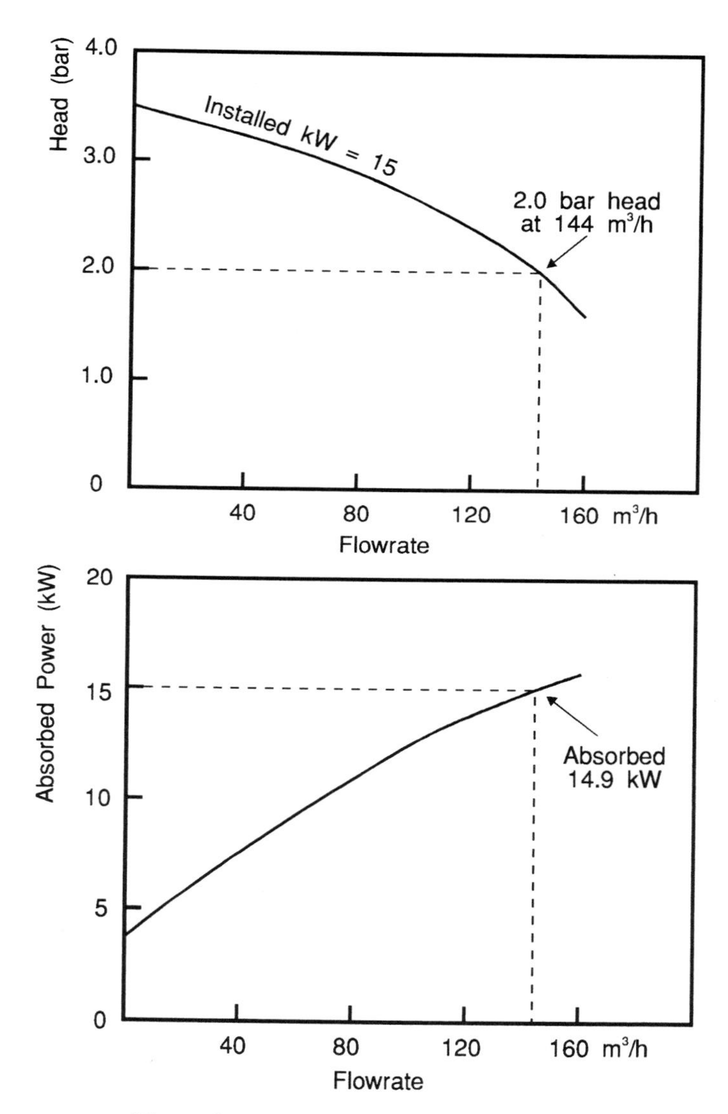

Figure 6. Pump Selection for a 144 m^3/h Flowrate.

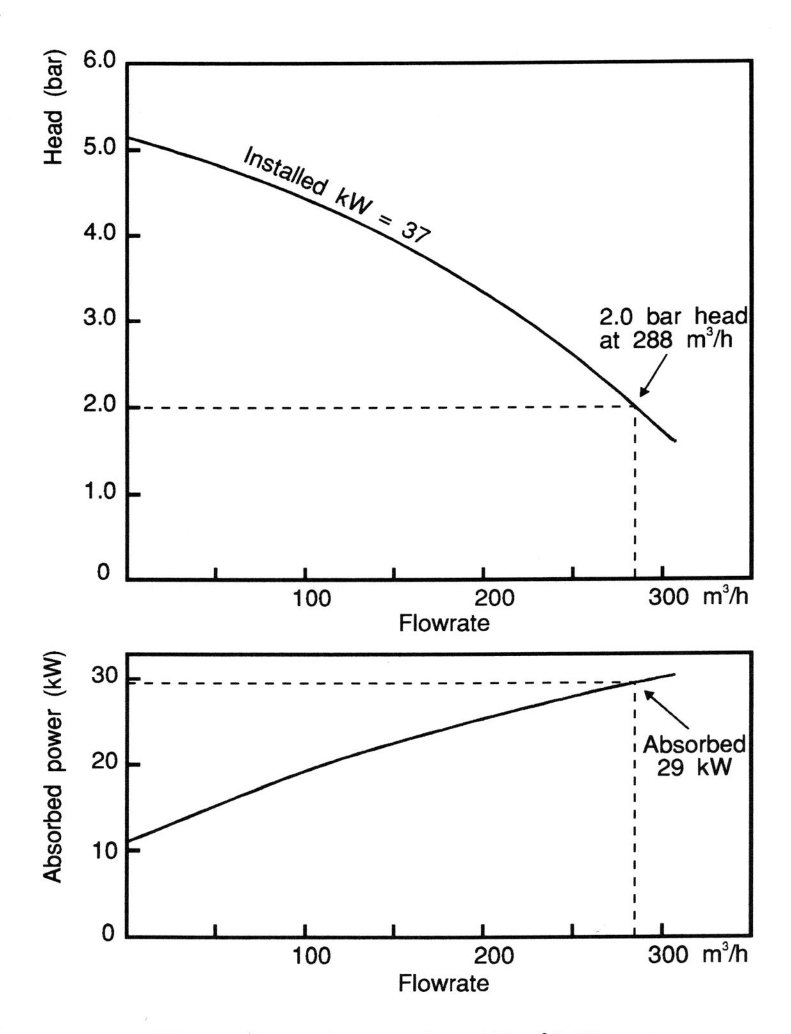

Figure 7. Pump Selection for a 288 m³/h Flowrate

Equipment Cost Savings

In addition to the substantial running cost savings, there is a considerable saving in equipment costs. For example:

Pumps
Pipework
Valves
Electrical Equipment
Overall Size

In the case of pumps and valves, the cost of these may be double that required for a system using the star-shape membrane.

Summary

From the example shown it can be seen that considerable savings in energy input can be made with many applications using the star-shaped membrane.

This energy input saving enables crossflow filtration to be applied to high volume - low value liquids including waste materials such as those damaging our environment - a topical subject, and an application where large installations of crossflow filtration is expected over the next decade.

Experimental Investigation on Enhancement of Crossflow Ultrafiltration with Air Sparging

Z. F. Cui
Department of Chemical Engineering
University of Edinburgh
King's Buildings, Edinburgh EH9 3JL, UK

Abstract: A new technique is proposed for enhancing the crossflow ultrafiltration process by injecting air into the system. The injected air provides a turbulence promoting effect, and can also increase the superficial velocity of the cross-flow, so effectively disturbing and suppressing the concentration polarisation layer, and hence enhancing the ultrafiltration process.

Experiments were carried out on the proposed gas-liquid cross-flow ultrafiltration process, with PVDF tubular membranes (PCI, MWCO 100kD) vertically installed and dyed dextran solutions (MW 87kD and 162kD) as the media. It was found that air injection could greatly suppress the effect of concentration polarisation and fouling, while maintaining a high ultrafiltration rate. Compared with the single phase operation, the permeate rate was increased by 70-250% even at a very low gas flowrate. At the same time, the rejection ratio of the membrane is increased considerably. An attempt was made to explain the mechanism for this enhancement. The limitations of the technique were considered and further research on this technique suggested.

Introduction

Membrane ultrafiltration plays an important role in many industrial separation processes, especially in biotechnological downstream separation (Jönsson and Trägårdh 1990a). Pressure driven systems are by far the most widely used and developed, and are of the greatest immediate interest.

The main problem with ultrafiltration is the decrease in flux caused by concentration polarisation due to boundary layer formation near the membrane surface, and by membrane fouling. These are considered to be the major barriers to the wider adoption of membrane processes (Jönsson and Trägårdh 1990b; Fell et al

1990). One way to control such fouling is to reduce the concentration polarisation by hydrodynamic methods; that is , using turbulence promoting techniques such as static mixers (Hiddink et al 1980), corrugated membrane surfaces (Racz et al 1986), fluidised beds (van der Waal et al 1977), pulsatile flow (Finnigan and Howell, 1989), surface rotation and vibration (Hermann 1982), electric fields (Brunner and Okoro, 1989), and so on.

The methods proposed for improving (or enhancing) the performance of ultrafiltration are similar to those used for augmenting convective heat transfer, which has been investigated extensively (e.g. Bergles 1985). One technique used to enhance convective heat transfer in a liquid involves the injection of air, to create gas bubbles and consequently a two-phase flow. This method was shown to be simple and effective (up to 400% increase in heat transfer coefficient) (Tamari and Nishikaa 1976), and was also used to suppress cake formation in the filtration of microparticles (Imasaka et al, 1989).

It is expected that a similar technique could be used to disturb the mass transfer boundary layer in ultrafiltration processes and to suppress concentration polarisation. In this study, experimental apparatus was designed to assess the performance of the proposed technique. Preliminary results indicate that this simple method could greatly enhance the ultrafiltration process and, at the same time, increase the recovery rate.

Experiment

Membranes

The membranes used in this study were PVDF tubular membranes (FP100) with an average molecular weight cutoff (MWCO) 100 kD supplied by PCI. The tubular membrane module was selected for its accurately determined surface area and its similarity to two-phase flow heat transfer equipment about which extensive information is available.

The inner diameter of the membrane tubes was 12.7 mm and two lengths were used in the experiments, 425 mm and 670 mm.

Media

A dyed dextran solution was used as the test medium to assess the performance of the two phase crossflow ultrafiltration process. It was chosen because of its easy analysis and because dextrans have been used by several research groups and some of the transport properties are known (Jonsson 1984; Goldsmith 1971).

Dextrans (industrial grade from Sigma) with average molecular weights of 87 kD and 162 kD and Procion dyes (from ICI) were chemically bound together. The concentration of the dyed dextran solutions, measured by the dry weight method, was correlated with the absorbance at a specified wavelength (blue dextran at 586 nm, red dextran at 525 nm), measured by a spectrophotometer. It was shown that a linear relationship exists between absorbance and concentration (Cui et al 1993).

Dextran solution proved to be Newtonian in behaviour (Jonsson 1984). It was also found that in the experimental concentration range, its viscosity (measured by a capillary viscometer) was proportional to the concentration of the dextran (Cui et al 1993).

Apparatus

The experimental apparatus used for this study is shown in Fig. 1. The tubular membrane was installed vertically with the medium ascending inside the tube. The pumping and control unit was based on the self-contained Bio-2000 system (BIO-FLO LTD, Glasgow), which has two peristaltic pump heads with different phase settings. Two pressure transducers are installed pre and post membrane to monitor the pressure in the feedline. The system allowed both the flowrate of the solution and the transmembrane pressure to be preset independently and controlled automatically. The pulsatile flow delivered by this system could also improve ultrafiltration performance (Finnigan and Howell 1989). The signals of the averaged flowrate, pre- and post membrane pressure and transmembrane pressure were output to the computer, where data logging and analysis were carried out. In the case of tubular membranes, the pressure drop along the tube was negligible, and therefore, the pre and post pressure were almost identical. The transmembrane pressure was then the pressure difference between the feed line and the atmosphere (outside of the membrane).

The permeate was weighed continuously with an electronic balance (METTLER PM4800), and the analogue signal was also input into the computer where the calculations of permeate flowrate and flux were carried out. The concentration of the feed and that of the permeate were continuously monitored by measuring the absorbance of the solution using a spectrophotometer. The rejection ratio could then be determined.

The desired flowrate of compressed air was set using a rotameter. The air was passed through an in-line filter, injected at the inlet of the membrane tube through a 1.5 mm orifice, and flowed concurrently with the liquid phase crossing the membrane surface.

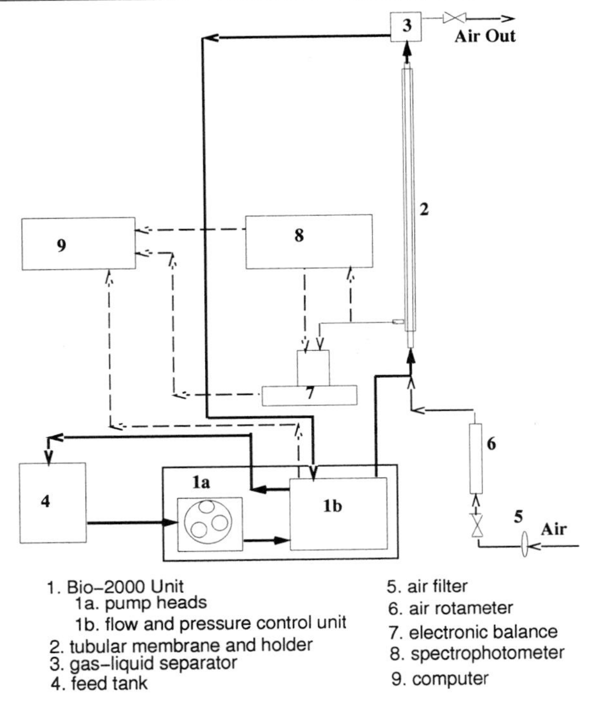

Figure 1: *Experiment Apparatus*

Procedure

The pre-cleaned membrane was washed with distilled water for half an hour. The flowrate of liquid and air (zero for single phase flow) and the feed line pressure were adjusted to the desired level. The water was replaced with a medium solution of known concentration and the computer began data logging and analysis to calculate permeate rate and flux as time progressed. The concentration of the permeate was also monitored continuously. Each experiment lasted for one hour. At the end of each run, distilled water was first used to flush out the residential solution. A cleaning solution was circulated through the system for 15 min, and this was followed by a distilled water wash to rinse the entire system.

Results and Discussion

Permeate flux

The effect of air sparging on the membrane permeate rate is shown in Fig.2. The operating conditions were: flowrate of the dextran (MW 87kD) solution 0.6 l/min, transmembrane pressure 0.75 bar, the feed concentration 1.2% (wt), liquid phase superficial velocity 0.081m/s with Re = 545 (based on the inner diameter of the membrane tube). The ranges of transmembrane pressure, crossflow velocity, and feed concentration used in the experiment were chosen according to those used in the practical operation of downstream separation.

From Fig.2, the polarisation phenomenon with single phase flow (air flowrate = 0) is clearly shown. For example, in the first 60 s, the average permeate flux was 0.005 kg/(m^2 s), but dropped to a relatively steady value of 0.0021 kg/(m^2 s) after 25 min. With gas-liquid two-phase flow, although the polarisation was not completely eliminated, it was significantly reduced even at a low gas flowrate (superficial velocity ratio V_G/V_L in the range of 0.06 - 0.32). The permeate flux was more than doubled and remained at a steady high level over the full experimental period. Over a wider range of operating conditions, using gas-liquid flow could increase the ultrafiltrate flux by 70% to 250% (see also Fig.3).

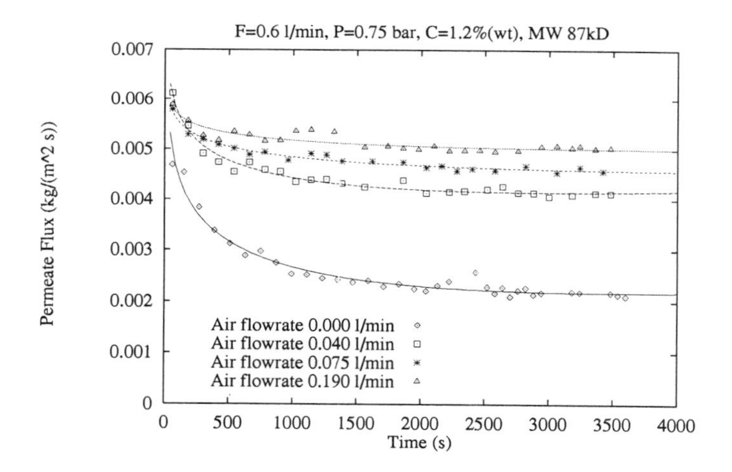

Figure 2: *Declines of the Permeate Flux*

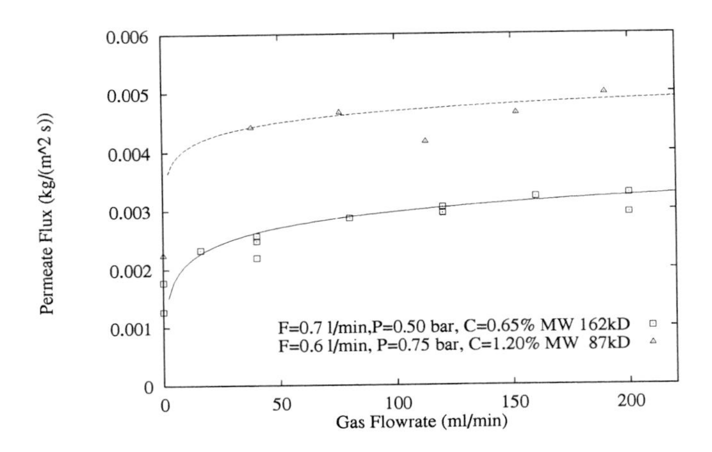

Figure 3: *Influence of Gas Flowrate on Permeate Flux*

the permeate flux is specified, the decreased local concentration results in fewer macromolecules passing through the membrane; and hence a higher membrane rejection ratio is observed.

Flow pattern

The flow patterns of the gas-liquid two-phase flow could not be determined by the usual methods (e.g. Butterworth and Hewitt 1977), because of the pulsatile nature of liquid phase. Although it was not directly measured, the gas phase probably adopted a slug flow pattern. In this case the frequency of slug generation would correspond to that of the pulse, with slug formation and release occurring when the fluctuating liquid pressure at the point of release was at one of its lowest. The principle effect of changing gas flowrate was to alter the size of the slugs. As long as their size was greater than a critical value (which needs to be determined), their effect of removing the polarised layer was substantially the same. This may explain why neither the enhancing effect nor the rejection ratio were sensitive to the gas flowrate in this series of experiments.

The influence of air flowrate on the permeate rate was examined and typical results are shown in Fig.3, where the results for single phase operation are indicated at zero gas flowrate. The volumetric gas flowrate at each operating pressure was used in this study instead of at standard condition, based on the assumption that the disturbing effect of the mass transfer boundary layer from bubbles would be related to the actual diameter of the bubbles at the operating condition.

From Fig.2 and Fig.3, it could be suggested that:

- air sparging could significantly enhance the ultrafiltration process, as a result of the disruption and suppression of the polarisation layer by gas bubbles;
- this enhancement was not sensitive to the gas flowrate in the range of the experiments, which indicates that a significant enhancement could be achieved even at very low gas flowrate. Low gas flowrates are important in minimizing the damaging effect of gas bubbles on the processed proteins or macromolecules. The detailed mechanism for this phenomenon was not very clear, but it might be related to the fluctuating nature of the pulsatile flow of the liquid.
- this enhancing effect was more pronounced with higher feed concentrations, and also at higher transmembrane pressures, under which conditions the polarisation effect was more severe.

Membrane rejection

It was also found that air sparging could also significantly increase the membrane rejection ratio, as shown in Fig.4. As mentioned earlier, the average molecular weight cutoff of the membrane was 100kD. The membrane rejection ratio, using gas-liquid two-phase flow, increased from 89% to 95% for the 162kD dextran solution and from 80% to 93% for the 87kD dextran. These results indicate that using gas-liquid two-phase flow could actually increase the recovery rate, which is a significant factor in the concentration of high value products.

This increase in membrane rejection ratio could be explained by the following mechanism. With two-phase flow operation, the gas bubbles effectively disturb the mass transfer boundary layer and promote a mixing process between the solution near the membrane surface and the main flow. By removing (or partially removing) the polarised layer, the local concentration of solution at the membrane surface is lower in comparison with the single phase flow situation. The membrane rejection ratio must somehow depend on the local concentration at the membrane surface. Assuming that the number of macromolecules crossing the membrane is dependent on the accumulation of macromolecules around the larger pores, if

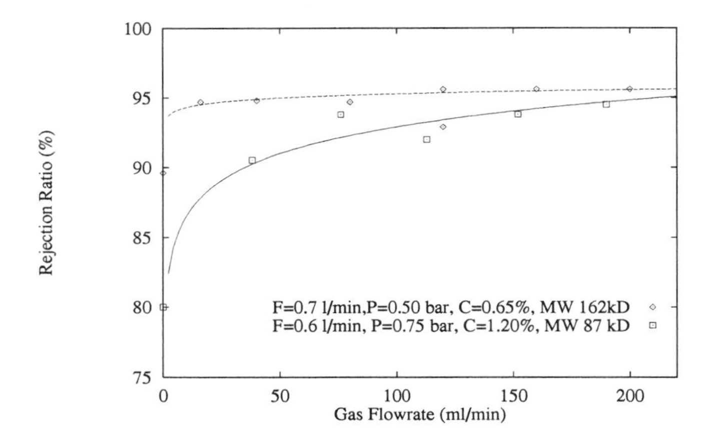

Figure 4: *Effect of Air Sparging on Membrane Rejection*

Further research

The main limitation of this technique is the effect of shear from the gas bubbles on the processed proteins or macromolecules, which indicates that it could not be applicable to shear sensitive media. However, this damage should be quantified and possible methods to minimize this damage should be sought.

On the assessment of this technique, more experiments are required to examine systematically the influence of operating conditions and a wider range of gas flowrates. Longer experimental duration and experiments with other membrane modules and other media are highly desirable.

The detailed mechanism of this enhancement on ultrafiltration needs to be clarified, such as the behaviour of air bubbles or slugs and the relation between bubble dynamics and permeate flux. Although the main effect of the injected air is to disturbe and reduce the concentration polarisation layer, the other effects, e.g. the 'flattening' of the fluctuating pressure profile in the pulsatile flow due to the compressibility of air, and their influence on permeate flux and rejection ratio need to be examined.

Conclusions

Air sparging to create a gas-liquid two-phase crossflow near the membrane surface proved to be a simple and effective technique for reducing polarisation phenomenon. By simply injecting air into the feed, increases of 70%-250% in permeate flux have been observed . This enhancement effect seems more pronounced at higher feed concentration and higher transmembrane pressures, but is insensitive to gas flowrate in the experimental range examined. Using gas-liquid two-phase flow can also increase the membrane rejection ratio, which indicates that a higher recovery rate could be obtained.

Acknowledgment: This project is supported by Science and Engineering Research Council (SERC, G/H/73752) and carried out in the Chemical Engineering Department, Edinburgh University. R. Nunn and I. Shiels were involved in part of the experiment in their Honours Degree research project. The author would like to thank Dr Colin Pritchard, the Head of Department, and Dr Donald Glass for their encouragement and support. He is also grateful to PCI Membrane System Ltd for providing the tubular membranes, and to Bio-Flo Ltd for providing some technical support.

References

Bergles, A. E. (1985), Techniques to augment heat transfer, IN: Rohsenow, W. M. et al (ed.) Handbook of heat transfer applications (2nd edition), McGraw-Hill, New York.

Brunner, G. and Okoro, E. (1989), *Berichte der Bunsen Gesellschaft fur Physikalische Chemie*, **93**, 1026-1032.

Butterworth, D. and Hewitt, G. F. (1977), *Two-Phase Flow and Heat Transfer*, Oxford University Press.

Cui, Z. F., D. H. Glass and K. Wright (1993), Membrane ultrafiltration studies using dyed dextran solutions, The 1993 IChemE Research Event, Birmingham.

Fell, C. J. D., Kim, K. J., Chen, V., Wiley, D. E. and Fane, A. G. (1990), *Chem. Eng. Process.*, **27**, 165-173.

Finnigan, S. M. and Howell, J. A. (1989), *Chem. Eng. Res. Des.*, 278-282.

Goldsmith, R. L. (1971), *Ind. Eng. Chem. Fundam.*, **1**, 113-120.

Hermann, C. C. (1982), *Desal.*, **42**, 329-338.

Hiddink, J., Kloosterboer, D. and Bruin, S. (1980), *Desal.*, **35**, 149-167.

Imasaka, T., Kanekuni, N., So, H. and Yoshino, S. (1989), *Kagaku Kogaku Ronbunshu*, **15**, 638-644.

Jönsson, A.-S. and Trägårdh, G. (1990a), *Desal.*, **77**, 135-179.

Jönsson, A. and Trägårdh, G. (1990b), *Chem. Eng. Process.*, **27**, 67-81.

Jonsson, G. (1984), *Desal.*, **51**, 61-77.

Racz, I. G., Wassink, J. G. and Klaasen, R. (1986), *Desal.*, **60**, 213-222.

Tamari, M. and Nishikawa, K. (1976), *Heat Transfer Jpn. Res.*, **5**, 31-44.

van der Waal, M. J., van der Velden, P. M., Koning, J., Smolders, C. A., and van Swaay, W. P. M. (1977), *Desal.*, **22**, 465-483.

Removal of hydrocarbons from waste water by pertraction

R. Klaassen[1], A.E. Jansen[1], J.J. Akkerhuis[1], B.A. Bult[2], F.I.H.M. Oesterholt[2], J. Schneider[3]

Abstract

Pertraction is a new membrane based solvent extraction process for the removal of organic components from industrial waste water.
Experiments in the laboratory with four different industrial waste water streams show that pertraction can treat a broad range of components including aromatic, halogenated and poly aromatic hydrocarbons. Removal of these components is possible below ppb level.
New special crossflow modules have been designed for pertraction. These modules give improved mass transfer and the possibility to install more surface area per module. In this way crossflow modules can treat much larger waste water flows than classical membrane modules used so far.
Pilot plant tests will be carried out to demonstrate the long term stability of pertraction.

1 Introduction

The TNO Institute of Environmental and Energy Technology, Tauw Infra Consult and the Separation Products Division of Hoechst Celanese Corporation (HCSPD), a subsidiary of Hoechst AG, are developing applications for pertraction: a new technology for the removal of organic components from industrial waste water.
Main objective of the study is the development of the pertraction process to a technical and economical feasible technology for the removal of a broad range of pollutants from waste water. Demonstration on pilot plant scale forms an essential part of the research

[1] TNO Institute of Environmental and Energy Technology, P.O. Box 342, Apeldoorn the Netherlands
[2] TAUW Infra Consult, P.O. Box 479, Deventer, the Netherlands
[3] Hoechst Celanese Corporation, Separations Product Division Rheingaustrasse 190, D-6200 Wiesbaden 1, Germany

programme. Besides this the development and testing of new module types is an important secondary object of the study.

The study is subsidized in the frame work of the Dutch Governmental Environmental Technology Program coordinated by NOVEM. It is carried out at the laboratories of TNO and TAUW Infra Consult. The pilot plant study will be carried out on the sites of two major chemical industries in the Netherlands.
Major phases in the research program are: selection of waste waters and extractants, laboratory experiments with waste water and different module types and finally pilot plant experiments. Laboratory experiments have been completed and a beginning has been made with the design and construction of the pilot plant.

2 Pertraction

2.1 Principle

Pertraction is a new, non-dispersive membrane based extraction process. Organic components can be removed from the waste water into an organic extractant. The interface between the waste water and extractant is immobilized using a hydrophobic microporous hollow fiber membrane by means of a small transmembrane pressure gradient (≈0.1 bar)(1). See figure 1. The membrane itself has no selectivity.

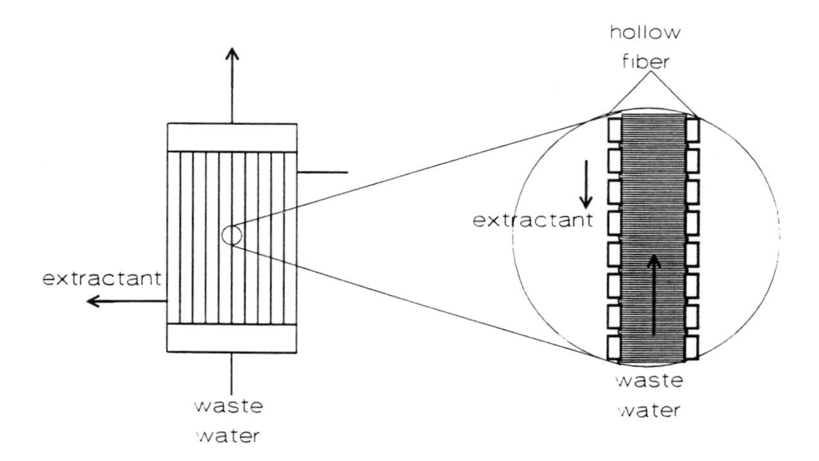

Figure 1 *Principle of pertraction*

The use of hollow fiber modules offers major advantages over conventional extraction processes, including:

- no emulsion formation in the water phase
- volume of extractant is relatively small
- process parameters are very flexible
- phase separation is not necessary
- large specific surface area when using hollow fibre membranes
- small energy consumption
- modular compact equipment
- no difference in density needed between waste water and extractant

Another advantage of pertraction over stripping or carbon adsorption is that both volatile and non-volatile components can be extracted in one step.

2.2 Applications

The current available techniques for the removal of hydrocarbons from waste water streams do have their limitations. Techniques such as air stripping or activated carbon filtration that are commonly used in these situations are only able to remove specific groups of compounds. By air stripping only volatile compounds can be removed and by activated carbon filtration only adsorbable compounds can be removed. In addition both methods are highly concentration dependent. For instance the physical process of carbon adsorption shows a decline of adsorption capacity at lower concentration levels.

Pertraction makes no distinction between volatile compounds such as halogenated and mono aromatic hydrocarbons with poor adsorption characteristics and good absorbable non-volatile compounds such as poly aromatic hydrocarbons. With pertraction all hydrocarbons can be removed even at low concentration levels. Because of this non-selectivity pertraction is believed to be a suitable technique for industrial waste waters containing low levels of hazardous hydrocarbons. Examples of these waste water streams can be found in different types of industries like chemical and petrochemical plants, tank cleaning firms, garages, dry cleaning firms and wood preservation companies.

3 Experimental

3.1 Experimental Set up

Pertraction experiments with real waste water samples have been carried out in the laboratory. Liqui-Cel^R(a registered trademark of Hoechst Celanese Corporation) membrane contactors, commercially available from HCSPD, have been used. A flow chart of the installation is given in figure 2. Both the extractant and the waste water are pumped through the membrane module and are then led back to a storage vessel.

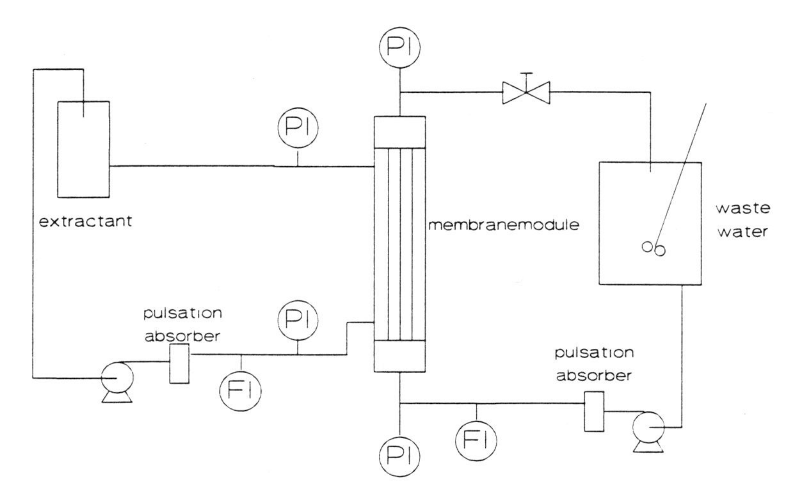

Figure 2 *Flowchart of laboratory installation*

The recirculation experiments make it possible to determine with a limited amount of waste water and one single membrane module the following facts:

- technical feasibility of pertraction with real waste water samples
- possible degree of removal from organics out of the water phase
- mass transfer in the tested modules
- pressure drop over the membrane module
- short term fouling effects

3.2 Mass transfer

The overall mass transfer coefficient characterizes the transport rate of the organic components from the waste water to the extractant.
When the mass transfer coefficient is known a design of a once through installation can be made. In an once through installation a certain number of modules are coupled in series to get a desired degree of removal of organic components from the waste water.

The determination of the overall mass transfer coefficient has been done with the simulation program Tutsim. Basically it works this way:

A mathematical model for mass transfer in the pertraction process has been formulated. The only unknown parameter in this model is the mass transfer coefficient Kw. With the simulation program the model is solved. The result is the change of concentration of a organic component in the waste water storage vessel over time. Kw is determined by variation of Kw in such a way that there is a good fit between the calculated and experimental curve of concentrations over time.

3.3 Results

Experiments have been carried out with four different types of waste water. The waste waters that have been investigated contained a broad range of different aromatic hydrocarbons, polycyclic aromatic hydrocarbons and chlorinated organic solvents with different concentration levels.

The waste water contains many components which seem not to influence each other. Degrees of removal of over 99% and end concentrations below 0.1 μg/l are possible. No problems with short term fouling effects have occurred.
A typical curve for the concentration versus time is given in figure 3.

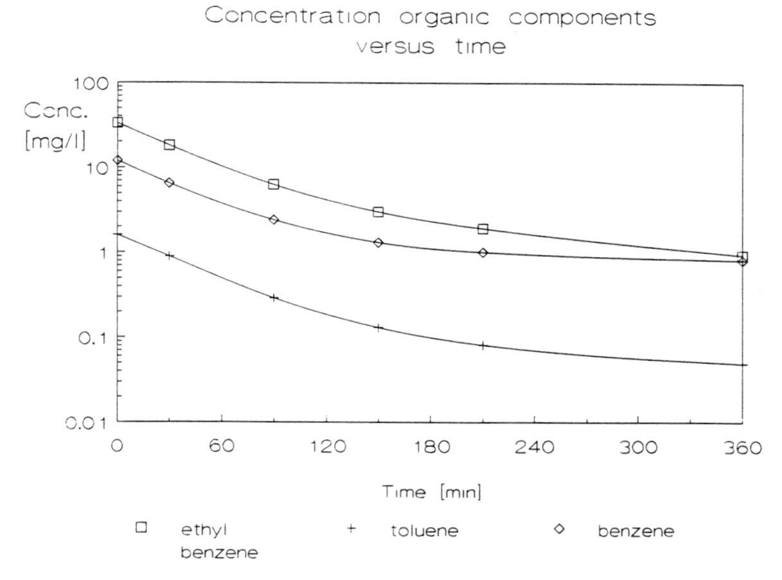

Figure 3 *Concentration benzene, ethylbenzene and toluene versus time*

A good fit between the model calculations and experimental curve was possible.

Mass transfer for components with a high distribution coefficient is limited by the transport in the water phase. Increase of water flow through the module increases mass transfer. The membrane wall and the extractant have no limiting resistance. Mass transfer coefficients are independent of the concentration level: concentrations in the waste water streams varied from 100 mg/l down to 0.1 µg/l (a difference of six orders of magnitude) and gave no significant change in mass transfer. Mass transfer coefficients are typically in the order of 10 - 20 E-6 m/s depending on the flow rate of the waste water through the fibers.

4 Module improvement

Modules available for pertraction so far have been originally designed using an unbaffled shell and tube (parallel flow) configuration.

Flow patterns on the shellside are not very desirable for mass transfer operations due to channelling and short circuiting. The dimensions of the module are limited; with larger bundles of fibers it is difficult to get a good fluid flow in the centre of the bundle. In order to improve module performance both TNO(2) and Hoechst Celanese have designed new module types. Schematic drawings are given in figure 4 and 5.

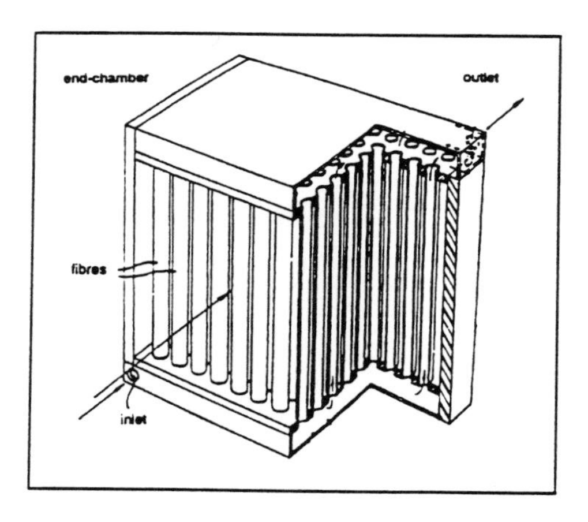

Figure 4 *TNO module*

CROSS FLOW HOLLOW FIBER CONTACTOR

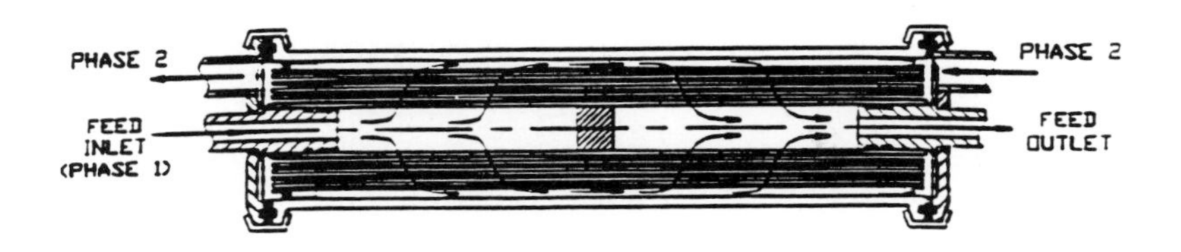

Figure 5 *Hoechst Celanese module (patent pending)*

Both new module types are based on cross flow: the shellside flow takes across the fiber bundle. The shellside mass transfer exceeds that of the tube side flow. The optimum way to operate the crossflow modules is to flow the stream with the limiting resistance on the shellside: for pertraction this is normally the waste water stream except for components with a very low distribution coefficient.

Experiments with prototypes of the crossflow modules and real waste water show a drastic increase in mass transfer. For the parallel flow modules with waste water on the tube side overall mass transfer coefficients are typically in the order of 10 - 20 E-6 m/s whereas for the crossflow module values of 40 - 80 E-6 m/s have been measured.

Due to improved mass transfer and the possibility to install more surface area per module crossflow modules can treat much larger waste water flows than parallel flow modules. Parallel flow modules used now can treat water flows in the order of 100 - 200 litre per hour whereas crossflow modules have been designed that can treat several cubic metres per hour.

5 Pilot plant

With the mass transfer coefficients obtained in the laboratory experiments it is possible to make a design of a once through installation. In an once through installation a certain number of modules are coupled in series to get a desired degree of removal of organic components from the waste water. In order to demonstrate the pertraction technology a once through pilot plant will be build into a container and operated on industrial locations.

A flowchart of the installation is given in figure 6.

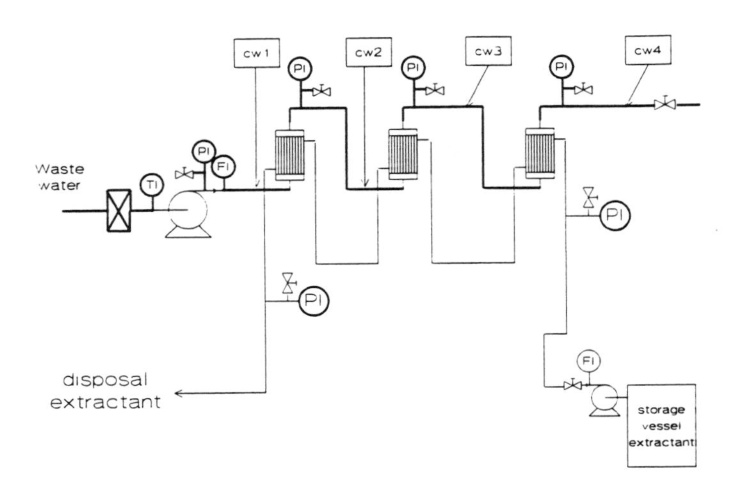

Figure 6 *Flowchart pilot installation.*

Model calculations for the pilot plant are shown in figure 7. In this figure the concentrations C_{w1}, C_{w2}, C_{w3} and C_{w4} as indicated in figure 6 are given as function of the mass transfer coefficient in the membrane module.
The calculations have been carried out with the following assumptions:

membrane area per module : 3.7 m^2
waste water flow : 150 litre/hour

In figure 7 the mass transfer region for the parallel flow and crossflow module are indicated. From the figure it can be seen that with three filtration type modules in series a removal of 90 to 99% is possible. The same degree of removal can be reached with almost one crossflow module. With three crossflow modules in series a degree of removal of 99.99% is possible.

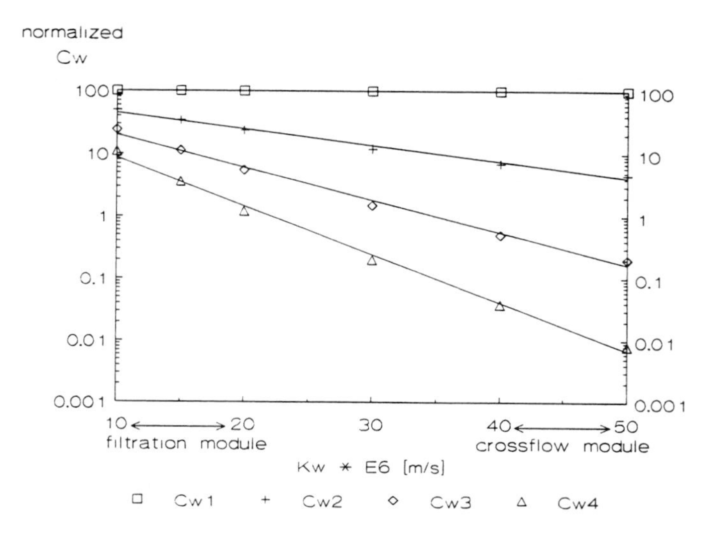

Figure 7 *Model calculations once through installation*

6 Conclusions

Labscale experiments with industrial waste water show that pertraction can treat waste waters with a broad range of components including aromatic, halogenated and poly aromatic hydrocarbons. Removal of these components is possible down to ppb level. The performance of the pertraction process characterized by the mass transfer coefficient is proven to be independent of the concentration from the components to be removed between 100 mg/l and 0.1 µg/l.

New special designed crossflow modules give improved mass transfer.
Due to the better mass transfer and the possibility to install more surface area per module crossflow modules can treat much larger waste water flows than classical filtration modules used so far.

A design for a pilot plant has been made: model calculations indicate that with three filtration type modules in series a removal of 90 to 99% is possible. The same degree of removal can be reached with one crossflow module. With three crossflow modules in series a degree of removal of 99.99% is possible. Pilot plant experiments on industrial locations have to demonstrate long term stability of the pertraction process.

7 Acknowledgement

The authors wish to thank NOVEM and Hoechst Celanese for their financial support.

8 References

1. Patents US# 4,789,468 - US# 4,997,569 - US# 4,966,707 - Hoechst Celanese

2. Meulen, B.Ph. ter - Transfer device for the transfer of matter and/or heat from one medium flow to another medium flow. International patent application PCT/NL91/00001

LABORATORY AND PILOT REACTOR EXPERIENCE FOR PHOTOCHEMICAL DEGRADATION OF ORGANIC CONTAMINANTS IN WASTEWATERS BY PHOTOCATALYTIC MEMBRANES IMMOBILISING TITANIUM DIOXIDE*

IGNAZIO RENATO BELLOBONO[a] and *ANNA CARRARA*[b]
[a]Department of Physical Chemistry and Electrochemistry, University of Milan
I-20133 Milano (Italy)
[b]Chimia Prodotti e Processi I-20124 Milano (Italy)

1. INTRODUCTION

Waste management problems are nowadays exacerbated by two factors: quantity and toxicity. On one side, waste quantities are steadily increasing due to population growth, and, perhaps more significantly, because the specific waste generation rate continues to increase, owing to the fact that polluting manufacturing processes are not being readily substituted by less-polluting technologies. On another side, pollution by hazardous organic wastes, by chloro-organics particularly, has become an important issue, due to their toxicity potential. Even removal of chloro-aliphatics from contaminated potable waters is causing progressive concern. An urgent need is thus felt for degradation procedures of hazardous wastes, possessing requisites of simplicity, acceptable economics, and general applicability.

The effectiveness of homogeneous and heterogeneous photocatalysis for the total oxidation of organic and halo-organic water contaminants have been thoroughly investigated in recent years, both scientifically and technologically [1-3]. Homogeneous photocatalysis, in the presence of oxidising agents, such as hydrogen peroxide or ozone, has begun to receive industrial attention [3]. It presents, however, several drawbacks, such as the low absorptivity of oxidisers and the necessity of cleaning continuosly the quartz walls of the UV lamps containers, due to frequent deposition of tarry and polymeric products on the highly irradiated parts of the reaction vessel. Heterogeneous photocatalysis is attractive, but it still retains burdensome inconveniences given by non-homogeneous

---

* Part 33 of the series "Photosynthetic Membranes"

light absorption by suspended semiconductors, leading also to difficulties in design of photoreactors, as well as by the necessity of a costly continuous filtration of photocatalyst. Even if the semiconductor could be immobilised by physical adsorption, as has been demonstrated [4-6], the thin layer of it would create problems of low absorption, the support would certainly be liable to interfere optically, cost of preparation and regeneration of photocatalyst would be high, due to frequent inactivation, and design of photoreactor would still involve many problems.

Greater ease of technological operation, on the contrary, has been found to be afforded by photocatalytic membranes [7-8] in which the semiconductor photocatalyst can be immobilised very conveniently, at high concentrations (up to about 40%) and without appreciable loss of its activity, because in these membranes surface area of immobilised species is not sensibly decreased, due to the very rapid process of photochemical grafting, which is employed for the immobilisation during the membrane manufacture [9-11]. Besides immobilisation of semiconductor in a reliable optically absorbing structure, which is able to remove all difficulties and drawbacks above, additional advantages may be provided by membrane processes employing photocatalytic membranes for waste degradation: *i)* the possibility of immobilising also other photocatalysts and/or photosensitisers, particularly when operating with semiconductors, such as titanium dioxide, with a fairly large band gap; *ii)* the possibility of continuously separating or concentrating the solutions being processed; *iii)* the possibility of modelling the most appropriate type of photoreactor by appropriately choosing the membrane geometry and taking benefit from the very high surface-volume ratio in membrane reactors.

A pilot membrane reactor has now been set up, and is described in the present paper, utilising a 12 KW high pressure mercury arc lamp, and able to operate either in a closed, or partially closed loop, or continuously, depending on concentration and kind of pollutants. In this same paper, laboratory data have also been obtained, examining systematically the influence of the most relevant parameters (radiant flux, concentration and chemical nature of pollutants, concentration of hydrogen peroxide, concentration of photosensitiser / photocatalytic system employed to increase reactivity, temperature effects). Comparison of laboratory data with performance and productivity of pilot plant, in several case studies, may thus be shown and discussed.

2. EXPERIMENTAL

2.1. Photocatalytic membranes

Photocatalytic membranes ("Photoperm" membranes CPP/313 by Chimia Prodotti e Processi, Milan), immobilising 23±2% of titanium dioxide (mainly anatase, by Degussa, P-25 grade), were prepared industrially by a pilot plant operating continuously. Controlled amounts of appropriate monomers and prepolymers, containing the semiconductor to be immobilised, and photoinitiated by a proprietary photocatalytic system, were photografted onto a non-woven polyester tissue. The final porosity of photosynthesised membranes was regulated at 2.5-4.0 μm, by controlling the rheological and photochemical parameters during membrane manufacture.

In some of the blends, varying amounts of proprietary photocalysts (I-VII) (Chimia Prodotti e Processi, Milan) were added, in order to sensitize the semiconductor dioxide outside its optical absorption range at wavelengths greater than that corresponding to its band gap. These photocatalysts were constituted essentially by stabilised preparations containing the following organo-metallic compounds, respectively:
(I)Peroxo-bis[N,N'-ethylene-bis-(salicylideneiminato) cobalt(III)]
(II)Peroxo-bis[N,N'-ethylene-bis(salicylideneiminato) dimethylformamide cobalt(III)]
(III) Triethylvanadate(V)
(IV) Oxo-(diquinolyloxo)vanadic(V) acid
(V) Tris-(phenylsilyl)vanadium(V)
(VI) Synergic mixture of tri-(*t*-butyl)- and tri-(*i*-propyl)vanadate(V)
(VII) Iron(III) potassium oxalate.
As the only information available on these photocatalytic systems was their nominal concentration of photoactive species, they were used as supplied, on the basis of this nominal concentration.

2.2 Layout of pilot plant

The reactor employed for pilot plant experiments, shown in Fig. 1, was substantially an annular photoreactor, made in stainless steel, in which two coaxial cylinders delimited the reaction zone, the inner one, made in quartz, containing the lamp. which was placed on the symmetry axis.

Fig. 1. General view of photodegradation plant. A points out the photoreactor, B the reservoirs, C the dosing pump of hydrogen peroxide.

This simple geometry allowed very convenient modelling and exploiting of the radiation field, as photocatalytic membranes (1.0 m^2 of apparent geometrical surface) were placed coaxially with the lamp as well as on both sides of suitable steel mesh baffles placed perpendicularly to the lamp itself. Distance between these baffles could be regulated, in order to increase or decrease their number and consequently surface area of membranes. The emission power of the lamp (Helios Italquartz, Milan) could be regulated in the range 8-12 KW. As in the reactor of Fig. 1 the outer shell acted as a metallic mirror, practically all photons emitted by the lamp reached the reaction medium and were integrally absorbed by the immobilised semiconductor and photocatalytic systems, when added. Reactor of Fig. 1 could be used either in a continuous process, or in a batch process, by entirely recirculating the processed solution, or with partial recycling of the same solution. When operating in a batch process, a cooling system was able to remove heat of irradiation, and to maintain solution temperature at a predetermined value within 1.5°C. Stirring of reservoirs was accomplished mechanically. Hydrodynamic conditions assured turbulence of processed solution, besides providing the pressure drop necessary for permeation of fluid through the membranes. Turbulence was needed to guarantee efficiency of adsorption and desorption processes of pollutants, onto and from catalyst immobilised within the membrane pores, which are among the mechanistic requisites of photocatalytic processes. All of the pilot plant experiments have been performed with "Photoperm" CPP/313 membranes, immobilising titanium dioxide alone, without any other photocatalytic system.

2.3 Apparatus and procedures of laboratory-scale studies

When carrying out laboratory-scale experiments, both with "Photoperm" membranes CPP/313 and with "Photoperm" membranes CPP/313(I-VII), at varying concentration of photocatalytic systems (I-VII) (see *2.1*) immobilised in the membranes, the same procedure described in previous work [7-8] has been followed. Disc membranes (6 cm diameter) were sandwiched by means of O-rings between optical Pyrex glass transmitting from 300-310 nm up to longer wavelengths, in order to cut off, in these laboratory studies, radiations shorter than about 300 nm, which could cause a direct photolysis or some direct photochemical transformation of organics tested, independent of the action and of the presence of photocatalytic system. The optical plates carried an inlet and an outlet tube, on one side and the other side respectively, so that solutions to be irradiated could circulate continuously from and to suitable reservoirs, and permeate trough the membrane, while being oxygenated by air in the reservoirs (if other oxidising agents were not added in the experiments). The distance between the optical plate and the membrane was 2 mm, and the ratio between the overall reacting volume (in the photoreactor cell and the reservoir) and the apparent, geometrical surface area of the irradiated side of the membrane was 1.1±0.1 mL cm^{-2}.

The cell was located at such a distance from a 2 KW high pressure mercury arc lamp (Helios Italquartz, Milan) as to have a mean polychromatic radiant flux, from 300 nm up to the higher wavelength edge of the absorption range of the semiconductor, by taking also into account absorption by other photocatalysts, if added, ranging from 2.5 to 480 mW cm^2, as measured actinometrically and as evaluated, given the emission spectrum of the lamp and the absorption spectrum of the photocatalysts.

The temperature was regulated, within 1°C of confidence, in the range 25-75°C, by thermostatting the reservoirs. The concentration of organics tested was determined by a Beckman 915A total organic carbon analyser. The initial rate r_0 of photodegradation of organics was evaluated from curves of their concentration *vs.* time in the linear range, where zero-order kinetics was apparent. Standard deviations of these values were obtained from four to six repetitions of experiments.

3. RESULTS AND DISCUSSION

3.1 Laboratory-scale experiments

As is known from the literature [1-2], and as has been done in previous work [7-8], the Langmuir-Hinshelwood kinetic rate law has been used to interpret experimental results of the laboratory runs, in which, owing to conditions employed, no degradation process of organic pollutants tested, other than the photocatalytic one, could occur:

$$r_0 = -dC_0/dt = kKC_0 / (1+KC_0) \qquad \text{(equation 1)}$$

where t is time, k is the reaction rate constant, which has been reported to be directly proportional to the square root of the absorbed radiation intensity [12], C_0 is the initial concentration of the organic compound, which is being photodegraded, and K is the equilibrium adsorption constant of this compound onto the photocatalyst. The linearised form of eqn. (1)

$$1/r_0 = (1/k) + (1/kKC_0) \qquad \text{(equation 2)}$$

clearly shows that initial rate of photodegradation equals the reaction rate constant for the process.

3.1.1 Influence of hydrogen peroxide

The influence of hydrogen peroxide was first studied in photodegradation of trichloroethene in aqueous solution, either in the presence or in the absence of photocatalytic membranes immobilising titanium dioxide alone ("Photoperm" CPP/313), without any other promoting photocatalytic system, on a laboratory scale. Results are reported in Fig.2, as a plot of initial rate of degradation, at 30°C, *vs.* the ratio N between the amount of hydrogen peroxide added and the stoichiometric quantity. In this same Fig., the datum relative to the rate measured in the absence of hydrogen peroxide, but only in the presence of dissolved oxygen, taken from previous work [7], is also reported.

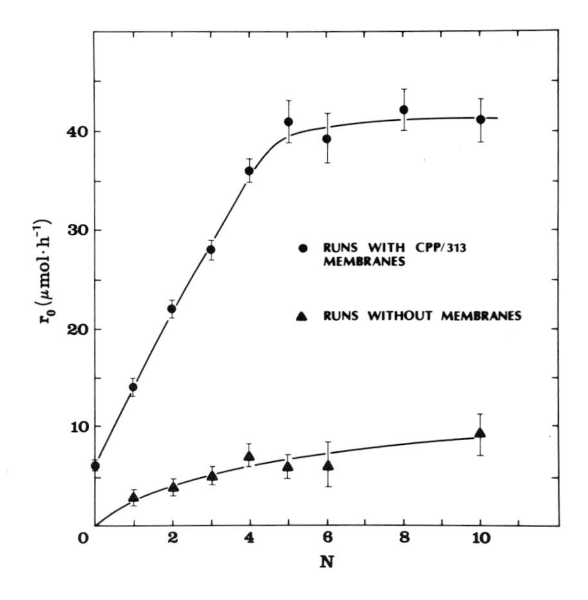

Fig. 2. Initial rate of photodegradation r_0 (μmol/h) (see eqn. 2) of trichloroethene in aqueous solution, at 30°C, by "Photoperm" CPP/313 membranes immobilising titanium dioxide, as a function of molar ratio *N* between hydrogen peroxide added and the stoichiometric amount. Uncertainties of values are expressed as standard deviations.

Two facts readily emerge from data of Fig. 2. In the absence of the membrane, the action of hydrogen peroxide stimulated by UV irradiation is rather scarce, at least in the wavelength range tested during the laboratory experiments. The corresponding reactivity (see Fig. 2) reaches a value practically equal to that of the photocatalytic process in the presence of dissolved oxygen alone, only if an amount of hydrogen peroxide at least five-fold or six-fold the stoichiometric one is added. This is not contradictory with reported literature data on the degradation effect of UV irradiation in the presence of hydrogen peroxide, since in the experimental conditions of these runs, owing to the wavelength range of irradiation spectrum and the absence of any catalyst, UV absorption by hydrogen peroxide and consequent reactivity were really very low or negligible. On the contrary, when the photocatalytic action of membranes is added, the presence of hydrogen peroxide increases markedly the photodegradation rate. A plateau effect of concentration is reached at a concentration of about five times the stoicheiometric one (see Fig. 2). The increase of rate is about seven-fold in these conditions.

All these facts point to the conclusion that the mechanism of action of hydrogen peroxide in the presence of the membranes is substantially given by a photocatalysed cleavage of the oxidiser molecule into .OH radicals, as has been pointed out otherwise [12] when considering this molecule as an intermediate in the possible chain reactions during photocatalysis.

Results substantially similar to those reported in Fig.2 were also obtained with other chloro-aliphatics such as those studied in previous work [7-8]. As a matter of fact, rates increase, at a stoichiometric concentration of hydrogen peroxide, with a factor ranging from about 2.4 (for trichloroethene; see Fig. 2) to about 3.0 (for perchlorinated compounds), while at the plateau factors range from about 7 (for trichloroethene; see Fig. 2) to about 9.1 [13].

3.1.2. Influence of promoting photocatalysts

Influence of chemical structure and concentration of some photocatalytic systems, such as those added during the manufacture of "Photoperm" CPP/313(I-VII) membranes (see *2.1* above), was tested both in the absence of hydrogen peroxide and with a five-fold amount of this oxidising agent respect to stoichiometry. Trichloroethene and tetrachloroethene were employed as reacting substrates, in the two series of runs respectively.

In Fig. 3 relative initial rates of photodegradation ***R*** (***R*** being the ratio between initial rates measured in the presence and in the absence of promoting photocatalysts (I-VII) respectively) are reported as a function of wt.% of (I-VII) in the photografted membranes, for runs carried out at 30°C without any addition of hydrogen peroxide. Advantages brought about by the different photocatalysts are more or less important, following the different central atoms, and, for a fixed central atom, following the nature of ligands. All behaviours show, anyway, a threshold effect: ***R*** values remain close to unity, if concentration of photocatalyst does not exceed a more or less significant minimum value. For some of the investigated systems, the influence of photocatalyst, such a for (VI), is already distinct and observable, beyond the experimental uncertainty, at concentrations as low as 0.01 wt.%, that is in the same range of concentrations in which these photocatalysts are active during photopolymerisation and photocrosslinking [11,14]. In general much higher concentrations than the latter are needed to reach practical effects and factors fairly greater than the unity (see Fig. 3). Finally, some kind of levelling results with increasing concentration, and a plateau is reached. Among the photocatalytic systems, the effect of (VI) is outstanding, leading in the plateau conditions (see Fig. 3), to an increase in rate of about 20 times.

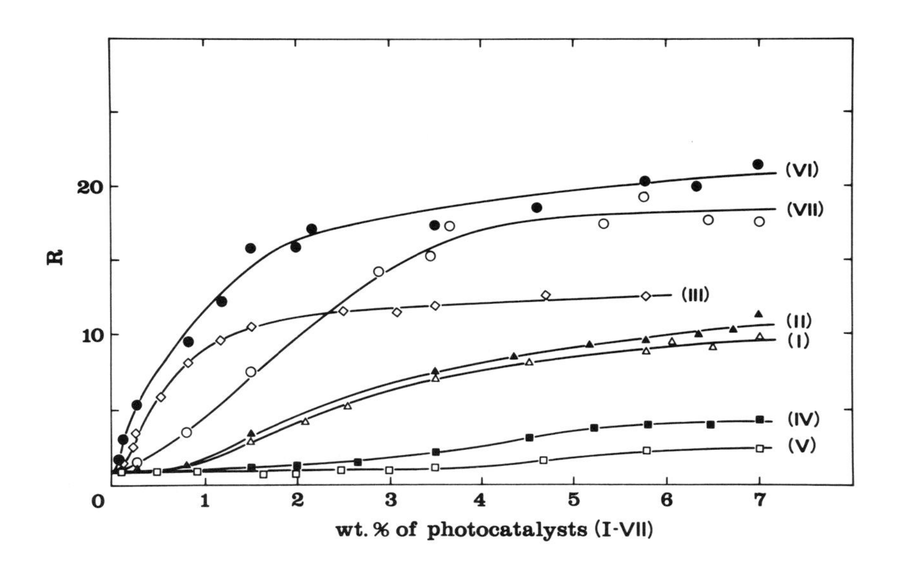

Fig. 3. Relative initial rates R (R is the ratio between rates (see eqn. 2) measured at 30°C in the presence and in the absence of promoting photocatalysts (I-VII) respectively), in the presence of dissolved oxygen alone, as a function of wt.% of photocatalysts (I-VII) in the photografted membranes.

The mechanism of action of these photocatalytic systems is not perfectly understood at the present moment, no one of them being a semiconductor. As a preliminary hypothesis, they may be considered as oxygen transporting agents, as in fact they are [11,14-16]. By this way, they may be able to intervene efficiently in the reversible redox processes involving oxygen itself or hydrogen peroxide, and consequently increase the rate of production of oxidising radical species. The threshold effect of Fig. 3 may also be the expression of this mechanism: depending on the physico-chemical nature of photocatalytic system, a minimum concentration is needed, in order to reach a measurable influence as oxygen transporter, in the complex radical reactions occurring during photodegradation of adsorbed species at the membrane interface.

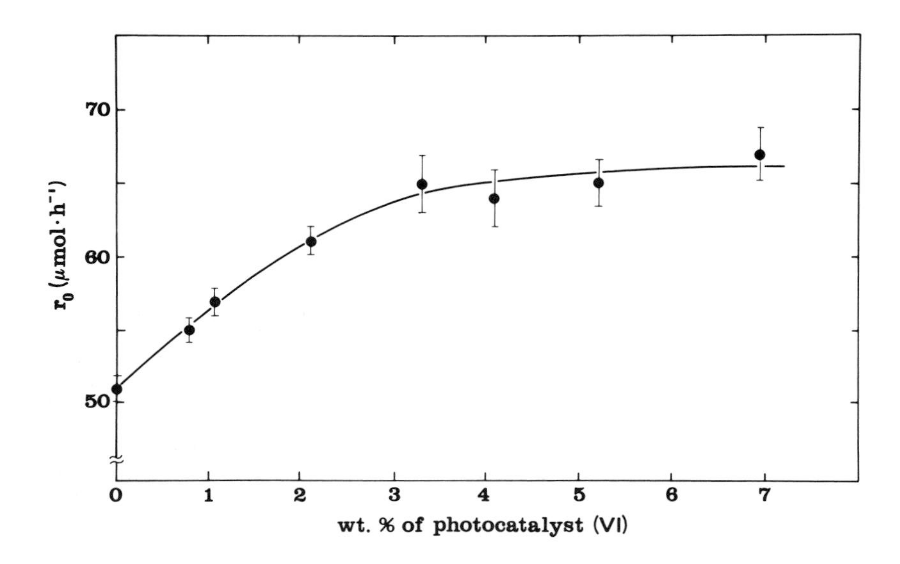

Fig. 4. Initial rate of photodegradation r_0 (μmol/h) (see eqn. 2) of tetrachloroethene in aqueous solution, at 30°C, in the presence of concentrations of hydrogen peroxide equal to five times the stoichiometric amount, as a function of wt.% concentration of photocatalyst (VI) in the photografted "Photoperm" CPP/313(VI) membranes. Uncertainties of values are expressed as standard deviations.

In Fig. 4, initial rates of photodegradation r_0 of tetrachloroethene in aqueous solution at 30°C, in the presence of hydrogen peroxide, at a concentration equal to five times the stoichiometric one, is reported as a funtion of concentration of photocatalyst (VI) in the membrane. The influence of (VI) as photocatalyst appears with clear evidence, even if its relative effect is not so marked as that observed in the absence of the added oxidising agent (see Fig. 3 for comparison). Again, a plateau effect is noticed. From the practical point of view, in conditions of Fig. 3 a further 27% increase may be gained in the otherwise already high value of rate yielded by the investigated substrate in the absence of (VI) but in the presence of hydrogen peroxide concentrations five-fold the stoichiometric one (51 μmol/h [13]).

As to the plateau effects observed in Figs. 3 and 4, they may indicate some kind of saturation of active sites available in the composite membrane structure: given a certain interface between membrane and solution, depending on structure and porosity of photografted membrane, further increase of photocatalyst concentration is not able to contribute further to photocatalytic sites, probably because the photocatalyst concentrates in the bulk or in unaccessible positions of the membrane itself.

3.1.3. Influence of temperature

Most photochemical reactions are reasonably insensitive to reactor temperature, because the excitation energy afforded by photon absorption is generally much greater than required to overcome ground state activation energy barriers. As a matter of fact, in a series of experiments, carried out on trichloroethene as substrate, in the absence of any oxidising agent other than dissolved oxygen, and with CPP/313 membranes immobilising titanium dioxide alone, it could be ascertained that, in the range of temperatures from 25 to 75°C, measured value of r_0, within experimental uncertainty (standard deviation of 0.9 μmol/h) substantially coincided with that obtained at 30°C (6 μmol/h [7]; see Fig. 2). This confirms that, even if adsorption of reacting species on photocatalyst should be unfavoured by increasing temperature, temperature activation of reacting radical species, on one side, and easier desorption of intermediates and products, on the other, together with excitation energy of optical absorption by photocatalyst, are able to more than compensate negative effects, on the light of the most accredited mechanism [12] involving the role of .OH radicals and related radical species.

The study of temperature effects, on the contrary, allowed it to be established that, in the runs of Fig. 2 carried out without membranes, reactivity of hydrogen peroxide with substrate (trichloroethene in this instance) under irradiation decreased by about 30-40% when temperature increased from 30 to 75°C and when the molar ratio *N* was 4-5. In these runs, in order to know exactly the concentration of hydrogen peroxide at the inlet of the photoreactor cell, the flowing solution was monitored spectrophotometrically at the inlet of the irradiated reactor. Conversely, no substantial variation in initial rate of photodegradation, outside the experimental uncertainty, was found, respect to values obtained at 30°C, when the same runs were carried out at 75°C, with the same molar ratios *N*, but in the presence of photocatalytic membranes. These effects may be explained by considering that thermal decomposition of hydrogen peroxide becomes progressively important at temperatures higher than 40-50°C. Hence, in the absence of any photocatalytic process, decomposition of hydrogen peroxide within the reactor cell decreases its concentration and consequently the whole reactivity. When photocatalyst is present, on the contrary, photocleavage of hydrogen peroxide predominates over its thermal decomposition in the photoreactor cell, even at relatively high temperatures.

3.1.4. Influence of radiation intensity

Initial rate of photodegradation of trichloroethene in aqueous solution at 30°C by "Photoperm" membranes CPP/313, immobilising semiconductor alone, was measured, in the presence of stoichiometric amounts of hydrogen peroxide, at various radiation intensities I, ranging from 2.5 to 480 mW/cm^2. Experimental results are collected in Fig. 5, in the form of plots of r_0 values, referred to apparent, geometric surface of membranes, as a funtion of both I and its square root. Experimental data clearly indicate a linear dependency of rate on square root of radiation intensity, thus confirming the general behaviour of this relationship, as reported in most literature data [1,2,12].

We may, consequently, infer that immobilised photocatalyst does not change sensibly and/or substantially alter its behaviour, as is known from physical properties of "free" (non-immobilised) species. This is otherwise a general conclusion [9-11] drawn from all studies on species immobilised by photografting procedures in a membrane structure. The fast method employed for the membrane manufacture allows it to maintain almost completely active the surface area of catalysts, sorbents, reagents, which are immobilised by this technology. Consequently, all physico-chemical properties of immobilised species remain practically unaltered.

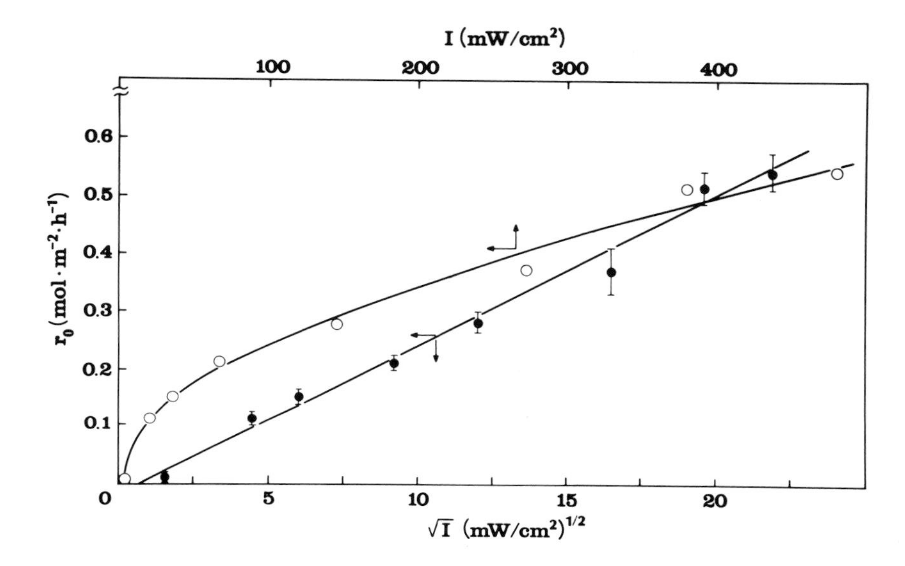

Fig. 5. Initial rates of photodegradation r_0 (mol/m^{2}h) of trichloroethene in aqueous solution, at 30°C, measured in the presence of stoichiometric amounts of hydrogen peroxide, as a function of radiation intensity I (mW/cm^2) (upper abscissa), or its square root (lower abscissa). Uncertainties are expressed as standard deviations. Rates are referred to unitary geometrical (apparent) surface area of membranes.

Finally, the linear dependence of r_0 on $I^{1/2}$ (see Fig. 5) (this may also be evinced from the curve of r_0 *vs.* I, though with less precision) shows that some kind of threshold value of intensity seems to exist in order to have a minimum observable effect on rate. This threshold may be evaluated as 0.6-0.7 mW/cm^2. If it really has a physical meaning, outside the possible uncertainty, it may represent, by hypothesis, the UV energy dissipated by the polymeric structure of photografted membranes, and consequently unavailable for the semiconductor activation.

3.2. Pilot plant experiments

Several photodegradation experiments have been carried out in the pilot plant, in order to compare results with those of the laboratory scale.

First of all, it should be remembered that irradiation conditions of the two devices were different. In laboratory experiments only photocatalytic degradation was possible, because all of the examined substrates did not show absorption in the UV range of irradiation. In the pilot plant, on the contrary, for substrates absorbing from 210 nm to about 300 nm, and for hydrogen peroxide, when used, direct photolysis could in principle occur, parallely with photocatalysed degradation. This must be borne in mind when comparing the two kinds of experiments.

Secondly, in the light of systematical information collected during laboratory experiments, particularly on the effect of irradiation intensity (see *3.1.4*) as well on the influence of hydrogen peroxide concentration (see *3.1.1*), results of pilot plant experiments may be conveniently compared, by interpolating laboratory data in the industrial conditions, even when these data are not available by direct measurements, given the ascertained relationships.

Four case studies will be presented in this paper, concerning pilot plant experiments with dichloromethane, trichloroethene, toluene, and phenol, as model substrates of chloro-aliphatic and aromatic, non-biodegradable compounds.

Dichloromethane may be considered as an example of non absorbing, or very low absorbing substrate, even at the high frequencies of the lamps employed. Initial rate of photodegradation during plant experiments with this substrate was measured as 0.40 mol/m^2h, with a standard deviation of 0.03 in these same units, at 30-45°C and at an irradiating power of 10 KW, corresponding to an absorbed radiation intensity, in the absorption range of photocatalyst, of 248 mW/cm^2 of apparent, geometric membrane surface. On a laboratory scale, at this same radiation intensity, a value of 0.41 mol/m^2h was obtained.

Trichloroethene may be considered as an example of low absorbing substrate in the high energy region of the UV range of interest for the emission spectrum of the lamps used. Its pilot plant degradation rate was measured as 3.3 mol/m^2h, at 30-45°C and radiation intensity as in the above case study, a value which is fairly greater than that obtainable from interpolation of the laboratory data of Fig. 5 (around 0.41 mol/m^2h, substantially the same as for dichloromethane, within the limits of experimental uncertainty). This occurs because direct photolysis and/or photocatalysed photolysis in the low wavelength UV range for this compound is kinetically more important than photocatalysed degradation in the high wavelength UV range, particularly at the remarkable values of radiation fluxes exploited in the pilot plant.

Toluene and phenol may be considered as examples of medium absorbing substrates in UV spectrum. Their photodegradation rates, in pilot plant experiments, in the same conditions of experiments cited above, were measured as 4.1 and 5.0 mol/m^{2}h respectively, as compared with the corresponding laboratory values, which have been the object of direct measurements, of 0.58 and 0.63 mol/m^{2}h respectively, with a standard deviation of 0.04 in these same units. Again, the explanation lies in the fact that long wavelength laboratory experiments represent a purely photocatalytic decomposition rate, while industrial experiments, in the conditions of the investigated layout, are undoubtedly more complex in nature, but fortunately allow to operate, particularly for absorbing substrates, at about seven- eight-fold photodegradation rates. Repetitions of laboratory-scale experiments with these same substrates, by using a quartz, instead of a glass plate, window, led, of course, to substantially the same values reported for industrial-scale experiments, within a confidence limit of about 0.3 mol/m^{2}h expressed as standard deviation.

To sum up, results of pilot plant operations are entirely satisfactory, and wholly promising for the photochemical degradation process carried out by photocatalytic membranes. All the values above, in fact, must not be considered as top values, since the membranes used in these experiments immobilised only titanium dioxide. If photocatalysts such as (III), (VI), or (VII) are added to the membranes in an industrial process, rates may be further and remarkably increased (see Fig. 3), as tests carried out in the present work (see *3.1.2*) have widely demonstrated.

Finally, also in conditions in which direct photolysis of pollutants is possible (even if photolysis is a very slow reaction, if compared with photocatalytic degradation [17]), and in which, in the absence of membranes, heavy formation of tarry products has been observed on the quartz container of irradiating lamps [17, 18], when photocatalytic membranes are present, no tarry deposits at all were observed on the windows. This is due both to the fact that, in the presence of membranes, the much faster photocatalytic reactions predominate kinetically, and that, in any case, a very limited and transient formation of tarry substances occurs on the adsorbing membranes, not on the windows. Furthermore, these optically absorbing tarry substances on the membranes disappear completely at prolonged irradiations. No cleaning device is thus necessary in the process described in the present paper.

4. CONCLUSIONS

Performance, flexibility, and productivity of photocatalytic membranes, prepared by photografting and immobilising titanium dioxide, as tested in the pilot plant reactor, for complete degradation of organic, non biodegradable water contaminants, are eminently conforming to a high standard.

1) The problem of light absorption in a heterogeneous medium, which presents a real challenge to establish working *a priori* models of the coupled hydrodynamics and radiation fields, is radically bypassed by the use of photocatalytic membranes.

2) These latter, indeed, allow not only to increase the surface-volume ratio enormously (this ratio was as high as 0.59 cm^2/mL in the pilot reactor, if computed with the apparent, geometrical surface of membranes: in terms of active surface area of semiconductor, it amounted to 340 cm^2/mL and may be increased further), but also to operate efficiently even where the most sophisticated traditional processes (such as carbon adsorption) fail due to low concentration of contaminants, or even where other traditional membrane processes (such as reverse osmosis) become ineffective owing to the low levels of pollution.

3) Photodegradation rates may be controlled and remarkably increased by immobilising appropriate photocatalysts, together with semiconductor, in the photografted membranes, and/or by employing oxidising agents, other than dissolved oxygen, such as hydrogen peroxide, in the solutions to be processed.

4) Adequately high values of photodegradation rates (up to several moles per hour and per square metre of geometrical surface area of membrane) may be reached by choosing an appropriate radiation intensity, as may be done in industrial plants, even without the aid of promoting photocatalysts, by operating with stoichiometric amounts of hydrogen peroxide. No excess of the latter, anyway, could remain in the processed solutions, as it is rapidly decomposed into fast reacting hydroxyl radicals by the photocatalytic membrane.

In conclusion, all the advantages brought about the "Photoperm" process described, as they have been evidenced both in laboratory and pilot plant experiments, and, last but not least, the relatively low cost of manufacture of these membranes, together with the general applicability of the investigated technology (from water potabilisation to treatment of wastewaters containing organic contaminants of practically any kind, as many preliminary tests have shown) are undoubtedly auspicious requisites of future successful developments.

5. ACKNOWLEDGEMENTS

Technical assistance by Mr. Franco Gianturco and Miss Barbara Barni, during pilot plant experiments, is gratefully acknowledged.

6. REFERENCES

1. Schiavello, M. (ed.): "Photocatalysis and Environment. Trends and Applications", *NATO ASI Ser.C*, vol.*237*, Kluwer, Dordrecht (1988).
2. Serpone, N. and Pelizzetti, E. (eds.): "Photocatalysis. Fundamentals and Applications", Wiley, New York (1989).
3. Tender,D.W. and Pohland, F.G. (eds.): "Emerging Technologies in Hazardous Waste Management", *ACS Symp. Ser.*, vol.*422*, ACS, Washington (1990).
4. Serpone, N., Borgarello, E., Harris, R., Cahill, P., Borgarello, M., and Pelizzetti, E., Photocatalysis over Titanium Dioxide Supported on a Glass Substrate, *Solar Energy Mater.*, 14, 121 (1986).
5. Matthews, R.W., Photooxidation of Organic Impurities in Water Using Thin Films of Titanium Dioxide, *J. Phys. Chem.*, 91, 3328 (1987).
6. Al-Ekabi, H. and Serpone, N., Kinetic Studies in Heterogeneous Photocatalysis. 1. Photocatalytic Degradation of Chlorinated Phenols in Aerated Aqueous Solutions over Titanium Dioxide Supported on a Glass Matrix, *J. Phys. Chem.*, 92, 5726 (1988).
7. Bellobono, I.R., Bonardi, M., Castellano, L., Selli, E., and Righetto, L., Degradation of Some Chloro-aliphatic Water Contaminants by Photocatalytic Membranes Immobilising Titanium Dioxide, *J. Photochem. Photobiol. A: Chem.*, 67, 109 (1992).
8. Bellobono, I.R., Composite Photosynthetic Membranes in Photochemical Treatment of Wastewaters, in "Wastewaters and Sludges" (Frigerio, A., ed.), CSI, Milan (1992), pp. 172-185.
9. Bellobono, I.R., Some Innovative Applications of Photografting Processes. Textile Dyeing and Printing. Sorbent and Reactive Composite Membranes, in "Grafting Processes onto Polymeric Fibres and Surfaces: Scientific and Technological Aspects", (Bellobono, I.R., ed.), *Conf. Proceed. Univ. Milan,* vol. 85 (1990), pp. 119-213.
10. Bellobono, I.R. and Selli, E., Photografting onto Polymers, in "Photopolymerisation and Photoimaging Science

and Technology" (Allen. N.S. ed.), Elsevier Applied Sci. Publ., London (1989), chapter 4.
11. Bellobono, I.R. and Righetto, L., Photochemical Production of Composite Membranes and Reinforced Plastics, in "New Aspects of Radiation Curing in Polymer Science and Technology" (Fouassier, J.P. and Rabek, J.F. eds.), Elsevier, in the press.
12. Turchi,C.S. and Ollis, D.F., Photocatalytic Degradation of Organic Water Contaminants: Mechanisms Involving Hydroxyl Radical Attack, *J. Catal.*, 122, 178 (1990).
13. Bellobono, I.R. *et al.*, unpublished results.
14. Bellobono, I.R., Selli, E., and Righetto, L., Influence of Photocatalytic Systems on the Photocrosslinking of Diallyl Oxydiethylene Dicarbonate Grafted onto Cellulose, *J. Photochem. Photobiol. A: Chem.*, 65, 431 (1992).
15. Bellobono, I.R., Kinetic Study of the Photocrosslinking of 1,6-Hexanediol Diacrylate Grafted onto Cellulose in the Presence of Photocatalytic Systems, *J. Photochem. Photobiol. A: Chem.*, 59, 91 (1991).
16. Bellobono, I.R., Muffato, F., Selli, E., Righetto,L., and Tacchi, R., Transport of Oxygen Facilitated by Peroxo-bis-(N,N'-ethylene-bis-(salicylideneiminato)-dimethylformamide-cobalt(III)) Embedded in Liquid Membranes Immobilised by Photografting onto Cellulose, *Gas Separ. Purif.*, 1, 103 (1987).
17. Bellobono, I.R., Carrara, A., and Schwienbacher, H., Pilot Plant Experience for Photochemical Degradation of Non-biodegradable Organic Contaminants in Wastewaters by Photocatalytic Membranes Immobilising Titanium Dioxide and Promoting Photocatalytic Systems, in " Acque Reflue e Fanghi" (Frigerio, A., and Faietti, L. eds.), CSI, Milan (1993), in the press.
18. Bellobono, I.R., and Schwienbacher, H., Pilot Reactor Potabilisation of Micropolluted Waters Containing Organic Contaminants by *PHOTOPERM* Process, in "Acque per Uso Potabile" (Frigerio, A., and Faietti, L. eds.), CSI, Milan (1993), in the press.

IMMOBILIZATION OF BACTERIA IN COMPOSITE MEMBRANES AND DEVELOPMENT OF TUBULAR MEMBRANE REACTORS FOR HEAVY METAL RECUPERATION

L. Diels, S. Van Roy, M. Mergeay
Laboratory of Genetics and Biotechnology

W. Doyen, S. Taghavi, R. Leysen
Membrane Group

Flemish Institute of Technological Research
V.I.T.O., Mol, Belgium

SUMMARY

Some bacteria like the heavy metal resistant Alcaligenes eutrophus CH34 strain are able to promote biomineralization, being the biologically induced crystallization of heavy metals. In presence of heavy metals, this strain may create an alkaline environment in the periplasmic space and outer cell environment after appropriate induction of heavy metal resistance mechanisms. In such an environment metal hydroxides are formed together with metal (bi)carbonates resulting from the (bi)carbonate production by the cells. Also metals bind to outer cell membrane proteins and the metal hydroxides and (bi)carbonates precipitate around these nucleation foci inducing further metal crystallization.

In order to keep cells viable for this biomineralization process, a Tubular Membrane Reactor was developed. A. eutrophus is immobilized in a composite membrane, based on polysulfone and ZrO_2. The membrane is casted on a tubular polyester support and separates a nutrient stream from an effluent stream. The ZrO_2 grains help in the linking of the cells to the membrane and the pores form a cavity to keep the bacteria immobilized. The nutrient is necessary to keep the bacteria viable to producecontinuously metal ligands. At the effluent side immobilized bacteria induce metal precipitation and metal crystals are recuperated on a recuperation column.

This column is filled with glass beads to which the metal crystals bind and grow continuously.

In this way metals can be recovered from the recuperation column by acid treatment without damaging the immobilized cells in the membrane. Metals can be removed for more than 99,9% (e.g. from 100 ppm Cd to < 0,1 ppm Cd).

INTRODUCTION

Recuperation of heavy metals can be realized by different techniques in function of their concentration. High concentrations (> 500 ppm) can be recovered by electrolysis and low concentrations (< 5 ppm) can be removed by biosorption or ion-exchange columns. At concentrations between 500 and 5 ppm precipitation with lime is possible generating high volumes of sludge with low metal/sludge ratios.
Also metal hydroxides, formed after addition of sodium hydroxide, can be removed via microfiltration. In this way metal concentrations can be reduced to ppm level but not lower. Therefore, in the above mentioned concentration range another system, the biologically induced crystallization of heavy metals called BIOMINERALIZATION becomes an important tool.

Biomineralization is a process in which bacteria actively induce the precipitation of heavy metals by the creation of a supersaturated environment around them and by the production of metal precipitation foci on their cell surface (Diels, 1990 ; Michel et al., 1986). The bacteria provide at the same time anions like carbonates, hydroxides (Diels, 1990) and biphosphates (Michel et al., 1986). This means that the bacteria must be kept viable. The addition of these nutrients to a Stirred Tank Reactor will lead to a too high nutrient consumption cost and to a secondary pollution of the effluent.
On the other hand, composite Zirfon® membranes (Leysen et al., 1989) were developed for ultrafiltration and the immobilization of enzymes. According to Kennedy et al. (1986) the inorganic filler ZrO_2 can help in the binding between the bacteria and the membrane.
The immobilization of metal precipitating bacteria in composite Zirfon® membranes led to the development of a Continuous Tubular Membrane Reactor for the removal of heavy metals from effluents. The principle of the reactor is based on the separation of two streams by a membrane in which the bacteria are immobilized. The bacteria remove the metals from one stream (effluent) and are kept viable via the other stream (nutrient) in which nutrient is provided. This combination makes

it possible to treat water, contaminated with heavy metals, with a minimal consumption of nutrients like carbon and phosphate.

MATERIALS AND METHODS

Heavy metal precipitating Alcaligenes eutrophus CH34

During the last years, bacteria resistant to a variety of heavy metals were isolated and identified (for review, see Silver, 1988). The mechanisms for these resistances are often controlled by plasmid encoded genes or by transposons. A remarkable example of those resistant bacteria is Alcaligenes eutrophus var. metallotolerans. The representative strain CH34 was isolated in sediments from a decantation basin of a zinc factory (Mergeay et al., 1978). Strain CH34 bears two large plasmids (Mergeay et al., 1985 ; Nies et al., 1987) controlling resistance against Cd^{++}, Co^{++}, Zn^{++}, Hg^{++}, Tl^{+}, Cu^{++}, Pb^{++} (pMOL30, 240 kb) and Co^{++}, Zn^{++}, Ni^{++}, Hg^{++}, CrO_4^{--}, Tl^{+} (pMOL28, 165 kb). As could be shown by Nies et al. (1989a, 1990) resistance to chromate is inducible and based on decreased net accumulation of the metal anion. Resistance to zinc, cadmium, cobalt and nickel are resulting from inducible, energy dependent cation efflux systems (Nies et al., 1989b ; Nies and Silver, 1989c ; Siddiqui et al., 1988 ; Sensfuss and Schlegel, 1988). In some physiological circumstances A. eutrophus can also accumulate and precipitate heavy metals (Diels et al., 1989, 1990). At increased concentrations of Cd or Zn ions, a removal of these metals from the solution is observed during the late log phase and the stationary phase. This accumulation and precipitation is correlated with the concentration and kind of carbon source (lactate or acetate) with the progressive alkalinization of the medium, with the concentration of phosphate and appears to be associated with the outer cell membrane. The alkalinization of the periplasmic space is due to the metal efflux system which works together with a proton antiport system continuously generating OH^- ions in the periplasm. The precipitation of $Cd(HCO_3)_2$, $CdCO_3$ and $Cd(OH)_2$ is proved by IR-spectroscopy. The interpretation of this feature is that the metal speciation will change at the cell surface due to the progressive pH increase, the steep pH gradient on this site, and the production of CO_2 by the cell metabolism (supersaturation of the microbial environment). The metal will in this way precipitate on the cell envelopes using a cell membrane component as a support. This trapping would occur at the uptake or the efflux of the metals or more likely during both processes and suggests a kind of biomineralization as reviewed by Mann (1988).

Immobilization and colonization of A. eutrophus in the composite membranes

Diluted bacterial cultures were filtrated over the membranes either vertical or tangential. Filtration is done from the open side (opposite to the skin side) where bacteria can penetrate and colonize the membrane.
After immobilization of A. eutrophus in the mentioned membranes, colonization happens by growth of single cells at random distributed in the membrane network. Nutrients are obtained via diffusion into the membrane from the nutrient solution.

Cd and Zn removal by immobilized A. eutrophus cells

A. eutrophus cells are immobilized in membrane M5 as described above. After filtration and rinsing, the membranes were incubated in minimal growth medium with 0.8% lactate as carbon source and 1 mM phosphate as phosphorous source. To each of the media ^{109}Cd (0.25 μCi) or ^{65}Zn (0.25 μCi) was added plus cold Cd or Zn to reach a final Cd or Zn concentration of 0.1 mM, 0.2 mM (only for Zn) and 0.5 mM, 1.0 mM and 2.0 mM (for Zn and Cd). After two days the radioactive metal content was measured on the membranes and calculated to give the final metal concentration per cm^2 of membrane surface.

Development of a Flat Sheet Reactor (FSR)

A Flat Sheet Reactor consists of two chambers with each an inlet and outlet and separated from each other by a composite membrane and a supporting frame as shown in Figure 1. In one chamber was placed a nutrient solution and at the other side synthetic effluent (in the presented experiments it was mostly a 1 mM of $CdCl_2$ solution) were pumped during reactor use.
Generally at each side 500 ml solution was pumped all the time and samples were taken from each chamber for measurement of optical density (caused by released cells), pH, metal concentration and lactate consumption. Parameters were evaluated by comparing two reactors under different experimental conditions.

Nutrient medium

A mineral medium according to Schlegel et al. (1961) was provided as nutrient to the bacteria. This medium contains only minimum amounts phosphate to keep the bacteria viable and used Tris(-hydroxy methyl)-aminomethane as a buffering agent. As carbon source 0.8% lactate and as a source of phosphorous 0.294 mg/l glycerophosphate was used.

Effluents

As synthetic effluent, tap water contaminated with 1 mM $ZnSO_4$ or 1 mM $CdCl_2$ are used.
A real effluent from a Portuguese surface treatment company was also tested. This effluent was the supernatant of a physico-chemical treatment and contained between 20 and 60 ppm of Zn and some traces of Cd, Cu, Cr and Ni.

Zirfon® membranes

The Zirfon® composite membranes, originally developed for electrochemical purposes have been developed in three different configurations : flat membranes with or without a reinforcing support, hollow fibres and tubes. However the hollow fibres are very sensitive to clogging especially in the presence of an effluent which contains suspended materials.

The tubular type of membrane which is probably the best choice for the bacteria immobilized reactor, corresponds to a bundle of tubes being mounted into a cylinder, the reaction chamber. Each tubular membrane is composed of a porous supporting tube covered with the active membrane at the inner tubular surface.

For the formation of suitable composite membranes $CaCO_3$, Sb_2O_3 (membrane type Zirfon M6 or M5) or polyvinylpyrolidone (PVP) (membrane type Zirfon M7, M8 and M10) or ZnO (membrane type Zirfon M13, M14, M15) were used as pore formers. Typical membranes for bacterial immobilization have a total pore volume between 60 and 80%, mean pore diameters between 1 and 5 µm and a thickness between 150 and 250 µm. They further possess a skin side with pores with a diameter smaller than 1 µm. In some cases, fingerlike structures are observed.

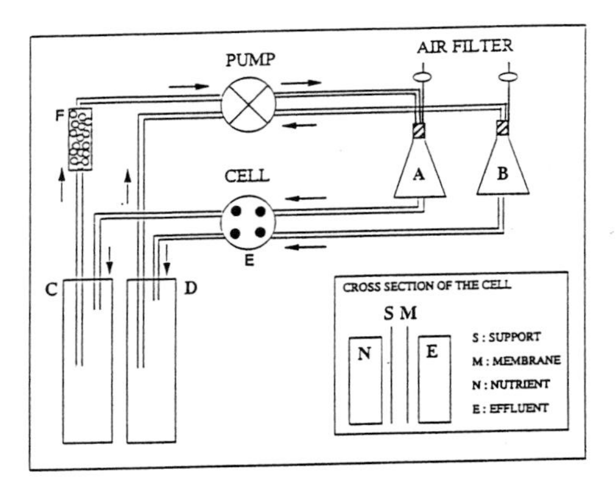

Figure 1. Schematic presentation of a Flat Sheet Reactor.
The general concept of a Flat Sheet Reactor is presented : A : nutrient ; B : effluent ; C : siphon ; D : siphon ; E : FSR ; F : glass bead column.

Development of a Tubular Membrane Reactor (TMR)

A first unit consisted of a stainless steel cylinder with a diameter of 20 mm with an inlet and outlet on the cyclinder wall. Inside was mounted a carbon pipe (inner diameter 6 mm and the outer diameter is 10 mm). The carbon pipe has a mean pore radius of 2 μm. At the inside, a membrane M5 was deposited with the open membrane side to the carbon pipe. The effluent flows through the lumen of the carbon pipe and the nutrient at the outside of the pipe (see Figure 2). In later experiments the carbon support pipes were replaced by polyester supports (with 100 μm thickness and a mean pore radius of 10 μm).
This TMR reactor was connected to an effluent and a nutrient flask (see Figure 2). Solutions were pumped at a flow rate of 40 ml/min for the nutrient and 5 ml/min for the effluent. At these flow rates, the ΔP over the membrane was very low. Bacteria were added from the nutrient side. After leaving the TMR reactor, the effluent to be treated, passes over a column, bearing glass beads, before it recirculates to the effluent flask. Effluent and nutrient solutions were stirred during the run.

a. Tubular Membrane Reactor (TMR)

b. Total flow sheet for metal removal with a TMR

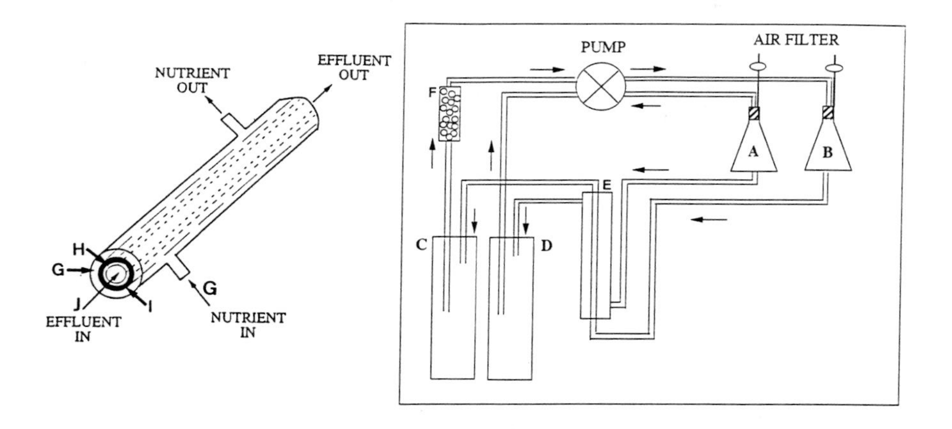

Figure 2. Schematic presentation of a Tubular Membrane Reactor.
A TMR is presented in detail (a) together with the general concept (b) : A : nutrient ; B : effluent ; C : siphon ; D : siphon ; E : TMR ; F : glass bead column ; H : membrane + supporting tube ; G : extralumen space.

Development of a Continuous Tubular Membrane Reactor (CTMR)

Figure 3 presents a continuous version of the TMR reactor system.
Effluent (contaminated with heavy metals) is pumped from tank A to B and from there circulated through the tubes of the reactor (CTMR) and the glass beads column (GBC) on which the metals are bound. After a certain residence time the effluent flows to C via two sand filtration units (SF1 and SF2).
Bacteria are immobilized in the composite membrane of the CTMR and are fed via the input of nutrient from D to E from where it circulates through the interlumen space of the CTMR. After use the nutrient is discarded into F.

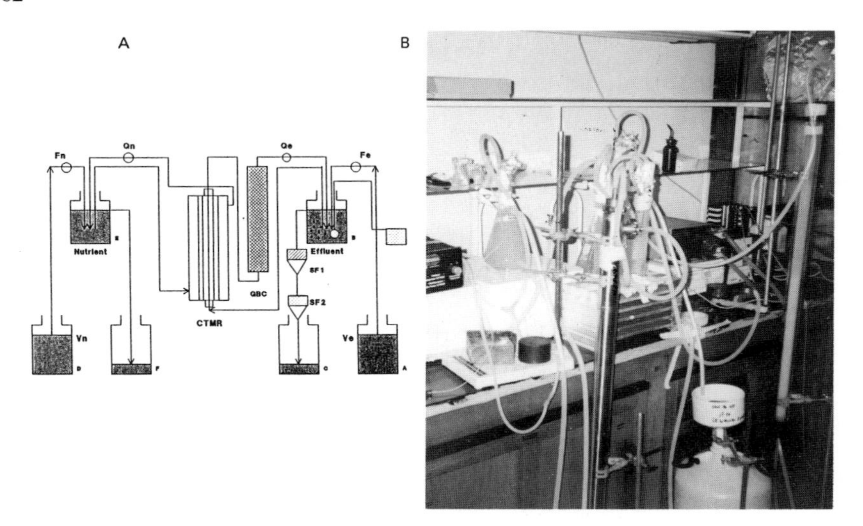

Figure 3. Continuous Tubular Membrane Reactor (CTMR).
Total flow sheet a) and a laboratory CTMR (b) are presented. A : effluent stock ; B : effluent flask ; C : effluent output ; D : nutrient stock ; E : nutrient flask ; F : nutrient output ; GBC : glass bead column ; CTMR : membrane module ; SF1 : sandfilter 1 ; SF2 : sandfilter 2.

RESULTS AND DISCUSSION

Zirfon ^{R} MEMBRANES

In order to obtain membranes with mean pore diameters between 1 and 5 μm, to the composite membranes, based on polysulfone and ZrO_2, some pore formers were added. These were inorganics like Sb_2O_3, $CaCO_3$ and ZnO, or organics like polyvinylpyrolidone.

Immobilization and colonization in the composite membranes

Figure 4 shows both sides of a membrane after a vertical filtration with a diluted A. eutrophus culture. Only few cells pass through the skin due to the small pore diameter of this membrane layer. A better distribution of the cells can be obtained by higher dilution of the cell suspension by tangential filtration.

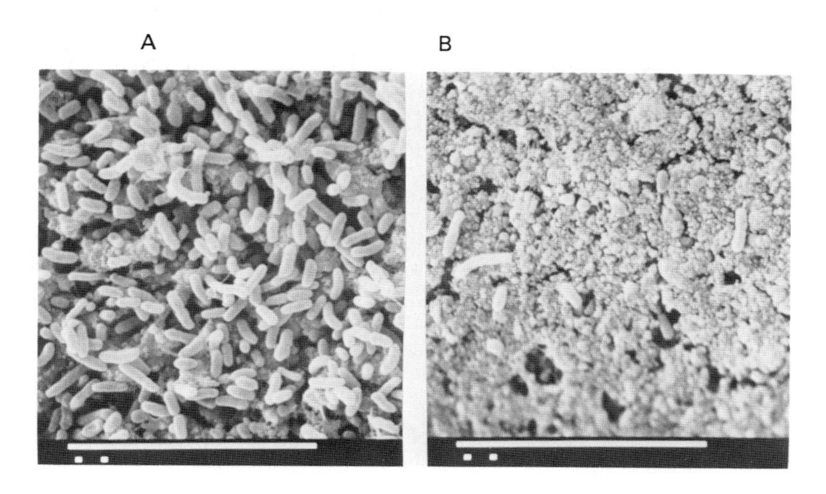

Figure 4. Microscopic pictures of a Zirfon M5 membrane after tangential filtration.
A diluted culture of A. eutrophus was filtered tangentially across the Zirfon M5 membrane. The filtration side or open side (A) and the opposite side or skin side (B) are shown. The bar presents 10 µm.

Figure 5 shows the colonization of the membrane at the open side in function of time. Immediately after the tangential filtration only a few cells are found in the matrix of the membrane. Afterwards bacteria tend to grow in the membrane at different spots producing a thin biofilm at the inside of the membrane and leading to the appearance of filament structures, typical for immobilized cells. This figure shows a complete colonization after about 4 to 6 days.

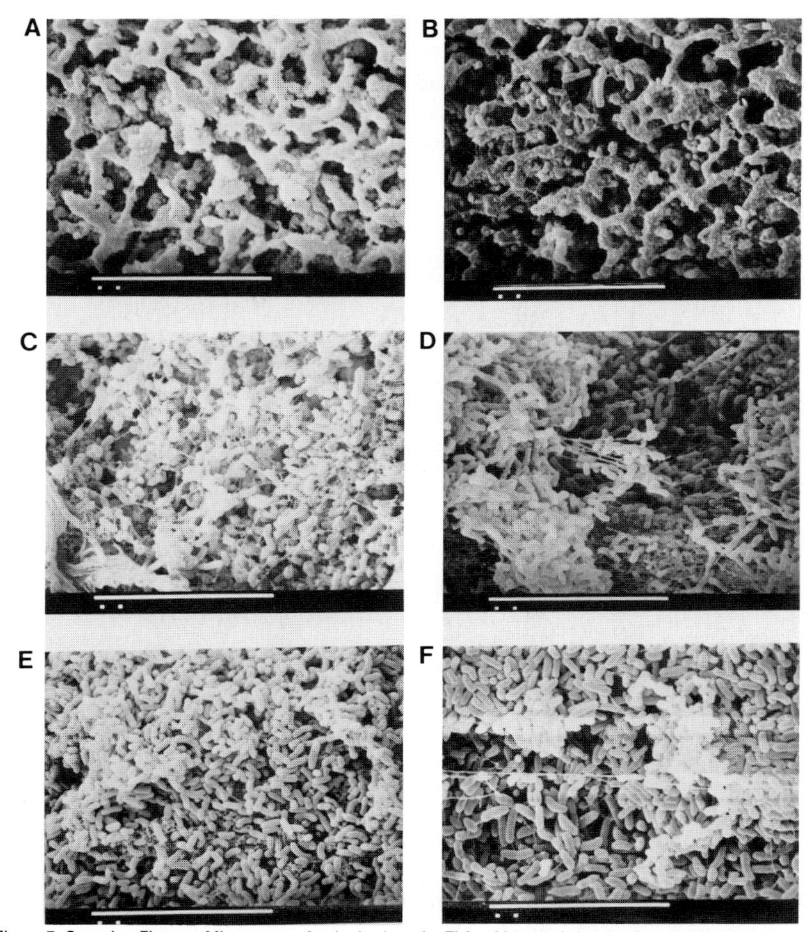

Figure 5. Scanning Electron Microscopy of colonization of a Zirfon M5 membrane by A. eutrophus in function of time.

The bacterial colonization of Zirfon M5 is shown before filtration (A), just after filtration (B), after 3 days (C), after 6 days (D), after 8 days (E) and after 10 days (F). The bar presents 10 μm.

Figure 6 presents the bacterial colonization of the cross section of a ZrO_2 membrane like M5 or M6 and a PVP membrane like M7, M8 and M10. The presented results show that composite membranes can be used for immobilization of bacteria. The Zirfon membranes M5 and M6 show also a very good immobilization inside of the membrane. Membranes with large fingerlike structures show a high release of bacteria and a bad colonization. In this case probably the fingerlike holes do not create a good niche for bacterial immobilization. Membranes with a more symmetric structure create better cavity like structures in which bacteria can bind and grow.

Membranes, casted with polyvinylpyrolidoneas pore formers showed the formation of biofilms only at their skin surface, but no good bacterial growth inside the membrane. Membranes with $CaCO_3$ as pore former created problems due to the CO_2 gas formation during the acid boiling procedure, necessary to remove the $CaCO_3$. Now membranes are casted with ZnO as pore former. In these membranes bacterial immobilization looks promising.

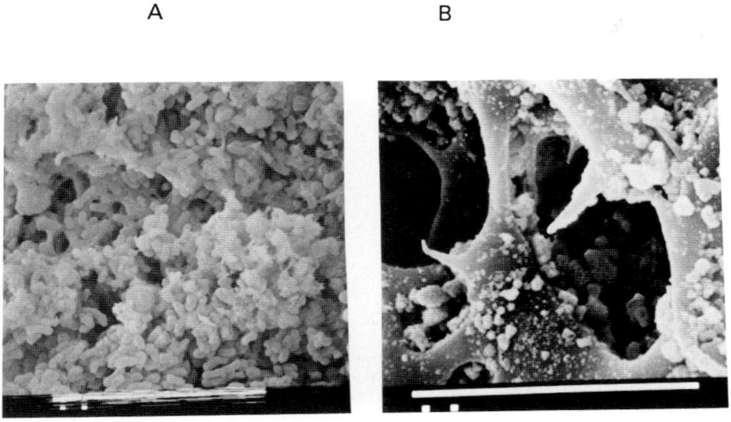

A B

Figure 6. Bacterial colonization in M5 and M7.
The bacterial colonization inside is presented for the membranes M5(A) with an inorganic pore former (Sb_2O_3) and M7(B) with an organic pore former (PVP).

Cd and Zn removal by immobilized A. eutrophus cells

Table 1 presents the Cd and Zn concentrations per square cm of membrane after one treatment cycle. The presented results are not obtained under membrane saturation conditions. They only show a first precipitation of metals on the membranes with immobilized bacteria. Other results (not shown) indicated that it was possible to remove high heavy metal concentrations from effluents via the immobilized bacteria. Concentrations of 100 ppm Cd or Zn could be reduced to below 1 ppm.

Original Metal Concentration	Metal Concentration/Membrane Surface
Cd	µg Cd/cm^2
0.5 µM	0.2
0.5 mM*	5.0
0.5 mM	169.2
1.0 mM	428.0
2.0 mM	900.0
Zn	µg Zn/cm^2
0.5 µM	0.2
0.1 mM*	2.0
0.1 mM	37.9
0.2 mM	52.9
0.5 mM	134.2
1.0 mM	287.5
2.0 mM	591.5

Table 1. Cd and Zn accumulated by Zirfon M5 membranes with immobilized A. eutrophus.
Cd and Zn concentration/membrane surface areas are presented for metal removal experiments with Zirfon M5 membranes colonized by A. eutrophus. Membranes with immobilized CH34 are incubated in 100 ml with different concentrations of Cd or Zn ions. (*) indicates experiments with membranes without immobilized bacteria

Flat Sheet Reactor

In the preliminary filtration experiments, bacteria were always immobilized in the membrane by perpendicular filtration which resulted in a fast clogging of the membrane. The nutrients were added to the heavy metal solution resulting in a wasting of unused nutrients. Therefore, to solve these problems, a flat sheet membrane reactor was developed. In this system bacteria were immobilized by tangential filtration in a Flat Sheet.

Figure 7 shows some results obtained with a Flat Sheet Reactor. The Cd concentration in the effluent and nutrient (due to diffusion) are shown.
Cd is removed from the effluent in a linear way from 140 ppm to 110 ppm after 28 hours. Between 28 and 36 hours a sharp decrease of Cd concentration from 110 ppm to 4 ppm was observed (in 5 hours). During this time the environmental

conditions around the bacteria are suitable to induce metal precipitation. Afterwards a small peak of 10 ppm (due to diffusion) appears to be followed by a second decrease to below 1 ppm after 80 hours. At the nutrient side some diffusion of Cd occurred to a concentration of 32 ppm after 72 hours, followed by a decrease to 25 ppm after 90 hours. Later on Cd was removed from the nutrient solution due to precipitation on the bacteria freely suspended in the nutrient. Larger membrane surfaces will accelerate the metal removal. The orientation of the skin side to the effluent prevents too high cell release but makes metal removal time longer due to the longer diffusion pathway.

In this system different parameters were studied. A membrane which was already colonized by bacteria could remove metals much faster. The choice of a membrane with a high bacteria immobilizing capacity is necessary. O_2 supply at or in the membrane is very important and can be increased by increasing the flow rate. pH adjustment of the effluent between 7 and 7.5 will accelerate the metal removal. Lactate turnover is measured to give the right dose in a right lapse of time without needless loss of this carbon source. Thick membranes will prevent cell release and will bind more metal.

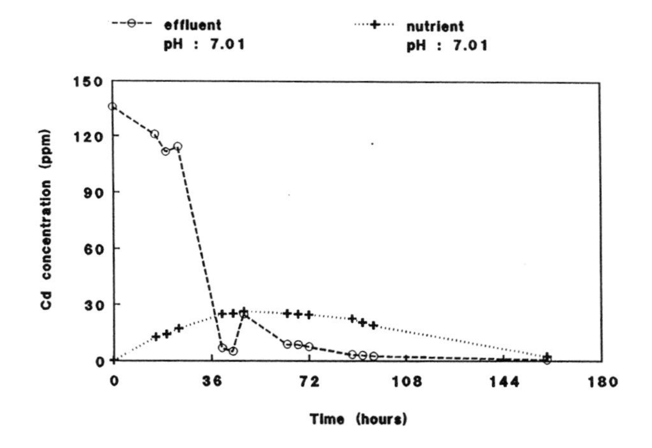

Figure 7. Cd removal in a Flat Sheet Reactor.
The results for Cd removal in a FSR reactor with Zirfon M5 are shown. Cd concentrations at effluent and nutrient side are presented.

Tubular Membrane Reactor

In order to make a stronger reactor structure and a less fouling sensitive membrane reactor, a Tubular Membrane Reactor (TMR) was constructed.

Figure 8a and b presents respectively Cd and Zn removal via a TMR system. In the beginning a long colonization period is necessary to colonize the whole carbon support pipe. Afterwards Cd removal goes faster and faster. Figure 8b shows the results for Zn removal for the same TMR. Zn removal started immediately because the C-pipe was already immobilized.

a. Cd removal

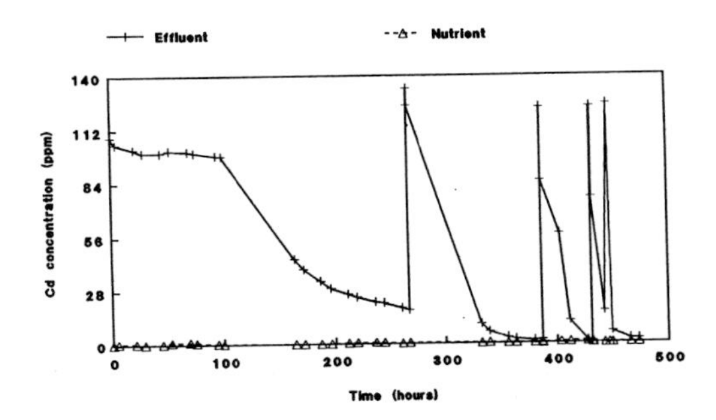

b. Zn removal

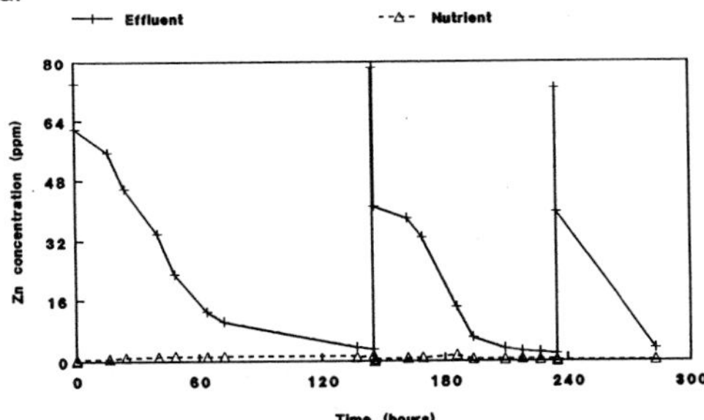

Figure 8. Cd and Zn removal with a TMR reactor.
Cd (a) and Zn (b) removal are presented in function of time. In (a) the first 100 hours are necessary for the colonization of the carbon-pipe support. Zn removal (b) was done with the same Zirfon M5 membrane on the same carbon-pipe support, while the colonization was still present.

Figure 9 shows the results for Cd removal in a TMR with a polyester support instead of a carbon support. The long immobilization period has now disappeared due to the thin polyester layer (only 200 μm), the high porosity and the large pore diameter. Bacteria can penetrate through the polyester support much faster than through the carbon support to reach the membrane where they can start the immobilization. Metal removal efficiency is the same in both systems (carbon or polyester support).

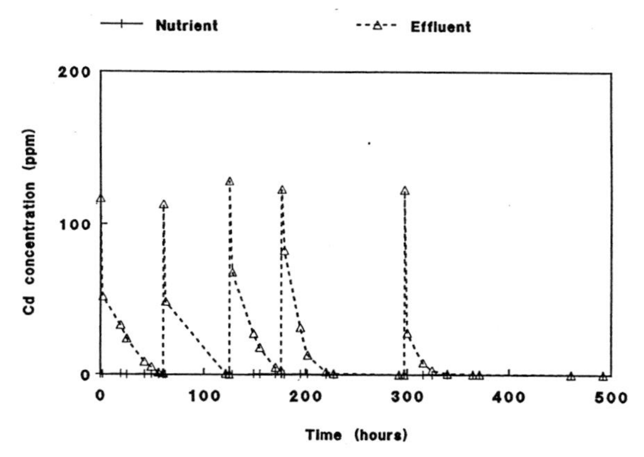

Figure 9. Cd removal with a TMR reactor with a Zirfon M5 membrane on a polyester support. Cd removal is presented in function of time.

Continuous Tubular Membrane Reactor

The tubular reactor was further optimized to a continuous system as shown in Figure 3a and b. In addition to the continuous operating system two sand filters (SF1 with coarse sand and SF2 with fine sand) were introduced at the output. The sandfilters help to do a final polishing of the metal content. Microbially formed crystallites are bound on the glass bead column which gives a reduction in metal concentration to 1 or 2 ppm. Very fine crystallites are removed by the sandfilters resulting in 0.1 ppm after SF1 and in 0.05 ppm after SF2, which is below the CEC norms. These results are presented in Figure 10. In this sytem a solution containing 120 ppm Cd is treated to provide an output below 50 ppb Cd at the output as presented in Figure 10. Similar results could be obtained with real effluent from a galvanisation factory. Such a real effluent containing 20 ppm of Zn gave after treatment with a CTMR a Zn concentration below 1 ppm.

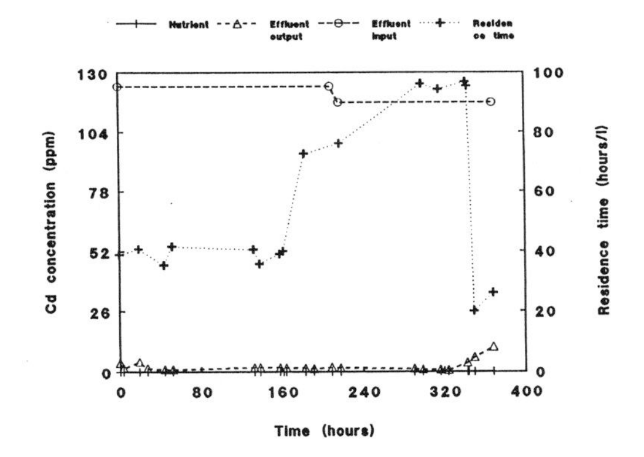

Figure 10. Results of Cd removal in a CTMR system.
The input and output Cd concentrations and residence time are presented in function of time.

Precipitation and crystallization of Cd and Zn

It was shown that bacteria immobilized on the membrane could induce, by the combination of their heavy metal resistance mechanisms (due to the metal efflux mechanism they create a supersaturated environment around the cell surface) and their metabolism, a precipitation of Cd and Zn hydroxides, bicarbonates and carbonates around the cells. Figures 11 shows Scanning Electron Micrographs of membranes with biomineralized Cd crystals.

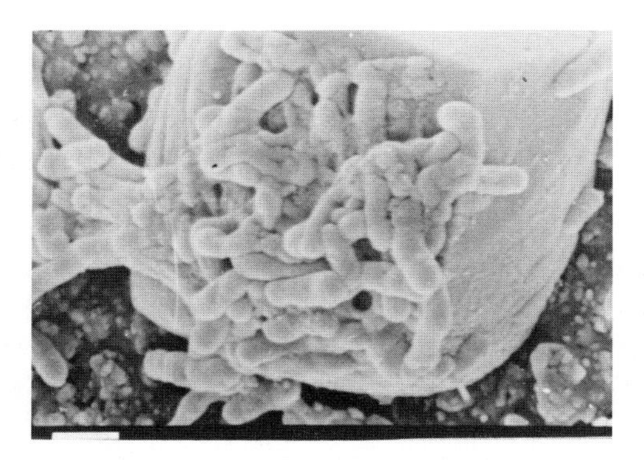

Figure 11. Bacterial induced formation of Cd crystals on Zirfon M5 membranes.
The figure presents the bacterial formed $Cd(HCO_3)_2$ and $CdCO_3$ crystals. The bar presents 1 μm.

From other results with glass vessels it could also be concluded that the Cd and Zn crystals stick very well to glass surfaces ; therefore a column of glass beads was introduced in the treatment cycle to trap the metal crystals. In fact, metal hydroxides are positively charged at pH 7 and will bind to the negatively charged silica beads. With the aid of this system regeneration becomes very easy. A column can be removed and replaced by another one and elution can be done by strong acids.

CONCLUSIONS

Different composite membranes were used with symmetric and asymmetric structures. Membranes with more symmetric structures (casted with the inorganic fillers, like e.g. Sb_2O_3 and $CaCO_3$) give a higher immobilization capacity.

The Zirfon membranes can clearly keep bacteria immobilized in their cavities. Bacteria stay in a kind of stationary phase due to their immobilization and to the limited nutrient provision at the open side of the membrane. This physiological situation allows the bacteria to induce a biomineralization process by creation of a supersaturation around the cell surface in which metal hydroxides and carbonates precipitate on crystallization foci created by metal binding to outer cell wall components. In this way small crystallites are formed that can be recuperated on a glass bead column. This system creates the possibility of the complete recuperation of the metals simply by elution of the column with acids. However very fine crystallites can only be retained by one of the two used sandfilter systems which results in a reduction of the Cd concentration from 150 ppm to below 0.05 ppm. With this concept the carbon source concentration necessary for Cd removal of 1 litre of effluent could be reduced from 8 g/l (Stirred Tank Reactor) to 0.08 g/l (Continuous Tubular Membrane Reactor).
However, until now, no clear distinction could be made between the efficiency of the membrane as a surface for adhesion of a biofilm, or as a volume which can be filled with active bacteria. Experiments are going on now to prove whether the tridimensional membrane structure is necessary or not.

In order to measure the performance of our composite membranes, not only an immobilization of the bacteria in the composite membranes could be realized, but also a new reactor concept was developed. The reactor is based on the separation of a process (effluent) stream and a nutrient stream (to keep the bacteria active). Both streams are separated by a membrane containing the immobilized bacteria. The bacteria catalyze in the effluent stream certain reactions (e.g. heavy metal precipitation, xenobiotic degradation, compound transformation) and they receive the necessary nutrients from the nutrient stream. First of all it seems to be clear that those kinds of reactors can be used for heavy metal removal. A. eutrophus cells are used for the removal of Cd and Zn. Moreover preliminary experiments with A. eutrophus and related bacteria for the removal of Cu and Ni are proved to be successful.

Secondly, the reactor can probably also be used for the recovery of certain precious metals like Pd. Some palladium resistant microorganisms could already be isolated.

Thirdly the degradation of xenobiotic organic compounds by specialized immobilized bacteria seems to be a good system for water treatment. Also due to the special design of the reactor : certain compounds can be used to induce cometabolism. Also a combination between heavy metal removal and xenobiotic degradation can be envisaged.

REFERENCES

- Diels L. et al. (1989) Large plasmids governing multiple resistances to heavy metals : a genetic approach.Toxic. Environm. Chem. 23, 79-89.
- Diels L. (1990) Accumulation and precipitation of Cd and Zn ions by Alcaligenes eutrophus strains. Biohydrometallurgy. Editors J. Salley, R.G.L. McCready, P.Z. Wichlacz, pp. 369-377.
- Kennedy J.F., Cabral J.M.S. (1986) "Use of titanium etc. species for the immobilization of bioactive compounds-enzymes. Transition Met. Chem. 11, 41-66.
- Leysen R., Vermeiren P., Doyen W. (1989) "Preparation and Application of Composite Membranes", In : Future Industrial Prospects of Membrane Processes", Edited by L. Cecille and J.C. Toussaint (CEC), Elsevier Science Publishers, London - New York (1989).
- Mann S. (1988) Molecular recognition in biomineralization. Nature 332, 119-124.
- Mergeay M., Houba C., Gerits J. (1978) Extrachromosomal inheritance controlling resistance to cadmium, cobalt and zinc ions : evidence from curing in a Pseudomonas. Arch. Int. Physiol. Biochim. 86, 440-441.
- Mergeay M., Nies D., Schlegel H.G., Gerits J., Charles P., Van Gijsegem F. (1985) Alcaligenes eutrophus CH34 is a facultative chemolithotroph with plasmid-bound resistance to heavy metals. J. Bacteriol. 162, 328-334.

- Michel L.J., Macaskie L.E., Dean A.C.R. (1986) Cadmium accumulation by immobilized cells of a Citrobacter sp. using various phosphate donors. Biotechnol. Bioeng. XXVIII, 1358-1365.
- Nies D., Mergeay M., Friedrich B., Schlegel H.G. (1987) Cloning of plasmid genes encoding resistance to cadmium, zinc and cobalt in Alcaligenes eutrophus CH34. J. Bacteriol. 169, 4865-4868.
- Nies A., Nies D., Silver S. (1989a) Cloning and expression of plasmid genes encoding resistance to chromate and cobalt in Alcaligenes eutrophus. J. Bacteriol. 171, 5065-5070.
- Nies D.H., Nies A., Chu L., Silver S. (1989b) Expression and nucleotide sequence of a plasmid-determined divalent cation efflux system from Alcaligenes eutrophus. Proc. Natl. Acad. Sci. USA 86, 7351-7356.
- Nies D, Silver S. (1989c) Plasmid-determined inducible efflux is responsible for resistance to cadmium, zinc and cobalt in Alcaligenes eutrophus. J. Bacteriol. 171, 4073-4075.
- Nies A., Nies D., Silver S. (1990) Nucleotide sequence and expression of a plasmid-encoded chromate resistance determinant from Alcaligenes eutrophus. J. Biol. Chem. 265, 5648-5653.
- Sensfuss C., Schlegel H.G. (1988) Plasmid pMOL28-encoded resistance to nickel is due to specific efflux. FEMS Microbiol. Lett. 55, 295.
- Siddiqui R.A., Benthin K., Schlegel H.G. (1989) Cloning of pMOL28-encoded nickel resistance genes and expression of the genes in Alcaligenes eutrophus and Pseudomonas spp. J. Bacteriol. 171, 5071-5078.

ACKNOWLEDGEMENTS

This work is supported by the CEC programme BRITE/EURAM (BREU 0084/C/SMA 1990).

OIL AND GAS

INTEGRATED MEMBRANE SYSTEMS WITH MOVING LIQUID CARRIERS FOR MULTICOMPONENT GAS SEPARATION

Beckman I.N.,[+] Bessarabov D.G.,[*] Teplyakov V.V[*].,
Teplyakov A.V.[+]

[+] *Chemistry Department, Moscow State University, 119899, MSU, Moscow, RUSSIA.*
[*]*A.V.Topchiev Institute of Petrochemical Synthesis*
Russian Academy of Sciences
117912, GSP-1, Moscow, Leninsky pr., 29, RUSSIA

ABSTRACT

The prospects for the gas separation membrane systems with moving liquid carriers are discussed in this paper. The combination of non-porous gas separation membranes with moving liquid carriers of different kinds allows to separate multicomponent gas mixtures on constituents.

The phenomenological theory of gas separation by membrane gas-liquid modules in circulating and flowing modes with non-specific and active carriers is considered. Computer simulation of the separation processes was carried out. The proposed technique for separation of gases was tested by separation of $CO_2/CH_4/H_2$, CO_2/CH_4 and He/O_2 gas mixtures by using membrane module with polyvinyltrimethylsilane (PVTMS) membranes.

INTRODUCTION

In many cases the development of membrane technology is connected with the improvement of productivity, selectivity and flexibility of the membrane gas separation systems. Recently the new types of the flowing liquid membranes in which a liquid solution flows along microporous (Sirkar, 1988) , (Teramoto, 1989) or non-porous membrane (Shelekhin and Beckman, 1989, 1990) have been proposed.

In this paper a new type of integrated membrane systems combining the membrane and absorbtion gas separation methods (selective membrane valve and permabsorber) are considered. One of the objects is a mathematical estimation for reasonable selection of the absorption liquid, its flow rate and approaches to module constractions. The liquid can be non-specific in relation to the gas mixture components; secondly, the solubility of the gas components in a liquid can considerably differ;

finally, a liquid can react with one or several gas components. Here we consider that gas A reacts with carrier C dissolved in the liquid with creation of AC-complex:

$$A + C \underset{k_2}{\overset{k_1}{\rightleftarrows}} AC \qquad (1)$$

where:
k_1- is the constant of a direct chemical reaction $((c^{-1})(t^{-1}))$; k_2- is the constant of a reverse chemical reaction (t^{-1}); the gas B does not react with the carrier C. Let the concentration of C is enough high so that constant $k_1^* = k_1 C_L^C$ and scheme (1) is a reversible process. The AC-complex does not diffuse through polymeric membranes.

SELECTIVE MEMBRANE VALVE

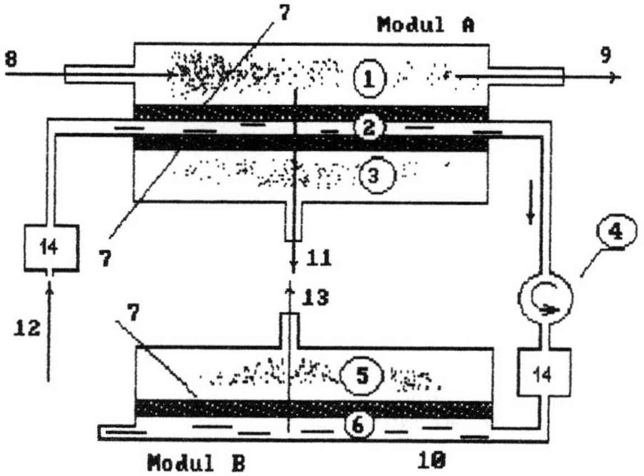

Fig.1. Schematic diagram of the selective membrane valve: flowing mode with desorber. 1,3- The gas chambers. 2- The thin liquid layer. 4- Pump. 5- The desorber gas chamber. 6- The liquid layer. 7- The polymeric non-porous membranes. 8- Feed gases ($CO_2/CH_4/H_2$). 9- Retentate (CH_4). 10- Dissolved gas (CO_2). 11- Permeate (H_2). 12- Liquid. 13- Degassing (CO_2). 14- Thermostates.

The selective membrane valve (SMV) consists of a permeation module (A) and a desorption module (B) (Fig. 1). SMV is the version of the integrated membrane system with a liquid flowing between two non-porous membranes. Three component gas mixture can be separated by SMV: first component (retentate) is insoluble in a liquid carrier; second one (permeate) diffuses through the membrane sandwich; third component dissolved well in a liquid layer is pumped to desorption module for degassing. Thus, the SMV has one inlet (Modul A) for the feed gas and three outlets for the products (Modules A and B). Desorption module (B) contains the same polymeric membranes and intends for degassing of a liquid. There are two thermostates for liquid in this system: on the inlet of modul A (cooling) and on the inlet of desorber (heating). For simplicity an one-membrane desorption module is considered. The possible modes of SMV operation are following: a flowing mode without or with a desorber; a circulation mode without or with a desorber.

MATHEMATICAL MODEL OF MASS-TRANSFER IN SMV.

The steady-state diffusion through SMV and co-current type module, in which well-mixed mode is maintained in gas phases over the membranes is considered; in a liquid layer the mixing occures in the transverse direction (x) while the mixing in the direction (y) does not take place. The three-layered membrane sandwich in Module A has a rectangular shape were d is a width, h is a membrane length, υ (cm/s) is a linear velocity of liquid. (Here subscript $_p$ - is permeator and $_d$ - is desorber). Let D_{1m}, σ_{1m}, ℓ_{1m}; D_{2L}, σ_{2L}, ℓ_{2L}; D_{3m}, σ_{3m}, ℓ_{3m} are diffusion coefficients, solubility coefficients of gases and layer thicknesses for the inlet membrane 1, the liquid layer and the outlet membrane 2, respectively (here subscript $_m$ is membrane; $_L$ is liquid)(Fig.1). The total sandwich thickness is $H = \ell_{1m} + \ell_{2L} + \ell_{3m}$. The partial pressure of the feed gas is p_{10}, and of the permeate is p_{HO}. The parameters of desorber are D_d, σ_d, ℓ_d respectively. Here $D_{2L} \gg D_{1m}$; $D_{2L} \gg D_{3m}$; $D_{2L} \gg D_d$.

In the case of chemical reaction (1) of carrier with component of gas mixture the gas diffusion through the sandwich being under steady-state conditions can be described by the equations:

$$\upsilon\ell_2 \frac{dC^A_{2L}}{dy} = i^A_1 - i^A_2 + k_2 C^{AC}_{2L} l_{2L} - k^*_1 C^A_{2L} l_{2L}$$

$$\upsilon\ell_2 \frac{dC^{AC}_{2L}}{dy} = k^*_1 l_{2L} C^A_{2L} - k_2 C^{AC}_{2L} l_{2L} \qquad (2)$$

where:

$$i^A_1 = -D_1 \frac{dC^A_{1m}}{dx}\bigg|_{x=\ell_1}, \quad i^A_2 = -D_3 \frac{dC^A_{3m}}{dx}\bigg|_{x=\ell_1+\ell_2} \qquad (3)$$

i^A_1 is the local flux of A component through the first polymeric membrane, i^A_2 is the local flux of A component through the sandwich; C^A_{2L}, C^C_{2L}, C^{AC}_{2L} are the concentrations of gas A, carrier C and AC-complex in the liquid. The complete permeate flux (I^A_2) of component A through the sandwich is:

$$I^A_2 = d\int_0^h i^A_2(y)dy, \quad \text{and} \quad C^A_3(y) = \frac{\sigma_3}{\sigma_2} C^A_2 \qquad (4)$$

In the steady-state the gas concentration profile along coordinate x in the polymeric membrane is known to be linear. Therefore the derivatives in eq.3 can be replaced by concentration differences. Hence according Fick's law the system 2 is equivalent to the differencial equation of the second order with the characteristic equation:

$$\alpha^2 + \alpha \frac{b + k_2 + k^*_1}{\upsilon} + \frac{bk_2}{\upsilon^2} = 0 \qquad (5)$$

where $b = \dfrac{\left(\dfrac{D_{1m}\sigma_{1m}}{\ell_{1m}\sigma_{2L}} + \dfrac{D_{3m}\sigma_{3m}}{\ell_{3m}\sigma_{2L}} \right)}{\ell_{2L}}$ and α_1, α_2 are the negative roots of (5). Here we consider the case when : $\ell_{1m} = \ell_{3m} = \ell \neq \ell_{2L}$; $\sigma_{1m} = \sigma_{3m} = \sigma \neq \sigma_{2L}$; $D_{1m} = D_{3m} = D_{md} = D$; $h_p = h_d = h$.

The flowing mode with a desorber (fig.1).

In this case the fresh liquid is fed on the inlet of permeator (p) and then pumped on the inlet of desorber module (d). When gas A reversibly reacts with carrier C the boundary conditions are:

$$C^{A}_{1m}\Big|_{x=0}=\sigma p_{10};\quad C^{A}_{3m}\Big|_{x=H}=0;\quad C^{A}_{2Lp}\Big|_{y=0}=0;\quad C^{AC}_{2Lp}\Big|_{y=0}=0;\quad C^{A}_{2Lp}\Big|_{y=h}=C^{A}_{2Ld}\Big|_{y=h};$$

$$C^{AC}_{2Lp}\Big|_{y=h}=C^{AC}_{2Ld}\Big|_{y=h}$$

In this case the complete permeate flux through the sandwich is :

$$I^{A}_{2}=\frac{P^{A}S}{\sigma_{2L}\ell}\left[\frac{a_{1F}}{h\alpha_{1}}\left(1-\exp(-\alpha_{1}h)\right)+\frac{a_{2F}}{h\alpha_{2}}\left(1-\exp(-\alpha_{2}h)\right)+\frac{p_{0}\sigma_{2L}}{2}\right] \qquad (6)$$

$$\text{where } a_{1F}=\frac{p_{10}\left(2D\sigma-\alpha_{2}\ell\ell_{2L}v\sigma_{2L}\right)}{2\ell\ell_{2L}(\alpha_{2}-\alpha_{1})v},\quad a_{2F}=\frac{p_{10}\left(2D\sigma-\alpha_{1}\ell\ell_{2L}v\sigma_{2L}\right)}{2\ell\ell_{2L}(\alpha_{1}-\alpha_{2})v} \qquad (7)$$

are the constants from boundary conditions (subscript F - is the index of Flowing mode).

When two modules consist of the same membranes and flowing liquid, the gas flux from one-membrane desorber is

$$I^{A}_{d}=\frac{P^{A}S}{\sigma_{2L}\ell}\left[-\frac{a_{1F}}{h\,\alpha_{1}}\left(1-\exp(-\alpha_{1}h)\right)^{2}-\frac{a_{2F}}{h\,\alpha_{2}}\left(1-\exp(-\alpha_{2}h)\right)^{2}\right] \qquad (8)$$

where the constants a_{1dF} and a_{2dF} are expressed by constants a_{1F} and a_{1F}.

Consequences.

1.For the flux of the permeate through the sandwich,

$I^{A}_{2}\rightarrow S\frac{D\sigma}{2\ell}\cdot p_{0}$ at $v\rightarrow 0$; $I^{A}_{2}\rightarrow 0$ at $v\rightarrow\infty$.

Thus in the flowing mode with a desorber the gas flux through the permeator can be completely closed. Hence, by variation of the liquid flux one can change the composition of the gas mixture after SMV. The gas flux after SMV can be closed also by increasing the value of k_{1}.

2.For one-membrane desorber flux $I^{A}_{d}\rightarrow 0$ at $v\rightarrow 0$; $I^{A}_{d}\rightarrow 0$ at $v\rightarrow\infty$. Thus the gas flux via SMV can be varied from the value $I^{B}_{2}=\frac{SPp_{0}}{2\ell}$ to zero.

The circulation mode with a desorber.

In this case a liquid from the module (B) is returned to the module (A) by pump operating in a circle. The liquid passed through modul (B) is degassed in some extent. The desorber parameters are the same as above. In the case when gas B does not react with the carrier C and gas A reacts with carrier C dissolved in the liquid with creation AC-complex (1) the boundary conditions are :

$$C^{A}_{1m}\Big|_{x=0} = \sigma p_{10}; \quad C^{A}_{3m}\Big|_{x=H} = 0; \quad C^{A}_{2Lp}\Big|_{y=h} = C^{A}_{2Ld}\Big|_{y=h}; \quad C^{AC}_{2Lp}\Big|_{y=h} = C^{AC}_{2Ld}\Big|_{y=h};$$

$$C^{AC}_{2Lp}\Big|_{y=0} = C^{AC}_{2Ld}\Big|_{y=h+h_d}; \quad C^{A}_{2Lp}\Big|_{y=0} = C^{A}_{2Ld}\Big|_{y=h+h_d}$$

MATHEMATICAL SIMULATION OF GAS SEPARATION

Separation of the CO_2 containing gas mixture

In the case of three components mixture $CO_2/CH_4/H_2$ (40/30/30) for given polymer (polyvinyltrimethylsilane (PVTMS)) under 298K solubility coefficients σ and diffusion coefficients D are $\sigma(CO_2)= 0,038$; $\sigma(CH_4)= 0,0I$; $\sigma(H_2)= 0,00II$; $D(CO_2)= 5\cdot IO^{-7}$; $D(CH_4)= I,8\cdot IO^{-7}$; $D(H_2)= I80\cdot IO^{-7}$ where $(\sigma)\left[cm^3 gas/cm^3 PVTMS\cdot cm.Hg.\right]$; $(D)\left[cm^2/sec\right]$ The pure water and water solutions of MEA (monoethanolamine) were used as a moving carriers of CO_2. The area of the membrane in one module was 75 cm^2. Thickness of the liquid layer was 0.2 cm. Thickness of the membrane 0,000I cm.

Figure 2 shows the dependences of permeate fluxes (I_2) on speed of liquid pumping under different values of constant k_1^*. In this case H_2 flux decreases smoothly with increasing of liquid speed (non-reactive gas); CO_2 passes away by liquid due to high solubility and reactivity; CH_4 is retentate. Figure 3 shows the dependences of desorption fluxes from desorber in flowing mode with consideration the chemical reaction between liquid carrier and gas component on speed of liquid.

PERMABSORBER.

In order to separate bicomponent gas mixture with very high selectivity it is possible to use the simplest integrated membrane system which is permabsorber. The principles and modes of its operating are the same as in SMV. In bimembrane permabsorber (Fig.1) the feed gas is fed in both gas chambers 1 and 3 (these chambers are connected each other). Thus, bimembrane permabsorber has one inlet for the feed gas and two outlets for the products (retentate and desorbate). In our experiments we used absorbtion and desorbtion modules consisting of 24 cells type of modules shown on Fig.1.

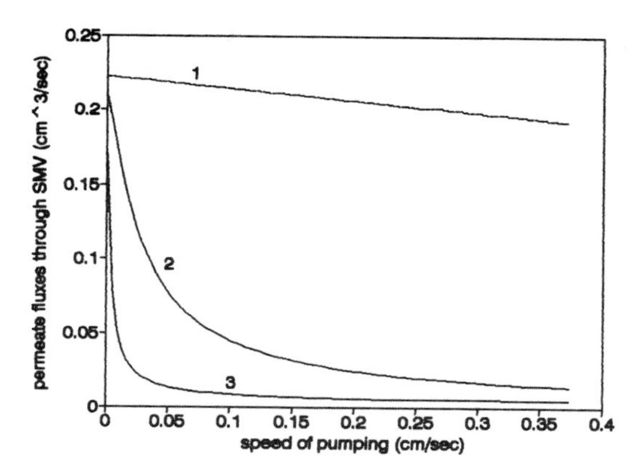

Fig.2. The dependences of permeate fluxes on speed of pumping in module A of SMV in flowing mode. Hydrogen is the most permeable component (1) (non-reactive gas), 2 - CO_2 (k_1^*= 0,05 k_2= 0,001), 3 - CO_2(k_1^*= 0,1 k_2= 0,001).

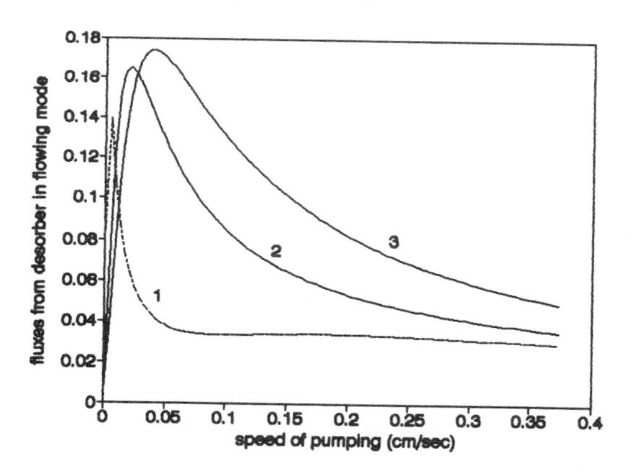

Fig.3. The dependences of desorption fluxes of CO_2 on speed of pumping in module B of SMV in flowing mode: 1- (k_1^*= 0,01 k_2= 0,001), (k_1^*= 0,01 k_2= 0,01),(k_1^*= 0,01 k_2= 0,1).

The co-current or countercurrent fluxes of gas and liquid in permabsorber are described by the following equations:

$$W_L h \frac{dC^A_{La}}{dy} = \frac{D_m S\sigma_m}{\ell_m \sigma_L}(\sigma_L C^A_{ga} - C^A_{La}) - V_{La}(k^*_1 C^A_{La} - k_2 C^{AC}_{La})$$

$$W_L h \frac{dC^{AC}_{La}}{dy} = \left[k^*_1 C^A_{La} - k_2 C^{AC}_{La}\right] \cdot V_{La} \qquad (9)$$

$$W_{ga} h \frac{dC^A_{ga}}{dy} = -\frac{D_m S\sigma_m}{\ell_m \sigma_L}(\sigma_L C^A_{ga} - C^A_{La})$$

where W_L is the volume speed of liquid; V_{La} is the volume of liquid layer; W_{ga} is the flow rate of feed. The boundary conditions in this case are the same as above. These equations can be solved by above mentioned procedure.

SEPARATION OF GAS MIXTURE CO_2/CH_4 BY PERMABSORBER

Table 1.

The separation of bicomponent gas mixture CO_2/CH_4 by PVTMS membrane permabsorber.

Liquid carrier	Temper. of desorber T°C	Gas flux ml/min	Liquid l/h	At the absorber's outlet,%		At the desorber's outlet,%	
				CO_2	CH_4	CO_2	CH_4
H_2O	I8	9.5	I.97	24.5	75.5	94.5	5.5
H_2O	I8	I.3	I.97	I5.9	84.I	73.I	26.9
H_2O	I8	4.2	3.66	25.5	74.5	75.2	24.8
K_2CO_3 (3M)	I8	4.2	3.66	25.6	74.4	92.2	7.8
K_2CO_3 (3M)	60	4.2	3.66	5.6	94.4	99.6	0.4

The proposed technique was tested on the separation of gas mixture CO_2/CH_4 with composition 46:54. Liquid flux was varied from 1.66 l/h to 8.24 l/h. The absorber's temperature was 18°C. The desorber's temperature was varied from 18 to 60°C. We used water and water solutions of monoethanolamine and carbonates of alkaline metals (Li_2CO_3, Na_2CO_3, K_2CO_3) as the absorbing liquid. The majority of experiments was carried out using K_2CO_3 solution with the concentration from 0 to 3 mol/l. The main results are shown in Table 1. As it is seen from Table 1 high purity of the both separated components can be achived by using permabsorber.

SEPARATION OF $CO_2/CH_4/H_2$ AND He/O_2 GAS MIXTURES USING SELECTIVE MEMBRANE VALVE.

The parameters of SMV construction are above mentioned. The speed of liquid pumping was varied in the range from 0.0013 cm^3/sec to 0.1 cm^3/sec. The pure water, water solutions of MEA and liquid $N(C_2F_5)_3$ were used as a carriers of CO_2 and O_2 respectively. The velocity of feed gas mixture was 0.66 cm^3/sec. The experiments were carried out with single Module A and SMV operating in the circulation mode with desorber (Module B). The main results are summarized in the Table 2.

Table 2.
The separation of $CO_2/CH_4/H_2$ and He/O_2 gas mixtures by PVTMS membrane SMV

FEED %	LIQUID MEMBRANE	SPEED OF PUMPING cm^3/sec	RETENTATE %	PERMEATE %	DESORBER GAS %
CO_2-40 CH_4-30 H_2 -30	water	0,04	CH_4- 75 CO_2- 20 H_2 - 5	CH_4- 2 CO_2- 7 H_2 - 91	CH_4- 24 CO_2- 75 H_2 - 1
	25% water solution of MEA	0,04	CH_4- 97 CO_2- 1 H_2 - 2	CH_4- 1 CO_2- 0 H_2 - 99	
He -70 O_2 -30	$N(C_2F_5)_3$	0,0I		He -98 O_2 -2	

CONCLUSION

The experiments with integrated membrane systems consisting of non-porous asymmetric PVTMS membranes and flowing liquid have demonstrated high effective gas separation for two and three components mixtures (for example, the selectivity $CO_2/CH_4 \sim 1000$ in comparison with $CO_2/CH_4 = 10$ for the PVTMS membranes). Multymembranes permabsorbers are very compact. The energy consumption for these systems is rather small due to the absent of the pressure compressor for gases and liquid. The application of polymeric non-porous membranes provides the stability of liquid phase for a long period of time and prevents the pollution of gas products by the liquid phase. The presence of two modules (desorption and absorption) provides the regeneration of liquid carriers. The mathematical simulation of above considered integrated systems allows to find the optimal gas separation conditions.

ACKNOWLEDGEMENT

The authors express their thanks to academician N.Plate who supported this study at the A.V.Topchiev Institute of Petrochemical Synthesis.

REFERENCES

Shelekhin, A.B.; Beckman, I.N.; Teplyakov V.V.; Gladkov, V.S. *Patent of RUSSIA*, No. 1637850A1, 1989.

Shelekhin, A.B.; Beckman, I.N. Ideal model for gas separation processes in membrane absorber. in: *Proceeding the 1990 Intr.Con.in Membr.Proc.*- Chicago, p.1419, 1990.

Sirkar, K.K. *US Patent* No. 4750918, 1988.

Teramoto Masaaki; Matsuyama Hideto; Yamashiro Takumi, Separation of ethylene from ethane by a flowing liquid membranes using silver nitrate as a carrier. *J.Membr.Sci.*,45, p.115, 1989.

SIMULATION OF IN-SITU ENRICHMENT OF NATURAL WELLS BY POLYMERIC MEMBRANES

by

Hisham M. Ettouney
Malik Al-Ahmad

and

Ahmed Helal

Department of Chemical Engineering
Faculty of Engineering
King Saud University
Riyadh, Saudi Arabia

and

R. Hughes

Department of Chemical and Gas Engineering
University of Salford
Salford, United Kingdon

ABSTRACT

Fracture of gas wells which is performed by the use of pressurized carbon dioxide gas results in an increase in the CO_2 content in the well. Reduction of the CO_2 in the gaseous mixture of the well is necessary before any handling or processing steps. In-situ enrichment by polymeric membrane modules has proved to be extremely effective in comparison with conventional chemical separation processes. The membrane separation process is characterized by its compact size, high separation efficiency, and low energy consumption. This simulation study aims at modelling the process of gas enrichment in the well by the use of cellulose acetate membrane modules. The mathematical model developed assumes a complete mixing flow configuration for both the module and the well, that the gas mixture contains only CH_4 and CO_2 with constant permeabilities, and ideal gas behaviour. Simulation results are presented in terms of the time required to perform the enrichment process as well as the final gas composition of the well and the removal efficiency for CO_2.

INTRODUCTION

Removal of CO_2 from natural gas is an essential step which precedes gas handling and conveyance. The mole fraction of CO_2 in the pipelines should be in the range of 2–3%. Since the removal process is performed on site, the removal system has to be compact, simple to operate, efficient, and utilize minimum energy.

The conventional removal process is based on the use of diethanolamine (DEA) to absorb the CO_2 gas. The absorption system is generally a permanent installation and is difficult to be moved from one location to another.

A novel technique for CO_2 removal is performed by the use of polymeric membranes. Membrane modules are compact and can be easily mounted on trailers, which facilitate their transport.

An economic analysis and comparison of the two systems shows that the use of the membrane system has a lower installed capital cost than that of the DA process (Schell, 1985). In addition, methane recovery by the membrane system is close to 90% compared to 95% methane recovery achieved by the DEA process (Schell, 1985).

This study concerns simulation of in–situ enrichment of natural gas by membrane systems. The study includes mathematical models for both the natural gas well and the membrane module. Analysis and results are presented in terms of gas composition as a function of enrichment time as well as other operating parameters, i.e., membrane area, operating pressure, and gas flow rate. Also, variations in CH_4 net recovery and CO_2 removal rate are presented as a function of the same parameter set.

Process description

Figure 1 shows a schematic of the gas well and membrane module arrangement. As shown in the figure the gas is fed to the module where it is separated into two streams. The permeate stream is rich in CO_2 and is vented to the atmosphere, while the second stream, the retentate, is rich in CH_4 and is recycled into the well.

The process continues for a long period of time, up to 50 days, to ensure reduction of the CO_2 content of the well from highs of 20–30 mole percent down to lows of 2–3 percent. The membrane module is operated at high pressures, ranging from 1000 to 10,000 kPa, with feed rates of 200 N m^3/day.

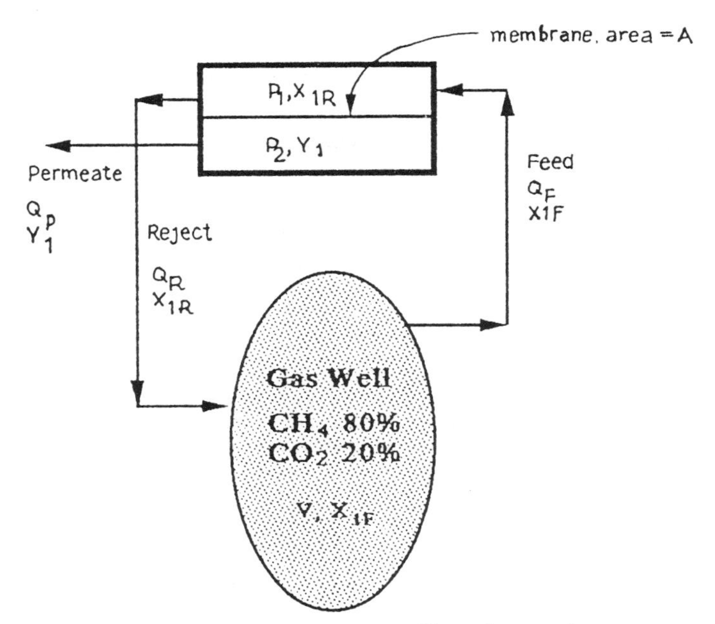

Figure 1: Schematic of gas well and membrane system.

Mathematical model

The enrichment model constitutes the membrane module and the gas well. A schematic diagram, figure 1, shows the two systems. As shown, the gas from the well is feed to the membrane module. The permeate stream, which is highly enriched in CO_2, is vented into the atmosphere. On the other hand, the retentate or unpermeated stream is recycled into the well. The retentate stream is of course highly enriched in CH_4.

The above process is unsteady because of variations in the composition and the amount of gas in the well during enrichment. Therefore, the material balance equations for the gas well are written in transient form. On the other hand, it is assumed that the material balances of the membrane module are for steady state conditions. This assumption is valid as long as the feed flow rate and feed pressure are maintained constant throughout the enrichment process. Of course the feed composition to the membrane module will vary during the process. As a result the model equations for the membrane module have to be solved simultaneously with the well equations.

The following is a summary of the model's assumptions.

- the gas mixture in the well is composed of CH_4 and CO_2 only,
- the well contents are perfectly mixed,
- the compartments of the membrane module are perfectly mixed,
- the permeabilities of the gas mixture are constant,
- the system is isothermal,

- the pressure in the well is constant,
- the feed flow rate from the gas well into the membrane module is constant,
- the feed pressure to the membrane module is equal to the pressure of the gas well,
- the feed and permeate pressure are constant, and
- the volumetric flow rates and well gas volume are normalized to standard conditions.

Clearly, if the effects of gas nonideality, permeability dependence on pressure and composition, and the folow condigurations are taken into account, this could result in a significant impact on the final results (see Fattah et al., 1991, and Ettouney and Hughes, 1991). However, in this initial study it was considered advisable to simplify the model in order to obtain a clearer pixture of the system behaviour, its dynamics, and the relations among the various operating and design parameters.

Use of the complete mixing flow configuration at this stage of model dvelopment has a number of advantages. Such a model requires a fairly limited number of physical parameters, which may require extensive and specialized measurements. Such parameters include a proper estimate for the initial volume of the gas well, the initial average composition of the gas well, and the permeability of various components in the membrane material. Other model parameters are those akin to oeprating and design conditions, e.g., flow rate, pressure, and membrane area.

The model equations for the gas well on invoking the perfect mixing flow configuration and the other stated assumptions are

- overall volume balance

$$\frac{dV}{dt} = - Q_p \qquad (1)$$

- volume balance of the CO_2 gas

$$\frac{dX_{1F}}{dt} = \frac{\left[X_{1R}Q_R - X_{1F}Q_F - X_{1F}\frac{dV}{dt}\right]}{V} \qquad (2)$$

Equations 1 and 2 are solved subject to the initial conditions

$$V = V_o \qquad (3)$$

and

$$X_{1F} = X_{1Fo} \qquad (4)$$

Based on the above stated assumptions the model equations for the membrane module do not have an explicit dependence on time. This is because the feed flow rate and the operating pressures are time independent. However, the equations have an implicit dependence on time because of the transient variations in the feed composition.

The complete mixing model equations which describe the behaviour of the membrane module (Weller, et al., 1954) are

- overall component balance of CO_2

$$X_{1F}Q_F = Y_1Q_P + X_{1R}(Q_F - Q_P) \qquad (5)$$

- flux of CO_2 across the membrane

$$Y_1Q_P = R_1A(X_{1R}P_1 - Y_1P_2) \qquad (6)$$

- flux of CH_4 across the membrane

$$(1 - Y_1)Q_P = R_2A((1 - X_{1F})P_1 - (1 - Y_1)P_2) \qquad (7)$$

The equations for the gas well are first order ordinary differential equations which are coupled with the algebraic equations of the membrane module. Accordingly, the gas well equations are solved numerically using Gear's method (Gear, 1971). The membrane module algebraic equations are solved simultaneously at each time step of the Gear's integration procedure.

Results and Discussion

The mathematical model is used to investigate the dynamic characteristics of the enrichment process and the effects of various operating and design parameters. The analysis is performed for a CH_4–CO_2 gas mixture and a cellulose acetate membrane.

The base case for the study of the system dynamic behaviour is made for the following set of parameters; feed flow rate of 600 m^3/day, feed pressure of 6000 kPa, and a membrane area of 3 m^2. The selection of these parameters is made to ensure a proper enrichment process which should have a final CH_4 recovery ratio of more than 90% and a final CO_2 mole fraction in the gas well of less than 0.03.

Also, constant permeabilities of 0.001 and 0.08 m^3/m^2 s kPa are used in all calculations for CH_4 and CO_2, respectively (Donohue, et al., 1989).

The results fo the base case study are displayed in figures 2 and 3. Figure 2 shows the transient variation in CH_4 recovery, which is defined as the ratio of the volume of CH_4 in the well. Also, the transient variation of CO_2 removal, which is defined as the ratio of the volume of CO_2 removed to the initial volume of CO_2 in the well, is displayed on the same figure.

As shown in figure 2, the CH_4 recovery ratio drops from an initial value of unity to a value of 0.9 after 50 days of enrichment. Similarly, the CO_2 removal ratio increased from an initial value of 0 to a final value of 0.86 after the same period of enrichment.

The variations in the CH_4 and CO_2 mole fractions in the well are shown in figure 3. As shown in figure 3, the CH_4 mole fraction steadily increases from an initial value of 0.8 to a final value of 0.965. On the other hand, the CO_2 mole fraction decreases from a high initial value of 0.2 to a final low value of less than 0.03.

Based on these results, it is evident that the selection of the operating and design parameters set for the enrichment process is successful. This is because the final enrichment properties are within the prespecified set points, i.e., 90% CH_4 recovery and a final CO_2 mole fraction of less than 0.03.

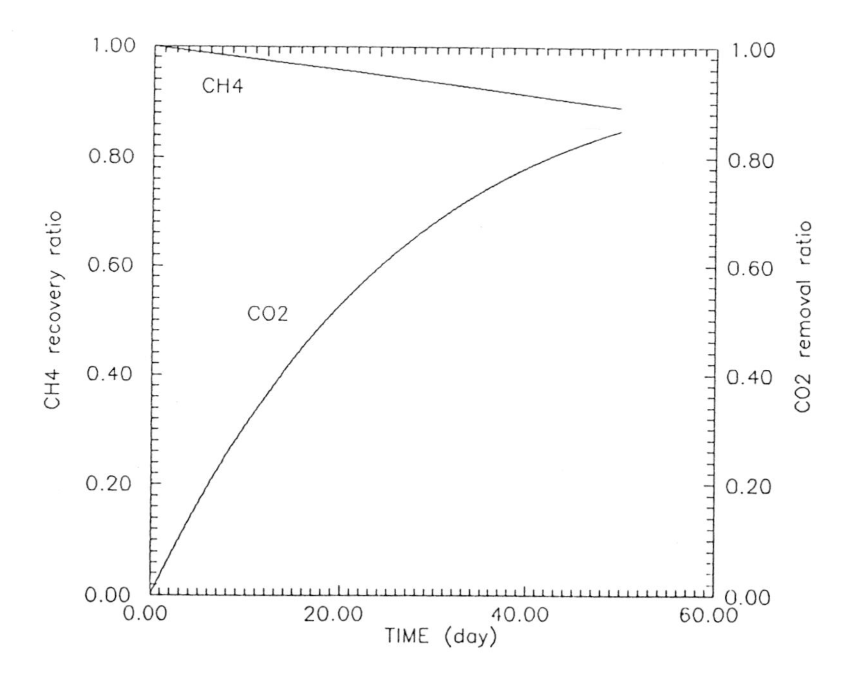

Figure 2: Variation in CH4 recovery and CO2 removal in gas well versus time, for feed pressure of 6000 kPa, initial well composition of 80% CH4 in CO2, feed flow rate of 500 m3/day, and membrane area of 3 m2.

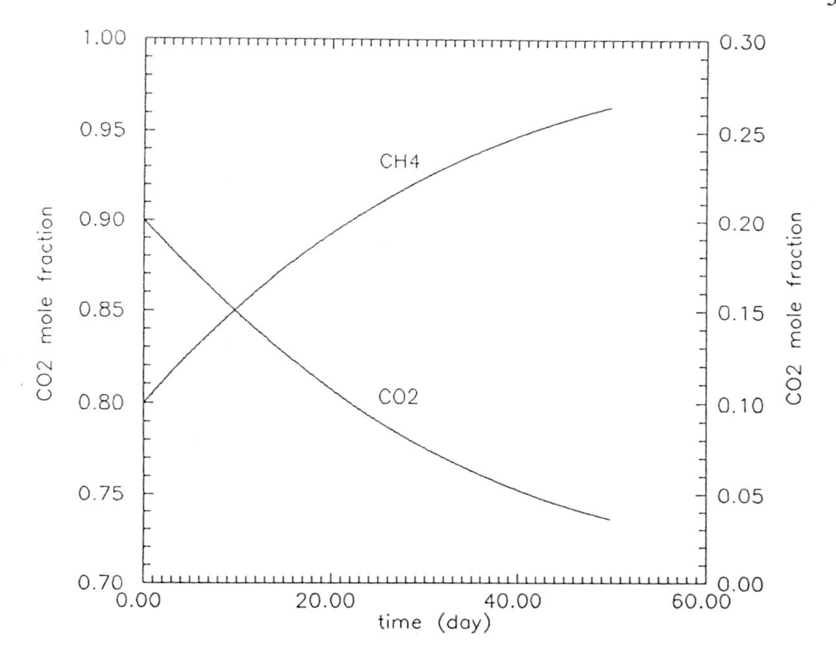

Figure 3: Variation in CH4 and CO2 mole fractions in gas well versus time for feed pressure of 6000 kPa, feed composition of 80% CH4 in CO2, and feed flow rate of 500 m3/day, and membrane area of 3 m2.

The selection of the above set of parameters is made in the light of the parametric analysis results. The parametric study includes variations in the feed flow rate and its pressure as well as the membrane area. All calculations are made for an initial well composition of 80% CH_4 in CO_2 and for an enrichment period of 50 days. The parameter ranges used in the caluclations are as follows:

- the membrane area is varied between 1 and 12 m^2,
- the feed flow rate is varied between 50 and 500 m^3/day,
- the feed pressure is varied between 1000 and 10,000 kPa.

Results are presented in terms of final CH_4 recovery, CO_2 removal, and CH_4 and CO_2 mole fractions in the gas well.

The effect the feed flow rate on the system behaviour is shown in figures 4 and 5. These calculations are made for a membrane area of 3 m^2 and a feed pressure of 6000 kPa. As shown in figure 4 the CH_4 recovery ratio has no dependence on the feed flow rate. On the other hand, the CO_2 removal ratio varies considerably with an increase of the feed flow rate, increasing over a range of 0.2 – 0.8 as the feed flow rate is varied from 50 to 500 m^3/day. This behaviour occurs because of the continuous increase in the driving force for CO_2 permeation due to the increase in the amount of CO_2 flowing in the membrane module. On the other hand, the permeation driving force of CH_4 does not increase with increase of the feed flow rate, being more dependent on the membrane area and feed pressure.

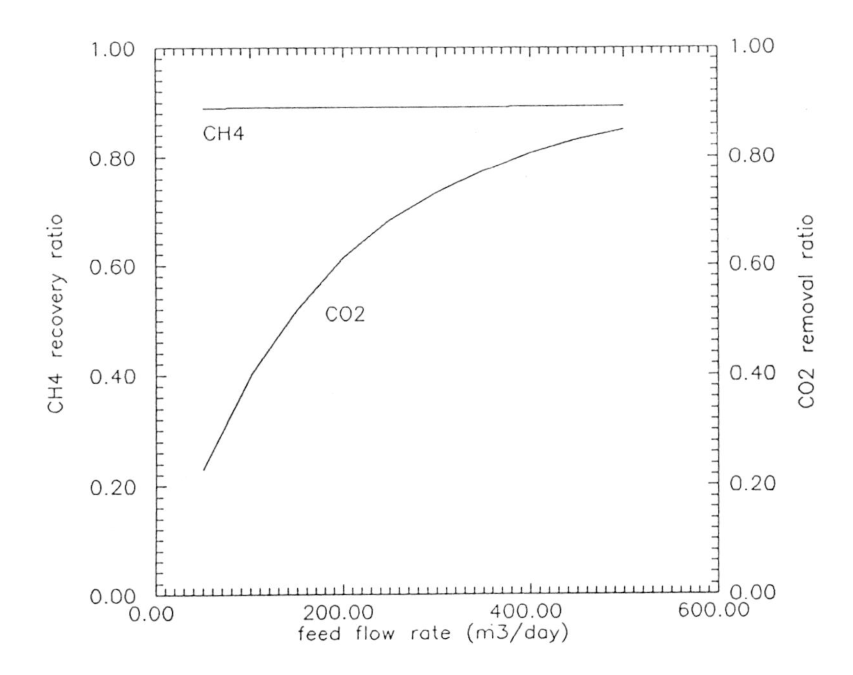

Figure 4: Variation in CH4 recovery and CO2 removal in gas well after 50 days of enrichment as a function of feed flow rate, for feed pressure of 6000 kPa, initial well composition of 80% CH4 in CO2, and a membrane area of 3 m2.

The effect of the feed flow rate on the CH_4 and CO_2 mole fractions in the gas well is shown in figure 5. As shown, the CH_4 mole fraction increases steadily with increase of the feed flow rate. On the other hand, the CO_2 mole fraction continuously declines with increase of the feed flow rate.

An analysis of the feed flow rate results shows that the best operating conditions for the membrane module are at the highest feed flow rate of 500 m^3/day. This is because the CH_4 recovery and CO_2 removal ratios at this condition are quite high, i.e., close to 90% for the recovery and more than 80% for the removal. Also, the CH_4 and CO_2 mole fractions in the well are at the required values of 0.96 and 0.03, respectively.

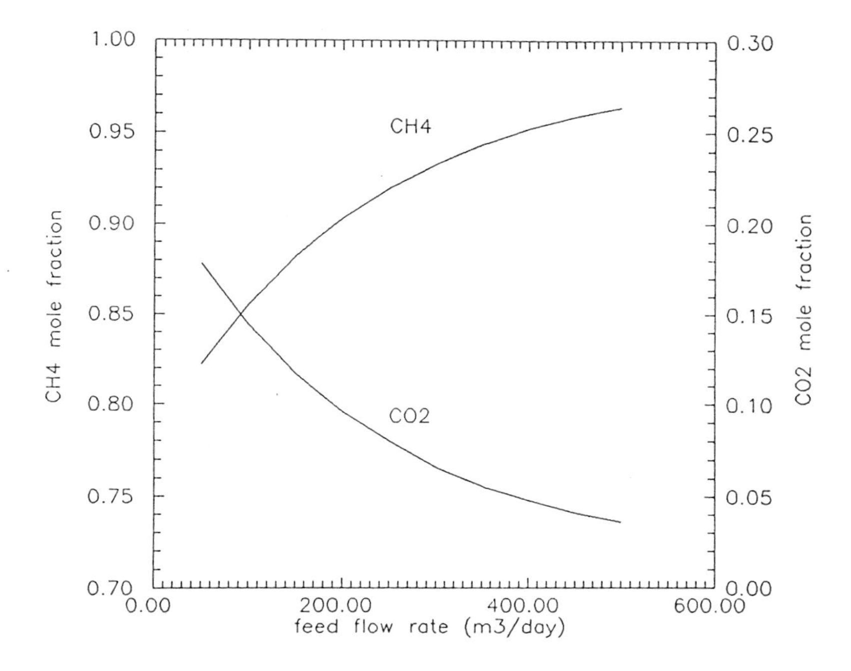

Figure 5: Variation in CH4 and CO2 mole fractions in gas well after 50 days of enrichment as a function of feed flow rate, for feed pressure of 6000 kPa, initial well composition of 80% CH4 in CO2, and membrane area of 3 m2.

The effect of varying the membrane area on the final properties of the enrichment process has been studied over a membrane area range of 1 – 12 m^2 and at a feed flow rate of 500 m^3/day and a feed pressure of 6000 kPa. Increase of membrane area reduces the permeation resistance for both CH_4 and CO_2. Results of this effect are shown in figures 6 and 7.

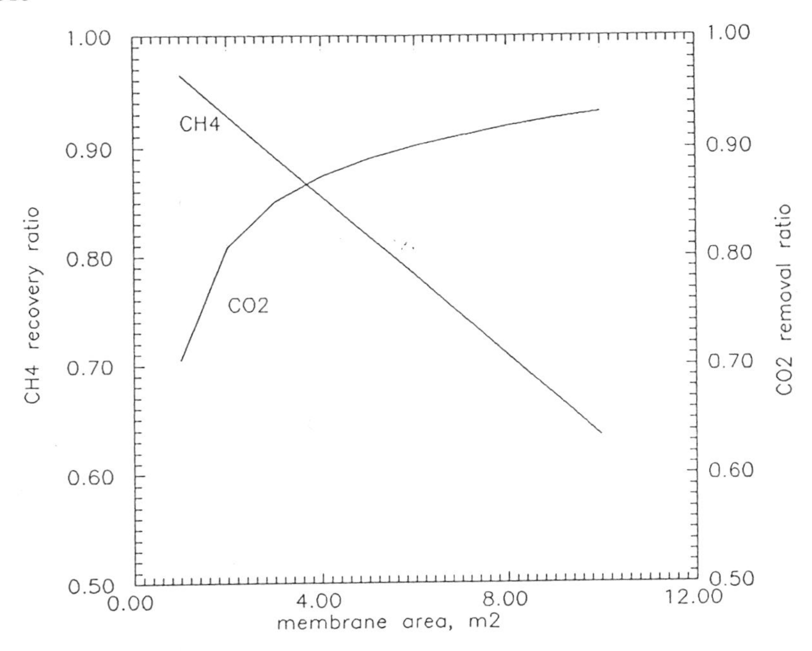

Figure 6: Variation in CH4 recovery and CO2 removal in gas well after 50 days of enrichment as a function of membrane area, for a feed pressure of 6000 kPa, initial well composition of 80% CH4 in CO2, and feed flow rate of 500 m3/day.

Clearly, an increase of the membrane area results in an improved CO_2 removal ratio and a decline of the CH_4 recovery ratio. Also, the CH_4 mole fraction in the well increases continuously with area enlargement. Similarly, the CO_2 mole fraction in the well drops continuously with increase of the membrane area.

Analysis of the above data (see figures 6 and 7) shows that enlargement of the membrane area above 5 m^2 results in depletion of both gases from the well. As a result, it is advisable to design the membrane module with an area in the range of 2 - 4 m^2 in order to achieve suitable enrichment. Under such conditions, the CH_4 recovery ratio is in the range of 0.9 and CO_2 mole fraction in the well is close to 0.03.

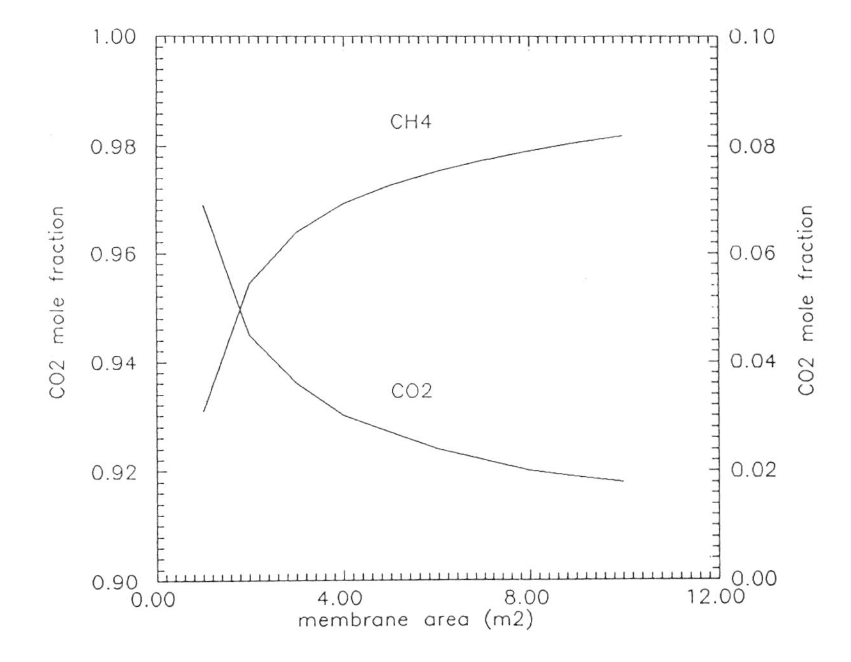

Figure 7 : Variation in CH4 and CO2 mole fractions in gas well after 50 days of enrichment as a function of membrane area for feed pressure of 6000 kPa, initial well composition of 80% CH4 in CO2, and feed flow rate of 500 m3/day.

The major effect of raising the feed pressure on the enrichment process is quite similar to that of enlarging the membrane area. Clearly, increasing the feed pressure increases the permeation driving force for both components.

Results of increased feed pressure are shown in figures 8 and 9. Figure 8 shows variations in CH_4 recovery and CO_2 removal. As shown the CH_4 recovery declines slowly with pressure increase, dropping from a value close to unity at a feed pressure of 1000 kPa to a value of 0.8 at a feed pressure of 10,000 kPa. On the other hand, the CO_2 removal increases from a low of 0.3 at a feed pressure of 1000 kPa to a high value of more than 0.9 at a feed pressure of 10,000 kPa.

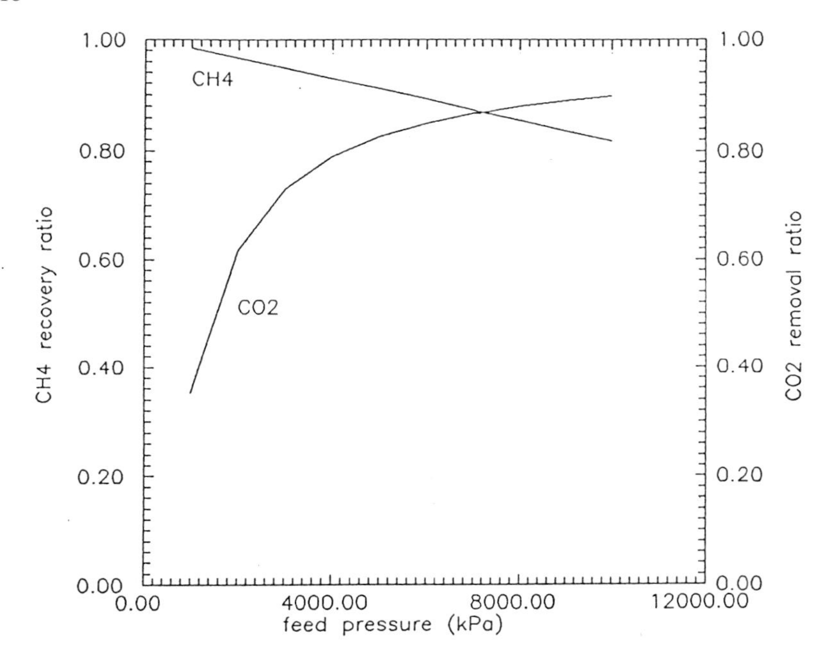

Figure 8 : Variation in CH_4 recovery and CO_2 removal in gas well after 50 days of enrichment as a function of feed pressure, for membrane area of 3 m^2, initial well composition of 80% CH_4 in CO_2, and feed flow rate of 500 m^3/day.

Variations in the mole fractions of the two species in the well as a function of the feed pressure are shown in figure 9. Operating the enrichment process at higher feed pressures is the likely, preferred option since the enrichment properties are within the prespecified operating conditions.

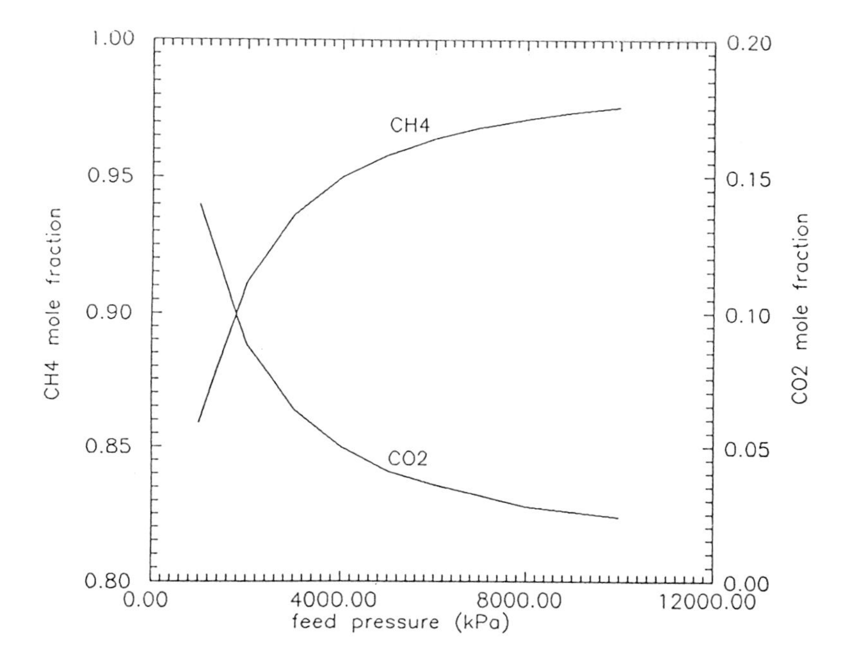

Figure 9 : Variation in CH4 and CO2 mole fraction in gas well after 50 days of enrichment as a function of pressure, for feed flow rate of 500 m3/day, initial well composition of 80% CH4 in CO2, and a membrane area of 3 m2.

Conclusions

A mathematical model has been developed to simulate the enrichment process of natural gas wells by membrane modules. The gas wells are characterized by a high initial concentration of the CO_2 gas, caused by the fracture procedure. The study aimed at characterizing the enrichment process and defining the proper operating and design conditions.

The mathematical models developed for both the gas well and the membrane module are those for the complete mixing flow configuration. The gas well model is developed for unsteady state conditions, because of the continuous decay in its total volume and CO_2 and CH_4 contents during the enrichment process. On the other hand, the membrane module equation do not include a transient term since the feed flow rate to the modules as well as the pressure are kept constant throughout the enrichment process.
Analysis of the simulation results shows that appropriate enrichment can be achieved at elevated feed pressures, low membrane areas, and large feed flow rates. For a cellulose acetate membrane and an initial well composition of 80% CH_4/20% CO_2, it is found that proper enrichment is achieved after 50 days of operation with a membrane area of 3 m^2, a feed flow rate of 500 m^3/day, and a feed pressure of 6000 kPa.

The results obtained by the mathematical model show that such a model is an extremely valuable tool for studying the dynamics of well enrichment, for defining appropriate operating and design conditions, and for developing a better knowledge and understanding of the enrichment process. As shown from the parametric study, the results can be used to select proper operating and design parameters to simulate the enrichment process within a prespecified set of final conditions.

References

Donohue, M.D., Minhas, B.S., and Lee, S.Y., "Permeation Behaviour of Carbon Dioxide-Methane Mixture in Cellulose Acetate Membranes", J. Membrane Sci., **42** (1989) 197-214.

Fattah, K.A., Hamam, S.M., Al-enezi, G., Ettouney, H.M., and Hughes, R., "A Nonideal Model of Separation Permeators", J. Membrane Sci, in print, 1991.

Ettouney, H.M., and Hughes, R., "Pressure and Composition dependence of Permeability on Gas Separation by Membranes", submitted, J. Membrane Sci., 1991.

Gear, C.W., "Numerical Initial Value Problems in Ordinary Differential Equations, "Prentice-Hall, Englewood Cliffs, New Jersey, 1971.

Schell, "Commercial Applications for Gas Permeation Membrane Systems", J. Membrane Sci., 22 (1985) 217 - 224.

Weller, S., and Steiner, W.A., "Separation of Gases by Fractional Permeation Though Membranes", J. Appl. Phys., **21** (1950) 279.

Symbols

A	Membrane area, m^2
P_1	Total pressure in feed compartment, kPa
P_2	Total pressure in permeate compartment, kPa
Q_F	Feed flow rate, m^3/day
Q_P	Permeate flow rate, m^3/day

R_1	Permeability coefficient of component 1, m^3/m^2 kPa day
R_2	Permeability coefficient of component 2, m^3/m^2 kPa day
V	Volume of the gas in well, m^3
V_o	Initial volume of the gas in well, m^3
X_{1F}	Mole fraction of component 1 in feed stream
X_{1Fo}	Initial mole fraction of component 1 in feed stream
X_{1R}	Mole fraction of component 1 in reject stream
Y_1	Mole fraction of component 1 in permeate side

POLYANILINE MEMBRANES FOR GAS SEPARATION AND ESR EXPERIMENTS

L. Rebattet [1], E. Geniès [1], M. Pinéri [2] and M. Escoubes [3],
[1] Laboratoire d'Electrochimie Moléculaire,
[2] Service d'Etudes des Systèmes et Architectures Moléculaires, Département de Recherche Fondamentale sur la Matière Condensée, Centre d'Etudes Nucléaires,85 X, 38041 Grenoble, France.
[3] Laboratoire d'Etudes des Matières Plastiques et des Biomatériaux, Université Claude Bernard, 43 boulevard du 11 Novembre 1918, 69622 Villeurbanne, France.

ABSTRACT

Conducting polymers have already been extensively studied for their numerous applications in electronic devices. Recent results have focused attention on the interest of this class of polymers and especially polyaniline for gas separation. Indeed interesting selectivity values between oxygen and nitrogen have been found. As the kinetic diameters of oxygen and nitrogen are very close to each other, only a very low selectivity can be expected from diffusion coefficient differences. In order to determine which kind of interaction may explain a selectivity, Electron Spin Resonance (ESR) measurements have been performed. A tenfold decrease in the apparent spin number is observed when the polymer is in presence of oxygen rather than nitrogen. This result is interpreted as being due to a specific interaction between oxygen and polarons of polyaniline.

KEYWORDS : electronic conducting polymer, polyaniline, polaron, oxygen, gas permeation, ESR.

NOMENCLATURE

A	membrane area	= 3 cm^2
$C_{(x,t)}$	gas concentration within the membrane at point x and time t	cc_{STP} cm^{-3}
ΔC	gas concentration difference accross the membrane	cc_{STP} cm^{-3}
C_1	gas concentration in the upstream compartment	cc_{STP} cm^{-3}
D	diffusion coefficient	cm^2 s^{-1}
e	membrane thickness	cm
F	permeation flux of gas	cc_{STP} cm^{-2} s^{-1}
$P_{(x,t)}$	gas pressure at point x and time t	cmHg
ΔP	pressure drop accross the membrane	cmHg
Pe	permeability coefficient	barrer
	1 barrer = 10^{-10} cc_{STP} cm cm^{-2} s^{-1} $cmHg^{-1}$	
P_2	gas pressure in the downstream volume	cmHg
Q	quantity of gas having crossed the film	cc_{STP}
S	solubility coefficient	cc_{STP} cm^{-3} $cmHg^{-1}$
t	time	s
V_2	downstream volume	= 144 cm^3
α	selectivity : coefficient of oxygen enrichment through the membrane	
θ	time-lag	s

INTRODUCTION

Potential uses of oxygen in medicine and combustion explain the interest in polymer films with large separation factors for oxygen and nitrogen. In addition to selectivity, a large permeability is necessary for the efficiency of the system.

Permeability of rubbery polymers can be increased by enhancing the chain flexibility and the side group bulkiness in silicone rubbers [1,2]. Permeability of glassy polymers such as poly(1-trimethylsylil, 1-Prop-1-yne) depends mainly on the free volume content [3,4]. Another substantial increase in permeability is due to the development of asymmetric membranes in which a thin skin of polymer with high gas selectivity is grown on a porous structural support [5,7].

Large differences in solubility can be obtained by incorporating within a membrane absorption sites for oxygen (cobalt, porphyrins) which act as fixed carriers for oxygen [8,9]. An O_2 / N_2 selectivity larger than 10 has been given for these membranes. A value of 16 has been reported for cellulose nitrate [10].

Recent experiments carried out with electronic conducting polymers such as polyaniline have shown promising results concerning gas separation [11,12]. The O_2 / N_2 pair was tested and provided large values of selectivity (30).

The doping, undoping and redoping process may induce permanent morphological changes in these polymers and constitute a way for tailoring membranes for gas separation. The authors present an interesting variation in permeability and selectivity during this treatment. Moreover the oxidation of polyaniline leads to the formation of defaults called polarons and bipolarons in the macromolecular chain [13-15]. ESR experiments have been performed on a polyaniline film and tend to evidence interactions between oxygen and polarons.

FUNDAMENTALS OF GAS SEPARATION

The method used is the transmission one [16,17].
Transport through the dense polymeric toplayer takes place by a solution - diffusion mechanism. This is governed by Fick's laws:

$$F = - D \frac{\delta C(x,t)}{\delta x}$$

$$\frac{\delta C(x,t)}{\delta t} = D \frac{\delta^2 C(x,t)}{\delta x^2}$$

Moreover the gas concentration $C(x,t)$ within the membrane at point x and time t is proportional to the gas pressure $P(x,t)$, which is expressed by Henry's law :

$C(x,t) = S \, . \, P(x,t)$ where S represents the solubility coefficient.

Concerning the quantity of gas Q which has crossed the membrane, it is determined by the second Fick's law when the stationary state is obtained.

$$Q = \frac{D\, C_1}{e} \left(t - \frac{e^2}{6\,D} \right)$$

The flux of gas F through the film can be expressed as :

$$F = - D \frac{\Delta C}{e} = - Pe \frac{\Delta P}{e}$$ where Pe is the permeability coefficient

By plotting Q versus t, the steady-state line enables the calculation of both the permeability coefficient (Pe) from the slope and the time-lag θ by extrapolation on the time axis.

$$Pe = \frac{dQ}{dt} \cdot \frac{1}{A} \cdot \frac{e}{\Delta P} \quad \text{and} \quad \theta = \frac{e^2}{6\,D}$$

The diffusion coefficient D is calculated from the time-lag value.

S is calculated from the relation : Pe = D S.

The selectivity of the O_2 / N_2 pair is calculated by the ratio of the respective permeability coefficients as follows :

$$\alpha = \frac{Pe\ O2}{Pe\ N2}$$

EXPERIMENTAL

Synthesis of polyaniline

Polyaniline in the emeraldine oxidation state was synthesized by chemical oxidation of aniline by ammonium peroxodisulfate in an acidic aqueous medium (1 M HCl) [18]. The polymer powder was deprotonated by immersion in a dilute aqueous ammonia solution under constant stirring and it was purified by washing cycles in acetonitrile. Then it was dried under dynamic vacuum for at least 24 hours before use. The emeraldine powder was dissolved in N-methyl 2-pyrrolidinone (NMP) at 2 g/l.

Materials

Asymmetric membranes used for gas permeation were prepared by the phase inversion process [19,20]. A polyaniline - NMP solution was cast on a glass plate and after partial evaporation of the solvent, the plate was immersed in a non-solvent coagulation bath. The resulting membrane was made up of a dense polymeric toplayer responsible for the separation and a porous sublayer which acts as a mechanical support. Following the method used by Anderson et al. [12], every cast film was doped by immersion in a 4 M chlorhydric acid solution Then the membrane was undoped by treating it with a 1 M ammonia solution and at last, it was slightly redoped by immersion in a 0.0175 M chlorhydric acid solution. Each treatment of this acid - base doping chemistry was performed during 3 hours.

Concerning ESR experiments, a little volume of the polyaniline - NMP cast solution was deposited on a gold wire (0.5 mm diameter) with a syringe. After most of solvent evaporation was performed at room temperature, the sample was put under dynamic vacuum for about 8 hours. The resulting film thickness was estimated as being close to 0.5 μm by microscopic observation. The deposit was doped by treating it with HCl 1 M and it was dried by curing it for a few minutes with a flow of warm air.

Methods

The thermogravimetric analysis (TGA) was performed on a Dupont 2000 apparatus with a TGA V5.1A software.

The permeation cell is made of two compartments (1 and 2) which are separated by the polymeric membrane. The cell is settled in a thermostatted chamber at 20°C.
A preliminary high vacuum desorption is realized in order to ensure that the static vacuum pressure changes in the downstream compartment will be much smaller than pressure changes due to permeation; indeed the static vacuum needs to be lower than 10^{-2} Torr/h. Then a 3 bars pressure is introduced in the upstream part. The gas pressure P_2 in the downstream volume V_2 is measured by a Datametrics pressure sensor (range 100 Torr).

The ESR apparatus is a Varian E4 and it is interfaced with a HP 9825 B computer to determine the spin numbers of the samples. A microwave frequency counter EIP 545 and a gaussmeter RMN2 are also used.

RESULTS AND DISCUSSION

Thermogravimetric analysis

As shown in Fig.1, every polyaniline film contains water that is lost below 100°C but the doped and the redoped samples seem to be more hygroscopic than the two others. According to the as cast film, the membrane still contains about 9% of NMP before acid-base treatment. For the doped sample, it is difficult to know whether the weight loss above 150°C is due to chlorhydric acid and NMP or only to the acid Nevertheless the undoped sample does not present significant weight loss above 100°C, which indicates it does not contain NMP anymore. Therefore we can suppose the weight loss of the doped sample above 150°C mainly corresponds to HCl. Concerning the redoped sample, it contains about 12% of doping agent, which is considerable regarding to the very low concentration in chlorhydric acid in the redoping solution.

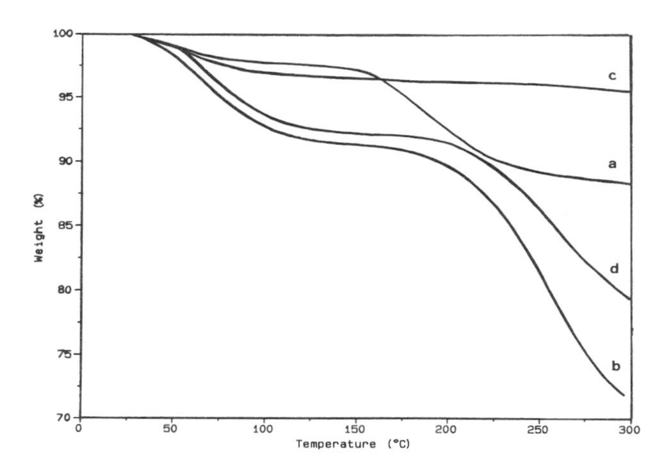

Figure 1. TGA of polyaniline films : (a) as cast; (b) doped in HCl 4M; (c) undoped in NH_4OH 1M; (d) redoped in HCl 0.0175 M.

Gas permeation

Fig. 2 presents typical permeation curves used for the calculation of permeability, diffusion and solubility coefficients.

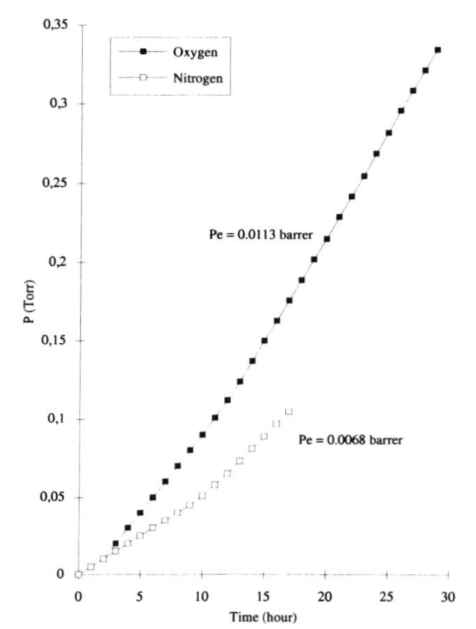

Figure 2. Gas permeation curves $P_2=f(t)$ for a polyaniline film doped in HCl 4M.

The results are presented in Table 1. It was not possible to calculate the diffusion and solubility coefficients for the undoped sample because the estimated time-lags were too short.

	Gas	Doped	Undoped	Redoped
Pe	O_2	0.0113	2.15	0.79
	N_2	0.0068	2.47	0.16
α		1.66	0.87	5.05
Time-lag (θ)	O_2	13,910	11.8	1056
	N_2	13,950	5.7	533
D	O_2	$2.7.10^{-11}$	–	$2.5.10^{-10}$
	N_2	$2.7.10^{-11}$	–	$4.9.10^{-10}$

Table 1. Gas permeation results for the doped, undoped and redoped samples.

The doping, undoping and redoping process obviously involves a large range of permeability coefficients. Moreover it provides an interesting increase in selectivity from 1.7 for the doped sample to 5 for the redoped one.

ESR experiments

In order to determine the role of polarons in polyaniline on the solubility of oxygen, ESR experiments were carried out by alternatively sending oxygen or nitrogen on the polymer coated on the gold wire.

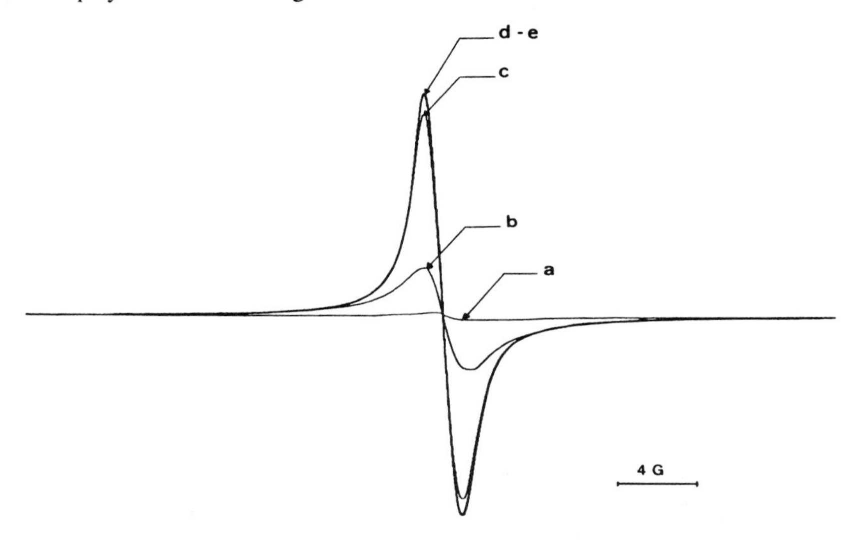

Figure 3. Variation of ESR spectra with the nature of gas which flows on polyaniline film: (a) under 1 bar O_2 - t=0; (b) after 1 mn under 1 bar N_2; (c) after 3 mn; (d) after 5 mn; (e) after 7mn. Magnetic field, 3396.2 G; frequency, 9.52 GHz; gain, 8.10^2; mod., 0.8 G; power, 10mW.

According to Fig. 3, there is a significant variation in the apparent spin number when a flow of oxygen replaces nitrogen gas in the cell. From 5 minutes of nitrogen environment, the apparent spin number stops increasing and remains constant. Indeed (d) and (e) merge into one. At that moment, the apparent spin number has reached a plateau. Further experiments have demonstrated that this is a reversible phenomenon.

There is another interpretation of Fig. 3 to explain the widening of the EPR signal with oxygen: indeed the introduction of oxygen in the polyaniline film considerably increases the signal half-width. Anyway this result demonstrates an interaction between the polarons of polyaniline and oxygen triplet states.

CONCLUSION

Specific interaction of oxygen with polarons in polyaniline may explain the selectivity of slightly doped polyaniline for O_2 / N_2 separation. Furthermore permeability flows can be changed by free volume modifications associated with doping-undoping-redoping process. Additional interest is due to the possibility of producing asymmetric membranes by thin film deposition on a microporous substrate. Further experiments are on the way to better understand the structural changes associated with the permeability variations. Moreover it would be interesting to know how varies the stability of polarons when other counter ions are used. Sorption measurements may also determine whether the selectivity between oxygen and nitrogen is due to a preferential solubility phenomenon or to a difference in diffusion coefficients.

REFERENCES

1. C. L. Lee, J. Membrane Sci., 38, 55 (1988).
2. S. A. Stern, V. M. Shah and B. J. Hardy, J. Polym. Sci., Part B, Polym. Phys., 25, 1263 (1987).
3. M. Langsam, M. Anaud and E. J. Karwacki, Gas Sep. Purif., 2, 162 (1988).
4. Y. Ichiraku, S. A. Stern and T. Nakagawa, J. Membrane Sci., 34, 5 (1987).
5. J. M. S. Henis and M. K. Tripodi, Science, 220, 4592 (1983).
6. C. Liu, W. J. Chen and C. R. Martin, J. Membrane Sci., 65, 113 (1992).
7. W. Lang and C. R. Martin, Chem. Mat., 3, 390 (1991).
8. H. Nishide, H. Kawakami, S. Toda, E. Tsuchida and Y. Kaniga, Macromolecules, 24, 5851 (1991).
9. H. Nishide, H. Kawakami, T. Suzuki, Y. Azechi and E. Tsuchido, Macromolecules, 23, 3714 (1990).
10. H. Yasuda and V. T. Stannett, in Polymer Handbook, 2nd edn., J. Bnandrup and E. H. Immergut, eds., Wiley, New York (1975).
11. M. R. Anderson, B. R. Mattes, H. Reiss and R. B. Kaner, Synth. Met., 41-43, 1151 (1991).
12. M. R. Anderson, B. R. Mattes, H. Reiss and R. B. Kaner, Science, 252, 1412 (1991).
13. E. M. Geniès and M. Lapkowski, J. Electroanal. Chem., 279, 157 (1990).
14. F. Devreux, F. Genoud, M. Nechtschein and B. Villeret, Springer series in Sol. St. Sci. 76, eds., Kuzmany, p.270 (1987).
15. A. Ray, G. E. Asturias, D. L. Keshner, A. F. Richter, A. G. MacDiarmid and A. J. Epstein, Synth. Met., 29, E141 (1989).
16. R. M. Barrer, in Diffusion in Polymers, J. Crank and G. S. Park, eds., Academic Press, New York, ch. 6, p. 165 (1968).
17. H. Zelsmann, M. Thomas, M. Escoubes and M. Pinéri, J. Appl. Polym. Sci., 41, 1673 (1990).
18. M. Angelopoulos, G. E. Asturias, S. P. Ermer, A. Ray, E. M. Scherr and A. G. MacDiarmid, Mol. Cryst. Liq. Cryst., 160, 151 (1988).
19. M. H. V. Mulder, Recent progrès en génie des procédés, Congrès Euromembrane 92, Paris, 6 (22), p. 17 (1992).
20. H. Strathmann, Materials Science of Synthetic Membrane, eds., Douglas and R. Lloyd, A. C. S. Symposium Series, 269, p.165 (1985).

Industrial Applications of Membrane Systems to Separate Hydrocarbon Vapors from Gas Stream

K. Ohlrogge, J. Wind, R.-D. Behling
GKSS-Forschungszentrum Geesthacht GmbH, Institut für Chemie, Geesthacht, Germany

Introduction

The removal and recovery of hydrocarbon vapors from gas streams by means of membrane technology is finding increasing acceptance in the chemical and petrochemical industry. Applications to treat off-gases generated from the handling, transportation and storage of gasoline membrane separation is a competitive technology to systems based on adsorption or cooling. Meanwhile 16 membrane plants with a capacity of 100 to 2000 m^3/h serving the operation of tank farms are installed or have been ordered [1]. Another technical approach is the recovery of valuable volatile solvents such as pure products at the emission source [2, 3]. The first applications in the chemical and pharmaceutical industry are small scale plants with a capacity of 1 to 25 m^3/h. The recovered solvents are Toluene, MEK (Methyl ethyl ketone), Hexane and 1,2 Dichloroethane.

Membrane and module configuration

The membrane which is used for the separation of volatile hydrocarbon vapors is a flat sheet composite membrane [4]. The selectivity of this membrane for various hydrocarbon vapors vs. nitrogen is shown in Fig. 1. These are ideal selectivities obtained by pure gas measurements.

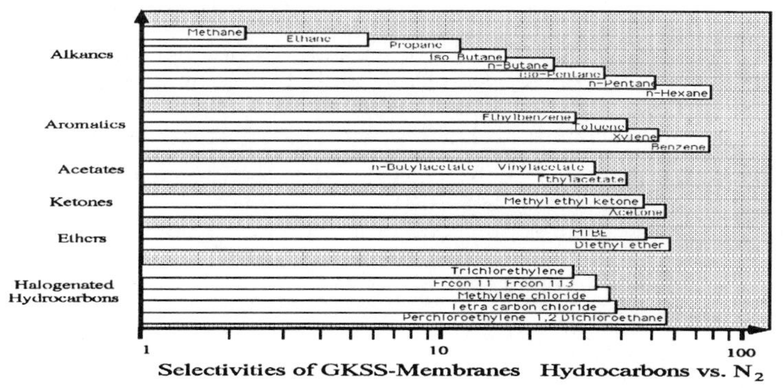

Figure 1: Membrane selectivity of various hydrocarbon vapors vs. nitrogen

The flat sheet membrane is manufactured as disc-like envelopes which are sealed by welding at the cutting edges (Fig. 2).

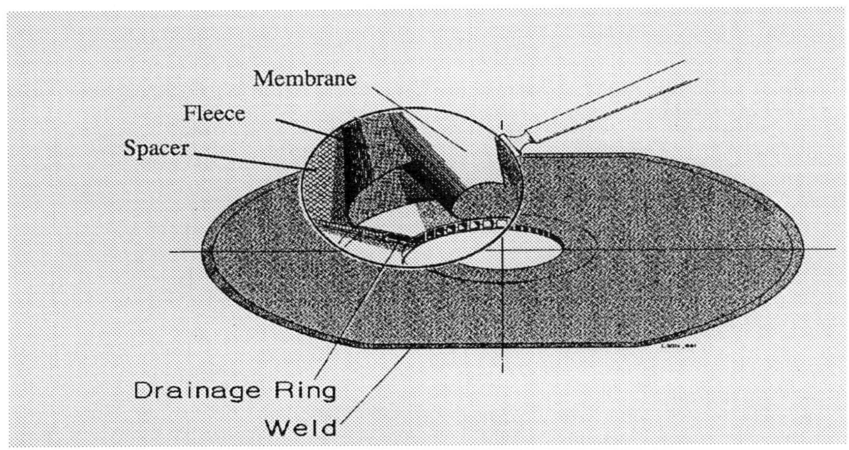

Figure 2: Membrane envelope

The membrane module is depicted in Fig. 3. The so-called 'GS-module' consists of the pressure vessel, a central permeate tube and a stack of membrane envelopes. This configuration allows a design of membrane compartments filled with a variable number of envelopes in parallel or in sequence arrangement to obtain a homogeneous flow across the membranes in dependence of the reduction of feed flow by gas permeation through the membrane.

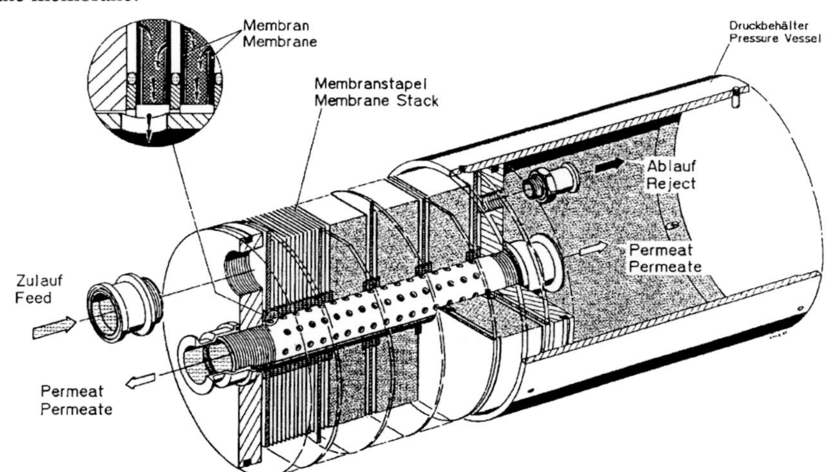

Figure 3: GS-Module

Basic considerations for the process layout

To design a membrane process a knowledge of the membrane/module performance, the operating conditions and the physical constants of the gas compounds is essential. Fig. 4 shows a simplified flow scheme with a list of the required input data for the process design.

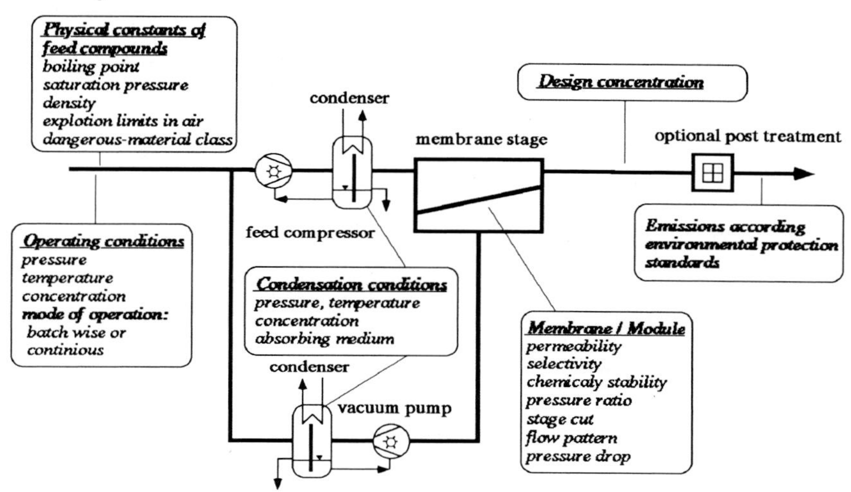

Figure 4: Flow scheme of membrane separation process

The positioning of the condenser for the recovery of the hydrocarbon vapors from the gas stream depends on the feed concentration and the vapor pressure of the gas mixture in the condenser. At high vapor concentrations it is more efficient to place the condenser in the feed line in front of the membrane module intake whereas at low vapor concentrations it is more economical to install the condenser in the permeate line after the vacuum pump.

The toluene recovery rate vs. membrane area is plotted in Fig. 5 depending on the location of the condenser and the intake concentration. The calculations are based on a given vacuum pump capacity. The input data for running the process simulation program are depicted in Figs. 6 and 7.

Table 1: Physical constants of Toluene

Formulae	Mol wt	Density	Boiling point	Saturation pressure			Saturation concentration		
	g/mol	kg/m^3	°C	mbar			g/m^3		
				20 °C	30 °C	50 °C	20 °C	30 °C	50 °C
C_7H_6	92.14	0.863	110.6	27.8	45.2	109	105	165	374

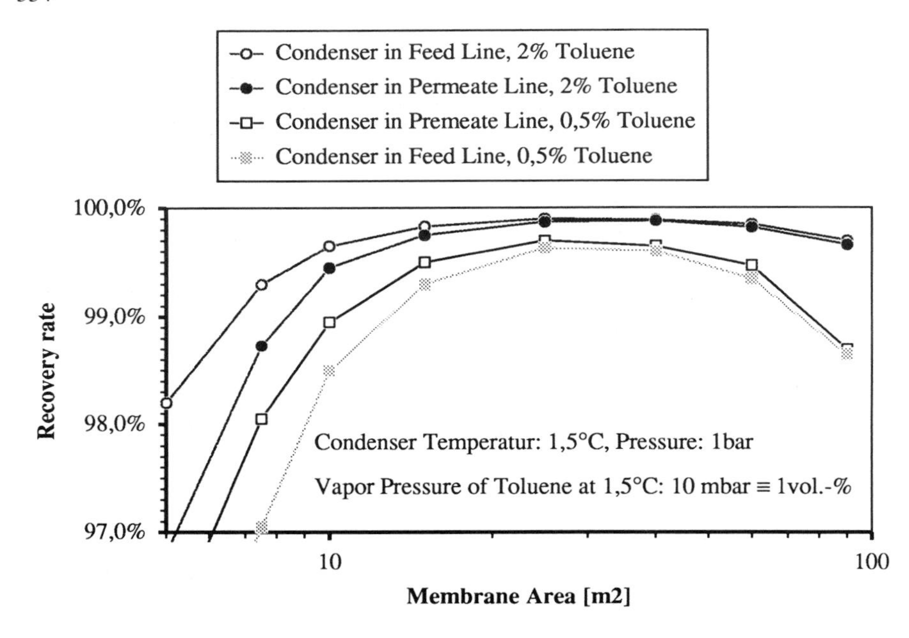

Figure 5: Toluene recovery vs. membrane area

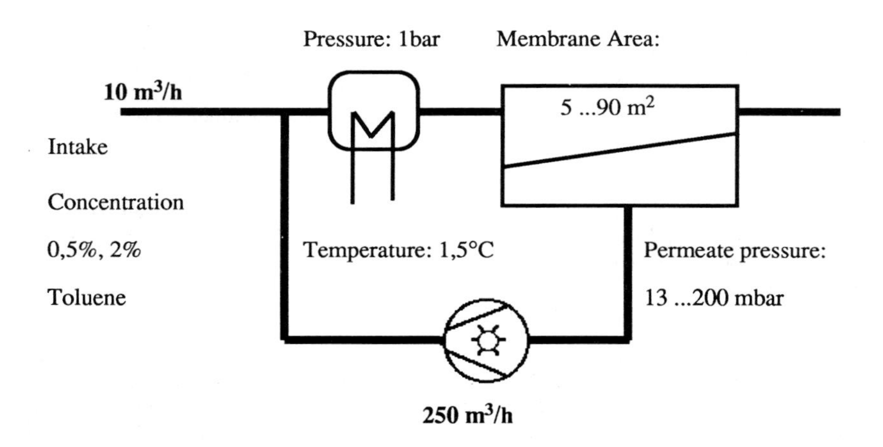

Figure 6: Input data condensation feed side

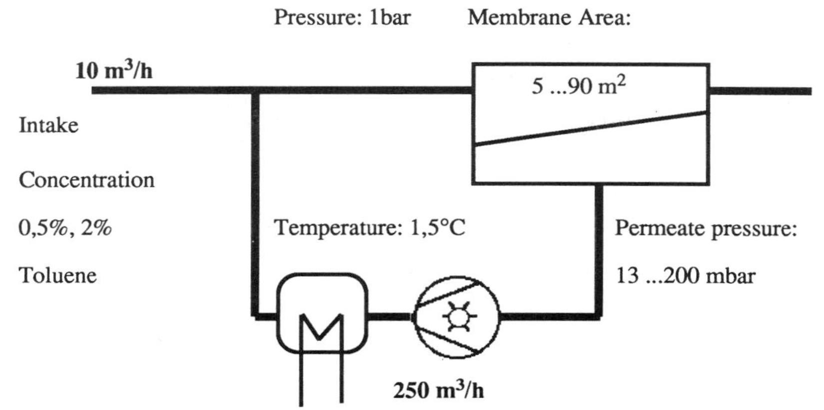

Figure 7: Input data condensation permeate side

Toluene recovery

A plant for recovering toluene vapor and for treatment of the vent gas conforming to TA-Luft-regulations was designed in accordance to the following details:

Design flow from emission source	1 m^3/h
min.	0,75 m^3/h
max.	1,75 m^3/h
Toluene intake concentration	0 vol% to saturation
intake temperature	ambient
condensation temperature	10 - 15 °C
vent gas concentration	TA-Luft = 100 mg Toluene/m^3 air
Type of pumps	liquid ring pumps
sevice liquid	Toluene.

The toluene recovery unit (Fig. 8) consists of two vacuum pumps (1) and (5), a combined separation and condensation vessel (2) jointly used for both vacuum pumps to separate the service liquid from the gas stream, to cool the feed gas and to liquify the condensable vapors and the membrane module (3).

A chiller (8) provides the cooling medium for the heat exchanger (7) installed in the service liquid line of the pumps. The condensed and recovered toluene vapors are collected in vessel (6). To achieve a higher vacuum at the permeate site a gas ejector is installed in the suction pipe of the permeate vacuum pump. The feed pressure is controlled by a pressure control valve (10).

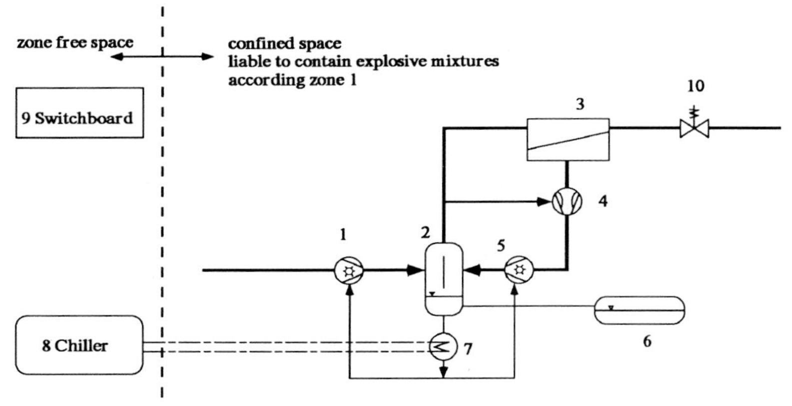

Figure 8: Flow scheme of the toluene recovery unit

Performance test

The performance test of the unit was executed under the following conditions:

- fixed membrane area
- variation of intake flow from 0.2 to 2 m^3/h
- intake concentration 0 %vol Toluene. Toluene saturation of the gas stream after the feed pump by evaporation of service liquid in the pump housing
- condensation temperature 13 °C
- Alternating permeate pressure, 61 mbar with gas ejector, 69.5 mbar without gas ejector.

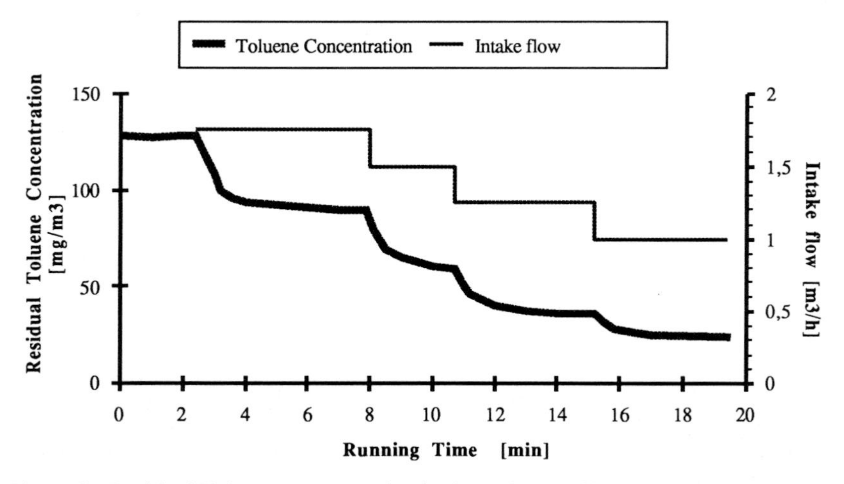

Figure 9: Residual Toluene concentration in dependence of intake flow

Fig. 9 shows that the aim of a toluene concentration in the vent stream < 100 mg/m^3 according to TA-Luft can be met at an intake flow from 0.75 to 1.75 m^3/h. During this test run the pressure ratio was kept constant.

The stage cut, the relation of feed flow to the module to permeate flow, decreases with an increasing intake flow.

The rise of internal circulated flow causes a higher flow velocity over the membrane surface. The recalculated membrane selectivity of toluene vs. nitrogen improves from approx. 45 % of the ideal selectivity at 0.75 m^3/h intake flow to 75 % at 2 m^3/h intake flow (Fig. 10). This selectivity is close to that measured for pure compounds and indicates a dependency of flow velocity and thickness of the boundary layer.

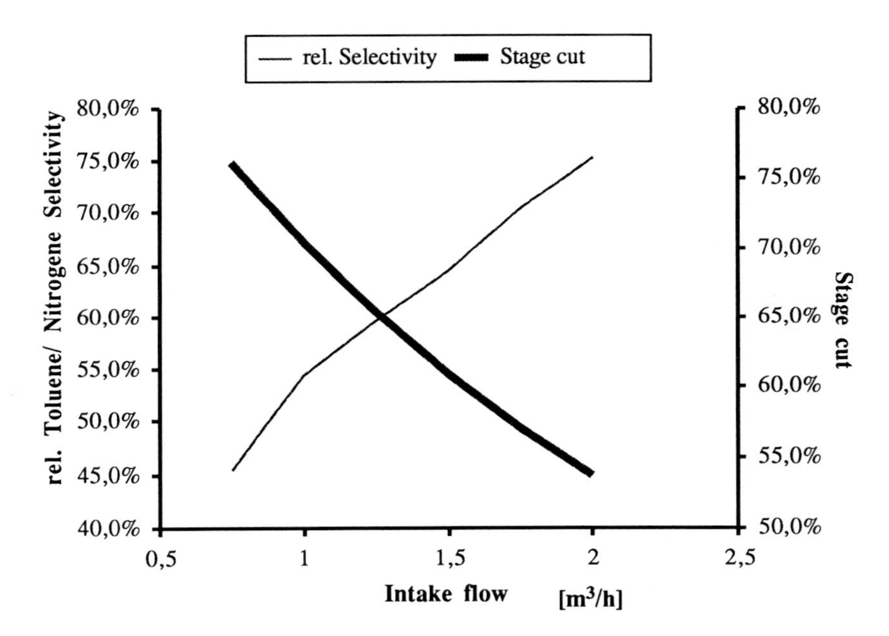

Figure 10: Selectivity and stage cut vs. intake flow

The effect of an alternating pressure ratio is shown in Fig. 11. The permeate pressure switches from 61 mbar to 69.5 mbar by turning on or off the gas ejector. The pressure ratio changes from 17.4 to 15.2 at a constant feed pressure of 1060 mbar. In dependence with the change in vacuum pressure the toluene concentration of the retentate stream changes immediately. The recalculated membrane selectivity varies from 45.5 (ideal selectivity) to 43.4.

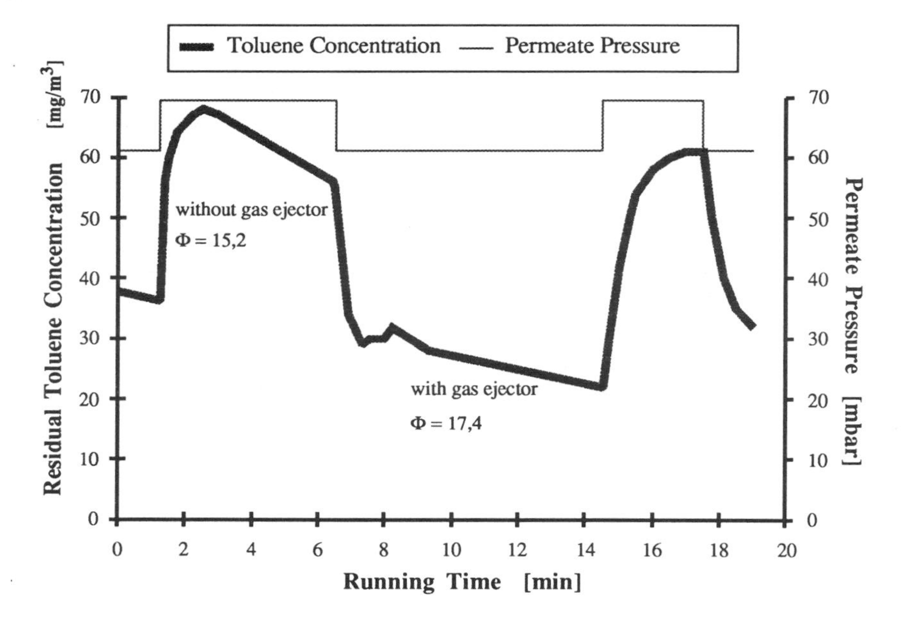

Figure 11: Effect of alternating pressure ratio

Conclusion

It is demonstrated that it is feasible to treat an off-gas contaminated with toluene according to the stringent TA-Luft specifications by means of membranes. Because of the small volume flow and the contamination of the air stream with only one compound (boiling point 110 °C) it is worthwhile to design a single stage unit. Process data and module layout have to be tailored very carefully conforming to the separation problem and the required vent gas purity.

References

[1] K. Ohlrogge; J. Wind; O. Roßhart and B. Schlicht: Membranverfahren in der chemischen und petrochemischen Industrie zur Abtrennung organischer Dämpfe. Aachener Membran-Kolloquium, 9. - 11.3.1993.

[2] R. W. Baker; N. Yoshioka ; J. M. Mohr; A. J. Khan: Separation of organic vapors from air. Journal of Membrane Science, 31 (1987) 259-271.

[3] J. G. Wijmans; V. D. Helm: A membrane system for the separation and recovery of organic vapors from gas streams.AIChE Symposium Series, 85 (1989) 272, 74-79

[4] K. Ohlrogge; J. Brockmüller; J. Wind; R.-D. Behling: Engineering aspects of the plant design to separate volatile hydrocarbons by vapor permeation. Separation Science and Technology, 28 (1-3) (1993) 227-240.

The Performance of a Membrane Vacuum Degasser

TorOve Leiknes and Michael J. Semmens, Ph.D.
Department of Civil Engineering
University of Minnesota
Minneapolis, MN 55455, USA

Charles Gantzer, Ph.D.
Membran Corporation
1037 10th Ave SE
Minneapolis, MN 55414, USA

1.0 Introduction

A study was conducted to evaluate the performance of hollow fiber membranes when used to vacuum degas water. Vacuum degassing may be used in combination with groundwater oxygenation for *in-situ* bio-remediation. Under conditions where oxygen enriched water is required without increasing the total gas pressure of the water, high concentrations of dissolved oxygen can be achieved by degassing a certain percentage of nitrogen and carbon dioxide prior to oxygenating the water. This technology is also useful for the removal VOC's from water. When membranes are treated with selective coatings, volatile organics can be separated from one another and from water; this is the basis of the pervaporation process.

This paper examines the kinetics of degassing and the influence of operating parameters on the performance of a membrane module that is specifically designed to treat high flowrates of water.

2.0 Theory of gas transfer.

When a vacuum is applied to one side of a gas permeable membrane that is immersed in water, a concentration gradient is created across the membrane. Dissolved gases in the water diffuse in the direction of decreasing concentration, through the membrane and into the vacuum. In the case of hydrophobic microporous membranes the interface between the vacuum and liquid phase is provided by the pores. If the pores are very small, $< 0.1\mu m$, the pores remain dry and gas-filled.

The kinetics of gas transfer may be modeled like the two-film theory with the addition of a membrane resistance. The mass transfer of gases across a membrane is therefore determined by three resistances in series as in Figure 1. These consist of a resistance across the liquid film layer created by the aqueous solution around the fibers, the resistance of the membrane, and finally the gas film layer created by the gases within the fiber. The sum of these resistances defines the overall resistance to gas transfer with a membrane system. The reciprocal of the overall resistance is the overall mass transfer coefficient of the system. The mass transfer characteristics and individual mass transfer coefficients for microporous gas-permeable hollow-fibers have been studied and the results indicate that the mass transfer coefficient is always controlled by the resistance in the liquid phase (Yang et al., 1986). This is due to the nature of the hydrophobic microporous membranes. Gas transfer through the membrane is by gas-phase diffusion (10^4 x faster than liquid phase diffusion) and the membrane resistance is negligible. As with aeration systems, the overall mass transfer can thus be estimated by evaluating the liquid film transfer coefficient, k_L.

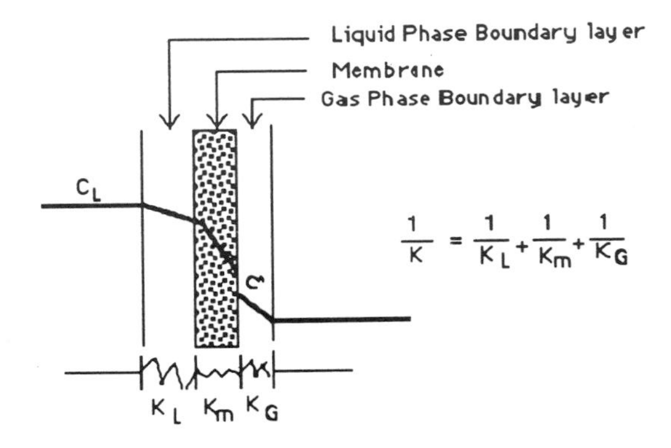

Figure 1 : Mass transfer Resistance in a Membrane System.

For a steady state condition the flux of gases across the membrane may be calculated using the formula:

$$N = K(C - C^*) \tag{1}$$

where:

N - is the gas flux.
K - is the overall mass transfer coefficient.
C - dissolved concentration of the gas in the liquid phase.
C*- concentration of gas in the fiber.

Looking at a single fiber, the gas transfer rate can be described by:

$$v_L \, {}^{dC}/_{dz} = K.a.(C - C^*) \tag{2}$$

where:

a - is the surface area of the fiber
v_L - is the water velocity past the outside of the fiber
z - is the length measured along the fiber axis

If the value of **C*** is assumed to be zero then the equation (2) becomes:

$$v_L \, {}^{dC}/_{dz} = K.a.C \tag{3}$$

The overall mass transfer coefficient can therefore be found by plotting ln **C** vs. time and measuring the slope of line of best fit.

Several investigators have developed mass transfer correlations for aeration using hollow fiber membranes. Since degassing is simply the reverse process of aeration the same correlations can be applied to the vacuum degassing system. The overall mass transfer coefficient may be determined by calculating the Sherwood Number (Sh) based on the correlations outlined in Table 1.

Reference	Correlation	Re Range
Knudsen & Katz	$Sh = 0.022\ Re^{0.6}\ Sc^{0.33}$	-
Yang & Cussler	$Sh = 1.25(Red_e/1)^{0.93}\ Sc^{0.33}$	5-3500
Prasad & Sirkar	$Sh = \beta[d_e(1-\phi)/1]\ Re^{0.6}\ Sc^{0.33}$	-
Cote et. al.	$Sh = 0.61\ Re^{0.363}\ Sc^{0.33}$	0.6-49
Ahmed & Semmens	$Sh = 0.0104\ Re^{0.806}\ Sc^{0.33}$	600-46000

Table 1: Observed dimensionless correlations for mass transfer across hollow fiber membranes in parallel flow with the liquid on the outside of the fibers and gas in the fiber lumen. $Re = {}^{vd_e}/_{\nu}$, $Sc = {}^{\nu}/_{D}$, $Sh = {}^{k_L d_e}/_{\nu}$ where:
d_e= equivalent diameter of the system; **D** = diffusivity of oxygen in water;
ν = kinematic viscosity

3.0 Model Development.

A preliminary model was developed to describe the degassing process. This model was used to predict the expected behavior of the membrane degasser and provide a frame of reference for interpreting the experimental results obtained under different operating conditions. The degassing model was developed in EXCEL and based on the above theory and several simplifying assumptions.

Only oxygen and nitrogen were considered. The overall oxygen mass transfer coefficient (**K**) in the system model was calculated assuming liquid film transfer control. The empirical correlation $Sh=0.018\ Re^{0.83}\ Sc^{0.33}$ (Ahmed and Semmens, 1990) was used to calculate **K**. The predicted effluent dissolved oxygen concentration at a given flow rate at any time interval was calculated using the equation:

$$C = Co.EXP(-a.k_L.\Delta t/V) \quad (4)$$

where

Co= DO concentration at time t_o

C = DO concentration at time t

a = total surface area of the fibers

k_L= mass transfer coefficient

Δt= time interval = $t-t_o$

V = volume of reservoir

The oxygen to nitrogen ratio was assumed to remain constant at 1:4 during the course of degassing and the mass transfer coefficients of oxygen and nitrogen were assumed to be equal. The influence of carbon dioxide and water vapor was neglected. The value of C^* in the fibers was assumed to be zero (i.e. pure vacuum) all along the fiber length.

4.0 Experimental Apparatus and Procedure.

A membrane module of microporous polyethylene was provided for testing by Membran Corporation (1037 10th Ave. SE, Minneapolis, MN, USA.). The membrane module contained 1500 microporous polyethylene fibers measuring 396μm in diameter and 1.55 m in length and provided an overall membrane surface area of $2.89 m^2$ for degassing. The fibers were individually sealed and they occupied only 5% of the crossectional area of the shell in which they were housed. When water was pumped through the module the fibers fluidized in the flow providing excellent contact between the fibers and the water.

The apparatus and equipment used in the study is shown in Figure 2. Water was drawn from a sealed water reservoir and pumped through the membrane module and returned to the reservoir. The performance was monitored by measuring the change in the DO in the reservoir.

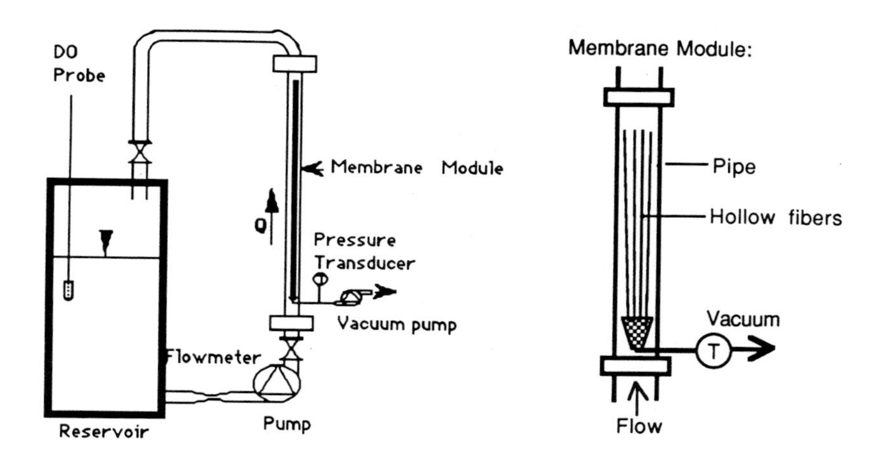

- Glass fiber water reservoir - volume 500 liters.
- PVC piping and fittings - diameter 5 cm.
- Celgard X10 membrane - 2000 fiber module
- MPE, EHF 300G membrane - 1500 fiber module
- Water pump - Oberdorfer Pumps, 3/4 Hp, 3450 rpm, Model # 6K581A
- Flow meter - Signet Scientific Company, Model # P51530-P0
- Vacuum pump - Edwards High Vacuum Pump E2M-1
- Pressure transducer - Contractor Instruments, Model # P8536.
- DO probe -Royce Dissolved Oxygen Analyzer, Model 9040

Figure 2. Experimental apparatus employed in this study.

Before each test the reservoir was aerated to ensure that the water was saturated with air. The reservoir was then sealed with a floating lid. The temperature of the water and air, the dissolved oxygen concentration, barometric pressure and the vacuum pressure were measured at the start of each test. The flowrate of water through the membrane module was then set to a value between 0.3 to 3.0 L/s. The DO was automatically measured every 20 secs, and the data was stored in a file, the vacuum pump was turned on. The vacuum exerted on the system ranged from 75 mtorr in the beginning to 48mtorr at the end of the experiment. The experiment was terminated when the DO dropped to 0.5 mg/l. The run time for the experiments varied from 1.5 to 3.5 hours.

5.0 Experiment Results.

Experiments were conducted at room temperature (23-26°C) to determine the influence of flow rate on the performance of the vacuum degassing configuration. The experimental data were then compared to the model predictions as shown in Figure 3. The experimental data compare well with the predicted behavior at low flow rates. At higher flow rates the experimental data fell below the theoretically predicted performance and the difference increased with increasing flowrate.

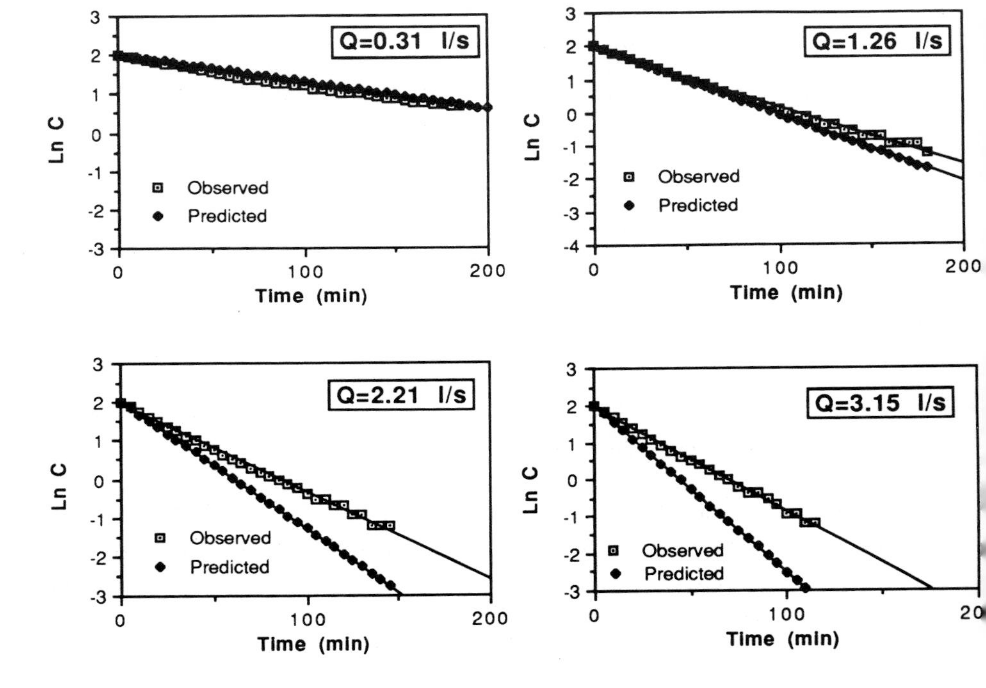

Figure 3: Experimental oxygen removal rate data are compared with the behavior predicted by the theoretical model as a function of flowrate; C*=0, Temperature = 25°C.

The overall mass transfer coefficients (**K**) may calculated from the experimental data as noted above. The predicted data are based on the rate of mass transfer across the liquid film, $\mathbf{k_L}$. Figure 4 compares the predicted values of $\mathbf{k_L}$ and the observed values of **K** and that were calculated by assuming that **C*=0**. The difference between the observed and the predicted coefficients may be explained in one of two ways: 1. the diffusion resistance in the membrane and gas phases are larger than thought (*i.e.* $\mathbf{k_G}$ and $\mathbf{k_M}$ are smaller than expected); or 2. the value of **C*** within the membrane is not equal to zero (*i.e.* the partial pressure of oxygen in the fiber is greater than zero).

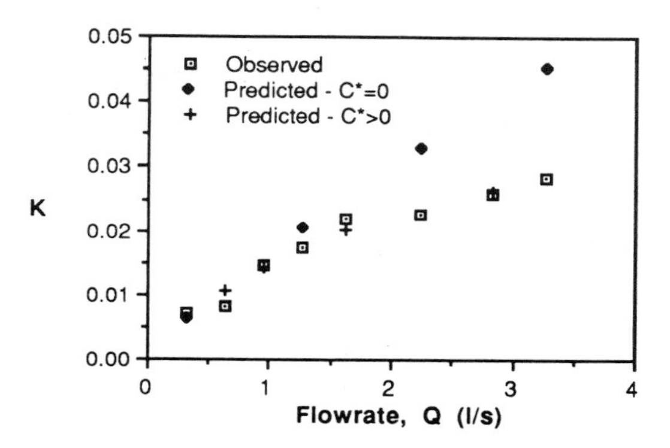

Figure 4 : A Comparison of the observed and predicted overall mass transfer coefficients as a function of water flowrate.

The values of $\mathbf{k_G}$ and $\mathbf{k_M}$ should not be affected by the water flowrate. Although the flux of oxygen and nitrogen across the membrane is faster when the water flowrate is higher, the gas phase diffusion rates are so much faster than rates in the liquid phase that it is very unlikely that there is a significant change in the resistance to mass transfer within the membrane. Since the model predicts the behavior well at low flowrates we conclude that the membrane and gas phase resistance to mass transfer remain negligible over the operating range investigated.

The assumed value of **C*** in the fibers provides a more likely explanation of the observed discrepancy. All the values of K are calculated assuming **C* = 0**. If **C* >0** the apparent value of K will be less than that predicted by the model.

Ahmed et al (1992) showed in studies on the transfer of oxygen into water with sealed hollow-fiber membranes that the effective value of **C*** varied along the length of the fiber as a result of changes in oxygen partial pressure. By modeling the oxygen profile within the fibers Ahmed et al. were able to demonstrate that an average value of **C*** could be used to model the gas transfer from the fiber. The value of **C*** was computed by multiplying the feed pressure of oxygen by a factor. The value of the correction factor was a function of the feed gas pressure and was determined empirically.

In a similar fashion we may reason that **C*** changes along the fiber during vacuum degassing. Furthermore, the absolute value of **C*** will be determined by the flux rates of gases across the membrane which will be strongly influenced by the water flowrate outside of the membrane. Following the approach of Ahmed et al (1992) a correction factor **Ø** may be calculated to provide an average driving force for mass transfer.
If we allow $C^* = Ø.C$ at any time then we may modify equation 3 and write the differential equation for mass transfer in the system as follows:

$$V \,.{}^{dC}/_{dt} = k_L.a.Ø.C \qquad (5)$$

integrating:

$$\ln.(Co/C) = Ø.k_L.a.t \qquad (6)$$

It is apparent from equation (6) that value of the overall mass transfer coefficient calculated from the data presented in Figure 5 is in fact the product $Ø.k_L$. (*i.e.* $K = Ø.k_L$ and therefore the value $Ø=K/k_L$)
The value of the factor was evaluated from the experimental data for the experiments conducted at room temperature and the results are shown in figure 5 as a function of flowrate.

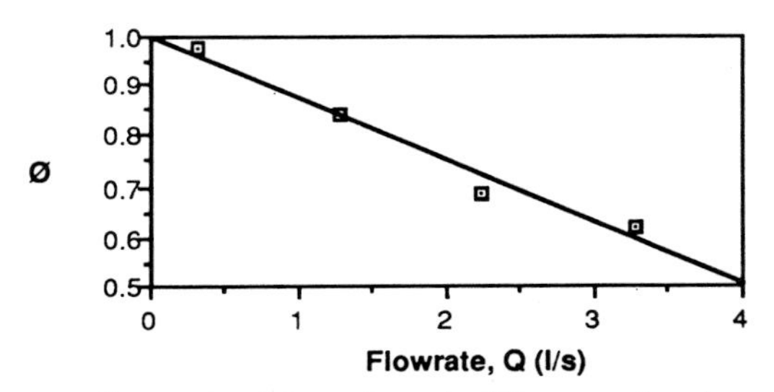

Figure 5: Values of **Ø** calculated from the ratio K/k_L as a function of flowrate.

A least squares linear regression through these data provide the following correlation between the flowrate and the correction factor **Ø**:

$$Ø = 1.00 - 0.124Q$$

When additional experiments were conducted at flowrates within the range the corrected predictions matched the observed data well as shown below in Figure 6.

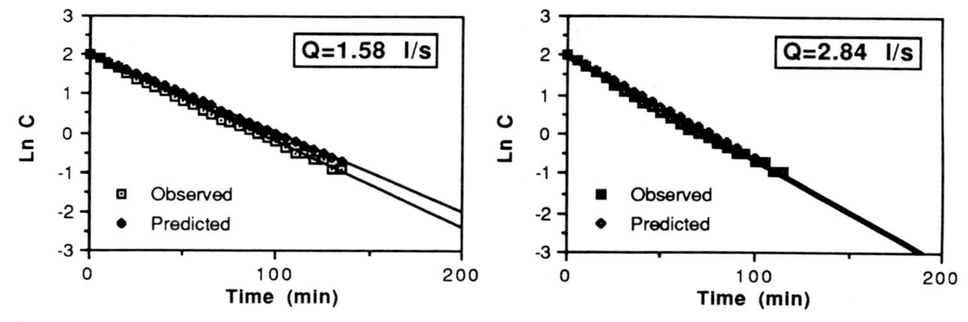

Figure 6: Additional gas transfer data showing the close agreement between the observed data and the model that incorporates the value of **Ø**.

6.0 Discussion.

This paper presents some initial findings on the kinetics of vacuum degassing with microporous hollow fiber membranes as a function of operating conditions. The type of module tested is very different from the more conventional shell and tube type modules in which the fibers are potted at both ends. By using smaller packing densities of fluidized fibers the contact between the fibers and the water can be achieved effectively without a large headloss penalty. These types of membrane modules can be designed to treat flowrates of 0.3 to 30 L/s with headlosses of less than 8 kPa per meter of fiber length. The membrane modules tested gave the expected performance at low flowrates but somewhat poorer performance as the flowrate increased and the residence time in the module decreased.

The degassing performance of the membrane has been analyzed by comparing the observed performance with the behavior predicted by a simple liquid film diffusion control model. This analysis suggests that the assumption that the partial pressure of oxygen inside the hollow fiber membranes (**C***) is zero is incorrect. When the average value of **C*** is assigned a value that is proportional to the external dissolved oxygen concentration, **C**, the proportionality constant **Ø** is found to be a linear function of the water flowrate past the fibers.

This analysis is preliminary and the factors that influence the value of the correction factor Ø have not been fully identified. What is surprising about the findings is the magnitude of the correction. For example, at the highest flowrate tested the corrected value of **C*** was approximately 60% of the dissolved oxygen concentration in the water outside of the fiber; far from zero. During these tests the pressure transducer coupled to the vacuum pump recorded vacuum pressures that were all very similar and very low, 5-7.5 kPa. It seemed therefore that the vacuum pump was adequate for the test apparatus. One would expect differences in the vacuum pressure if the pump was inadequate for the higher gas flux rates expected at higher flowrates. It would appear that at high flowrates the mass flux of gases into the membrane is great enough to cause a significant change in pressure along the length of the fiber. This must be verified experimentally.

In this study we have only examined the influence of water flowrate. However, water temperature and fiber number, length and diameter, are also expected to influence degassing performance. The influence of these parameters will be incorporated in a complete mass balance analysis of gas flow within the fiber. The model will be similar to that developed by Ahmed et al. (1992) for oxygen transfer at higher pressures. Such a model would allow design correlations for vacuum degassing equipment to be developed such that modules could be tailored to suit the application.

7.0 Conclusion.

The use of sealed-end hollow fibers of gas permeable membrane shows great potential for the removal of dissolved gases from water. The use of fluidized fibers ensures good fiber - water contact and effective use of the total fiber surface area. However, to be able to predict the system performance under a variety of operating conditions a more sophisticated mass transfer kinetics model that incorporates gas transport out of the hollow fibers is required. Studies are in progress to develop such a model and to determine the influence of fiber length, surface area, temperature, degree of vacuum and membrane coatings and other operating conditions on the rate of dissolved gas and VOC removal from water.

References:

Yang, M.C. and *Cussler, E.L.*, "Designing Hollow-Fiber Contactors", AIChE Journal, 32(11): 1910-1916, 1986.

Cote, P.L., et al., "Bubble-Free Aeration Using Membranes: Mass Transfer Analysis", J. Membrane Science, in press (1989).

Knudsen, J.G., and *Katz, D.L.*, Fluid Dynamics and Heat Transfer, McGraw Hill, NY (1958)

Prasad, R. and *Sirkar, K.K.*, "Dispersion-Free Solvent Extraction with Microporous Hollow Fiber Modules", AIChE J., 34(2), 177-188 (1988)

Ahmed, T. and *Semmens, M.J.*, "Use of Sealed End Hollow Fibers for Bubbleless Membrane Aeration: Experimental Studies", J. Membrane Science, 69 (1992) 1-10

Ahmed, T. and *Semmens, M.J.*, "The Use of Independently Sealed Microporous Hollow Fiber Membranes for Oxygenation of Water: Model Development", J. Membrane Science, 69 (1992) 11-20

CORRELATIONS BETWEEN STRUCTURAL FACTORS AND PERMEATION PROPERTIES IN POLYIMIDES

M. Escoubes[1], J.Y. Dolveck[1,2], M. Pineri[2], P. Moser[2] and R. Avrillon[3]

[1] Laboratoire Etudes Matériaux Plastiques et Biomatériaux-URA CNRS 507
Université Claude Bernard Lyon I, 69622 Villeurbanne Cedex, France
[2] PCM-CENG, 85X, 38041 Grenoble Cedex, France
[3] IFP, 1 Avenue du Bois Préau, 92500 Reuil Malmaison, France

SYNOPSIS

Diffusion is the main parameter controlling the gas separation properties in polyimides and this parameter depends on both size-concentration of the elementary free volumes and chain movements. This study is focused on the H_2/CH_4 separation. Transport properties are studied in a classical permeation cell. Segment chain mobility is evaluated through NMR and mechanical spectroscopy. Characteristics of free volumes are determined by positron annihilation technique. Three factors are analyzed : chemical structure, physical alterations through annealing and quenching, experimental conditions (for instance water sorption influence). The results allow to conclude that selectivity is due to methane diffusion which is mainly linked to the size and concentration of free volumes. A threshold of 6.3 Å is evidenced.

KEYWORDS : gas permeation and separation, polyimides, free volumes, positron annihilation technique, molecular movements

NOMENCLATURE

A : membrane area :	= 3 cm^2
C_1 : gas concentration on the upstream face of the membrane :	cc_{STP} cm^{-3}
C_2 : gas concentration on the downstream face of the membrane :	cc_{STP} cm^{-3}
D : diffusion coefficient :	cm^2 s^{-1}
e : membrane thickness :	cm
I_3 : long intensity of the positron life time spectrum :	%
J : permeation flux density :	cc_{STP} cm^{-2} s^{-1}
P_1 : gas pressure in the upstream compartment :	cm Hg
P_2 : gas pressure in the downstream compartment :	cm Hg
P_e : permeability coefficient 1 Barrer = 10^{-10} cc_{STP} cm cm^{-2} s^{-1} cm Hg^{-1}	Barrer
Q(t) : quantity of gas that crossed the membrane at time t :	cc_{STP}
S : solubility coefficient :	cc_{STP} cm^{-3} cm Hg^{-1}
t : time :	s
V : downstream volume :	144 cm^3
T : temperature :	K
α : selectivity coefficient :	without dimension
θ : time-lag :	s
τ_1,τ_2,τ_3 : lifetime constants of the positron spectrum : 1 ps = 10^{-12} s	ps
ϕ : free volume diameter	Å

INTRODUCTION

Polyimides are synthesized from diamines and aromatic dianhydrides following a condensation reaction.[1] These materials offer interesting properties : they exhibit a very good resistance to temperature and chemical attacks. Their excellent dielectric and mechanical characteristics provide them with various uses in electronics[2] and mechanics for the making of composite materials.[3,4] For a few years, the membranes obtained from polyimides have been arising a special interest for gas separation because they present a more interesting ratio permeability/selectivity than most of polymers. At the moment, numerous investigations are undergone in order to explain these particular properties.[5,6,7,8]

The solution-diffusion mechanism of gas transport through a dense membrane is based on the following hypothesis : molecules diffuse induced by a concentration gradient in the preexisting (or not) free volumes between polymer chains.[9,10,11] As a result of chain movements, these free volumes may recombine. The chemical structure has a sensible incidence : it can favour or not the emergence of chain segment movements[12,13] and it determines the network compactness as well as the free volumes size. The chemical structure also interferes in solubility that is to say in the gas concentration within the membrane. The aim of this study is to simultaneously increase the membrane permeabilities and selectivities thanks to the choice of an appropriate chemical structure.

Concerning polyimides, selectivity is basically due to diffusion at ambient temperature. Previous studies have demonstrated that permeability increases when the d spacing is improved by a side-group addition in the macromolecular chain.[14] Such an addition does not involve a variation in selectivity.

In this work, the authors propose to correlate the permeability properties of polyimides with the free volumes size and the chain movements.

EXPERIMENTAL

Materials

Three different polyimides have been used for this study, these were prepared from the hexafluorodianhydride 6FDA :

Dianhydride : O O CF_3 CF_3 O O O O 6FDA

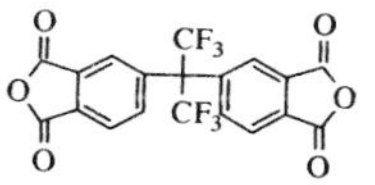

and a monoaromatic diamine which comprises zero, one and three methylated groupements :

Diamines :
A NH_2 NH_2 meta-phenylene diamine : mPDA
B NH_2 CH_3 NH_2 2-6 toluene diamine : 2-6 TDA ou MmPDA
C CH_3 NH_2 CH_3 CH_3 NH_2 tri-methyl-meta-phenylene diamine : tMmPDA

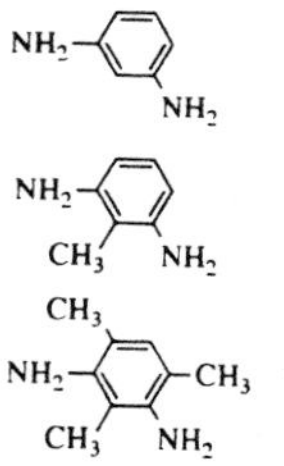

Polymides A, B and C were synthesized in metacresol or 1-methyl 2 pyrrolidone (NMP) at a temperature close to 200°C.

The polymer was precipitated and washed with a non-solvent such as ethanol or water. After being crushed, the obtained powder still contains about 5-6 % residual solvent.

Cast polyimide films were prepared by phase inversion process with NMP or metacresol solvent. Then these were annealed during 4 hours at

180°C or 300°C under vacuum. The films' thicknesses were about 20 µm. The residual solvent content depends on the annealing temperature (Table I).

Table I-Solvent contents of the different polyimide films

Sample	Δm %	
	annealing 180°C	annealing 300°C
A	4.7	1-1.7
B	9	2
C	5	4.5-6

Only sample C which is the most methylated keeps its solvent content after the annealing at 300°C.

An annealing at higher temperature can involve a beginning of degradation. We were brought to study the case of an heating at 400°C during one hour under vacuum and at atmospheric pressure. The weight losses are given in Table II and they show a considerable increase of the degradation for the product C under atmospheric pressure.

Table II-Weight losses of the different polyimides during an heating at 400°C

Sample	Δm % at 400°C during 1 hour	
	under vacuum	under atm press.
A	0,.6	4.3
B	0.7	6.8
C	0.4	19.3

The determination of glass transition temperature was carried out by thermomechanical analysis and DSC and it reveals that Tg significantly increases with the methylation degree (Table III).

Table III-Glass transition temperatures

Polyimide	Tg (°C)
A	300
B	335
C	385

Wide Angle X-ray Scattering (WAXS) shows that films are amorphous ; they nevertheless allow to calculate a value of d-spacing using the Bragg relation at the top of the amorphous bump ; this value also increases with the methylation degree (Table IV) :

Table IV-d-spacing determinated by WAXS

Polyimide	d (Å)
A	5.6
B	6.1
C	6.7

Methods

The correlations study between the polyimides permeation properties and structural characteristics is composed of the following points :

- the measurements of permeability, diffusion and selectivity coefficients in a classical permeation cell ;
- the chain movements can be studied by mechanical or dielectric spectroscopy and by Nuclear Magnetic Resonance (NMR). The mechanical spectroscopy[15,16,17] is particularly interesting because it allows a good resolution of phenomena at low frequency ;
- the determination of free volume size is carried out with the positron annihilation technique PAT.[18,19] This technique allows the measurements of free volumes diameters of a few Å[20] whereas electronic microscopy has a resolution limit of 20 Å.

Permeation measurements

The permeation cell is made up of two compartments (upstream : 1 and downstream : 2) which are separated by the polymeric membrane. The cell is thermostated in a chamber heated at 20°C. A preliminary high vacuum desorption is realized in order to ensure that the static vacuum pressure changes in the dowstream compartment will be much smaller than pressure changes due to permeation : indeed the static vacuum needs to be lower than 10^{-4} torr/min. Then a 3 bars pressure is introduced in the upstream part. The pressure variation in the downstream volume V_2 is measured by a Datametrics pressure sensor and is very low compared to P_1. By plotting the measured pressure P_2 versus t, the study-state line enables the calculation of both the permeability coefficient (P_e) from the slope and the time-lag θ by extrapolation on the time axis.

Indeed, if transport process is governed by Fick's laws, the gas quantity Q(t) (in cm^3 STP) that crossed the membrane (surface A and thickness e) at time t is given by the following equation :

$$Q(t) = \left[\frac{Dt}{e^2} - \frac{1}{6} \frac{2}{\pi^2} \Sigma_1^\infty \frac{(-1)^n}{n^2} \exp\left(\frac{-D\pi^2 n^2 t}{e^2} \right) \right] C_1 \, e \, A$$

when the stationnary state is obtained, this relation can be simplified :

$$Q = \frac{DC_1 A}{e} \left(t - \frac{e^2}{6D}\right)$$

In these relations C_1 and C_2 are respectively the upstream and downstream face concentrations.

$C_1 = SP_1$ (Henry's law where S is the solubility coefficient)

The time-lag θ is obtained for $Q = 0$:

$$\Rightarrow \qquad \theta = \frac{e^2}{6D}$$

The slope $\frac{dQ(t)}{dt}$ (which is proportional to $\frac{dP_2}{dt}$) allows the determination of the permeability coefficient P_e through the flux J :

$$J = P_e \frac{(P_1 - P_2)}{e} \approx \frac{P_1}{e}$$

$$J = \frac{1}{A}\frac{dQ}{dt} \quad \text{with} \quad Q = \frac{V_2 P_2}{76} \quad \text{and} \quad \frac{dQ}{dt} = \frac{V_2}{76}\frac{dP_2}{dt}$$

$$\Rightarrow \qquad P_e = \frac{e}{P_1}\frac{1}{A}\frac{V_2}{76}\frac{dP_2}{dt}$$

The solubility coefficient S can be calculated through the relation $P_e = DS$ assuming a fickian diffusion

$$\Rightarrow \qquad S = \frac{P_e}{D}$$

Molecular movements analysis

The NMR performed on proton and fluorine is carried out at 20°C on a Brücker CXP90 spectrometer which enables to get the spectra on a width of 125 KHz. The characteristic time of NMR is 10^{-4} s, which allows to analyze the influence of molecular movements within a frequency superior to 10^3 Hz ($\frac{1}{2\pi\nu} < 10^{-4}$).

The mechanical spectroscopy of films or plates is achieved thanks to two techniques : Torsion pendulum and Rheometrics. The measurements are carried through between 77 K and 350 K for frequencies around 1 Hertz. For the tg δ maxima, the relation $\omega\tau = 1$ provides us with the relaxation time t of the movement at the given temperature. It is possible to extrapolate this value through the activation energy E_a in order to obtain the relaxation time at 20°C which is the permeation measurements temperature.

Free volumes analysis by positron annihilation technique

The positron was born from the ^{22}Na disintegration. The lifetime between birth and annihilation characterizes the material. The spectrum acquisition needs 10^6 lifetime countings.

In the polymeric materials, the positron can either be annihilated right away by an electron or be linked to this electron in a free volume in order to form the positronium (ortho or para according to the parallelism or antiparallelism of the spins). These two positronium are annihilated in a different way. They are therefore three annihilation mechanisms and the spectrum can be divided into three exponential components :

$$N(t) = I_1 \exp{-t/\tau_1} + I_2 \exp{-t/\tau_2} + I_3 \exp{-t/\tau_3} + Bg$$

I_i are the intensities ; τ_i are the lifetime constants ; Bg is the background noise.

The lifetime τ_1 of the parapositronium is very short (125 ps).

The lifetime τ_2 of the free positron varies between 200 and 500 ps.

The lifetime τ_3 of the orthopositronium depends on the size of the free volume in which it is trapped between 10^3 and 10^4 ps. Eldrup[21] showed that there is a linear relation between this lifetime and the holes volumes with diameters between 3 and 15 Å. The intensity I_3 corresponding to τ_3 characterizes the concentration in free volumes and the product $I_3\tau_3$ can be related to the total free volumes.

The samples were dimensioned to 5x7 mm and piled up on each side of the source according to the two following processes : i) process 1 : the piling up is 0.6 mm wide : all the positrons are absorbed - ii) process 2 : the piling up is 0.2 mm wide with a golden sheet reflecting the unabsorbed positrons. Nevertheless, the measured intensities might be lower than in the first process. The τ_3 value should not be wrong. The spectrum studying is obtained thanks to the «Positron FIT» program.

RESULTS AND DISCUSSION

Three parameters are likely to have an effect on molecular movements, on free volumes, and consequently on transport properties :

- The chemical structure, hence the choice of the 3 polyimides
- The physical alteration through annealing processes around Tg, followed by a rapid quenching. Two processes were tested : i) 330°C during 1 hour in N_2 with quenching in liquid argon - ii) 400°C during 1 hour in air with quenching in iced water.
- The experimental conditions in particular the water influence.

Gas permeation

The figure 1 exhibits the typical curves of $P_e = f(t)$. For methane, the time-lag allows a fine calculation of D. For H_2, the time-lag is always very short (< 15 s) and it can only be ensured, according to the films thicknesses, that D is $\geq 10^{-7}$ cm^2/s.

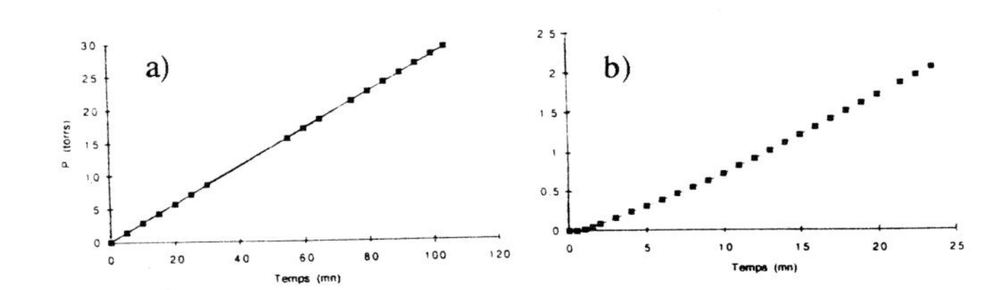

Figure 1-Typical permeation curves for H_2 (a) and CH_4 (b)

The following Table V displays the extreme values obtained on a large range of samples of the 3 polyimides.

Table V-Permeation characteristics on initial films

Samples	Pe (H_2)	P_e (CH_4)	D(CH_4).10^{-10}	S(CH_4)	$\alpha(H_2/CH_4)$
A	25-31	0,11-0.18	2.7-3.3	0.04-0.067	167-280
B	66	0.39-0.43	6.5-7.8	0.055-0,060	151-166
C	70-180	2-18	23-95	0.08-0.18	13-40

The permeability and diffusion coefficients increase with the degree of methylation (a lower homogeneity is observed in the trimethylated samples C). The C selectivity is by far inferior to the ones of A and B (factor > 10). This selectivity is mainly due to the methane diffusion factor.

The annealing and quenching effects are given by the values in the Table VI which are to be compared with those of Table V.

Table VI-Permeation characteristics on treated films

Samples	Pe (H_2)	P_e (CH_4)	D(CH_4).10^{-10}	$\alpha(H_2/CH_4)$
A treated 330°C	50	0.33	3	150
A treated 400°C	55	0.28	4.4	195
B treated 400°C	480	150	≥ 1000	3.2
C treated 330°C	400	23	140	17
C treated 400°C	800	180	≥ 1000	4.6

For the non methylated samples A, the characteristics, in particular the H_2/CH_4 selectivity remain unchanged whereas the 2 heating processes are above Tg. For both B and C samples, the 400°C annealing leads to a downfall in selectivity. The characteristics of B move then towards those of C.

The water influence on the gas permeation is studied by equilibrating the samples with different water partial pressures before the permeation experiments. The permeability Pe decreases linearly with the water sorption for both gases (Fig. 2).

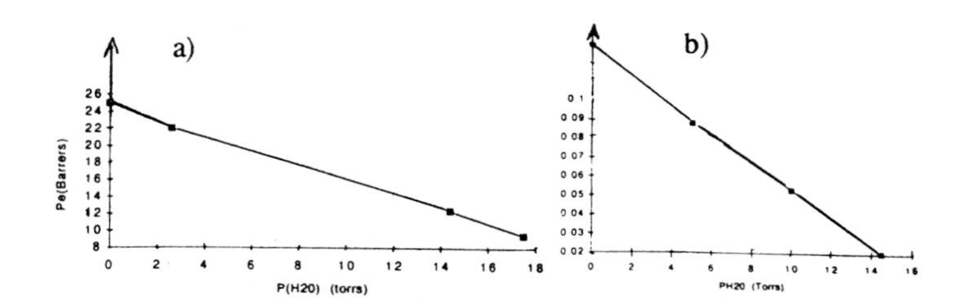

Figure 2-Influence of the water vapor pressure on H_2 permeabilities (a) and methane permeabilities (b)

Molecular mobility

The preceeding diffusion coefficients D enable us, by means of equation $D = \frac{1}{6}\nu\lambda^2$, to calculate the jump frequency ν (or jumptime $\tau = \frac{1}{2\pi\nu}$) assuming as known from the literature the jump distance λ :

For $CH_4 : 10^{-10} < D < 10^{-8}\ cm^2/s \rightarrow 10^{-6} < \tau < 10^{-4}\ s \quad (\lambda \cong 30\ Å)$

For $H_2 : D > 10^{-7}\ cm^2/s \rightarrow \tau < 10^{-8}\ s \quad (\lambda \cong 10\ Å)$

In NMR analysis, the figure 3 shows that the half width of the fluorine band remains the same for the 3 samples whereas the proton half width decreases from A to C. Simultaneously the permeability coefficients rise as previously described (Table V). Owing to the fact that the NMR characteristic time is in agreement with the methane jumptime, it seems possible to consider that the local movements of the methyl groups on the diamine play a role in the methane diffusion.

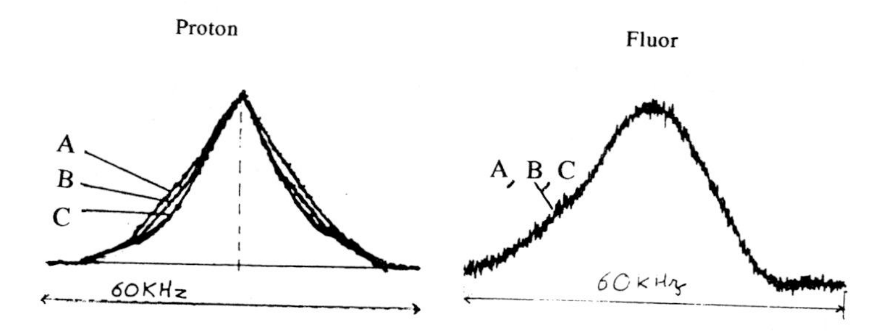

Figure 3-Proton and fluorine NMR experiments on the non (A) mono (B) and tri (C) methylated polyimides

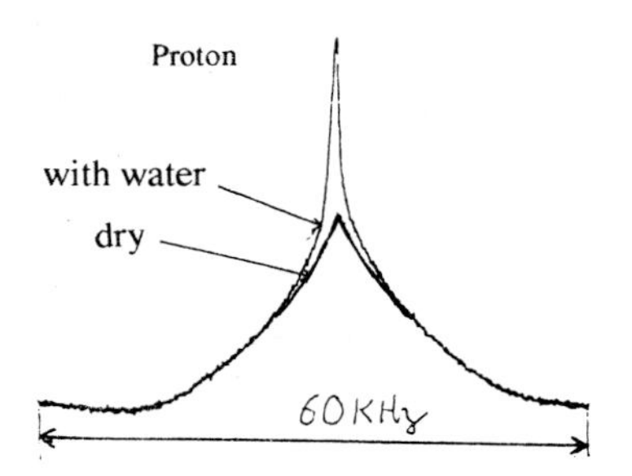

Figure 4-NMR spectra of dry and hydrated monomethylated polyimide

According to the figure 4, the presence of water does not modify the proton band width (a thin band is simply superposed on the large one without modifying the latter) whereas there is a linear decrease in the gas permeation (Figure 2). On this subject matter, the preceeding correlation is not confirmed.

In internal friction analysis (torsion pendulum and Rheometrics), the figure 5 displays 2 relaxation peaks : γ near 150 K due to water and ß near 270 K due to the residual solvent.

Rheometrics : film (< 5% solvent) Pendulum : sheet (>10% solvent)

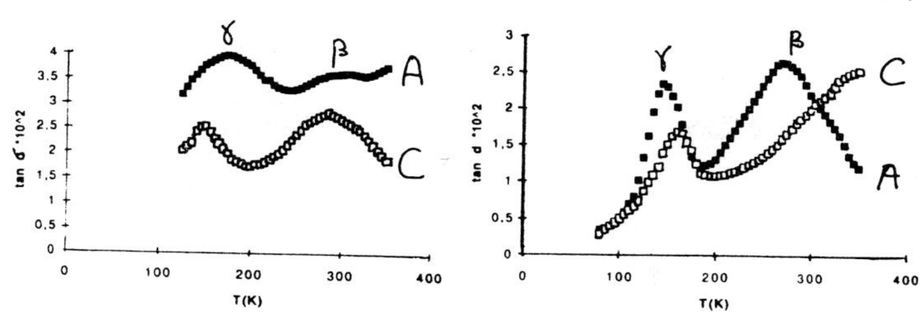

Figure 5-Internal friction analysis on non (A) and tri (C) methylated polyimides

Once extrapolated to 293 K (E_a = 50 kJ/mole), the relaxation time of the γ peak is around 10^{-9} s and a correlation can be assumed with the transport properties. After dehydration, though, the γ peak disappears and is replaced by a smaller one around 130 K (antiplasticizing effect of water). The permeability coefficients are yet higher in dry state. The molecular movements evidenced by the γ peak and the transport properties are therefore not correlated.

Free volumes

The PAT results concerning the orthopositronium lifetime data on the as cast films are given in the Table VII. (the two piling up processes are distinguished for the I_3 values ; the diameter ϕ is calculated from the τ_3 = f(volume) curve of Eldrup assuming a spherical hole).

Table VII-P.A.T. data on initial films

Samples	I_3 %		τ_3 (ps) ± 100	ϕ (Å) ± 0.2
	process 1	process 2		
A	5.7-6	3.8	2100-2400	5,4-6.1
B		3	1836	5.1
C	17.9-18.4	12	3150-3650	7.3-8.1

Concentration and size of free volumes increase in proportion with the degree of methylation. It is to be noticed that our τ_3 values are very close to the ones given by Tanaka and coll.[19]

It is shown on the figure 6 that I_3 and τ_3 are temperature sensitive in the glassy state.

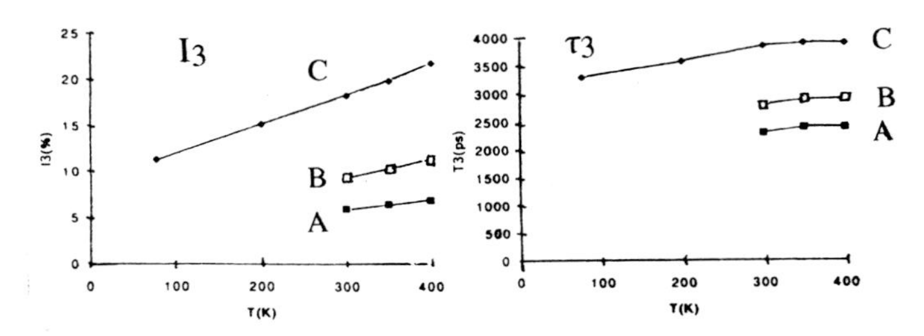

Figure 6-Variations of I_3 and τ_3 with temperature

After the annealing treatments, the Table VIII indicates the variations observed on a given sample under the same measurement conditions.

Table VIII-P.A.T. data on treated films

Samples	I_3 %		ϕ (Å)	
	before	after	before	after
A annealing 330°C	3.8	5	6	6.3
A annealing 400°C	5.7	5.9	5.9	5.9
B annealing400°C	3	2.4	5.1	5.1
C annealing 330°C	18.4	20.5	8	8.3
C*annealing 400°C	12.1	3.5	7.3	5,8

* *important damaning (weight loss : 19.3 %)*

The 330°C annealing treatment results in a slight increase in I_3 and ϕ, whatever the methylation degree may be. The 400°C annealing treatment induces the same effects on the non methylated sample A. On the opposite the same treatment applied to B and C reveals a reduction in I3 and even in the mean diameter ϕ

According to the figure 7, I_3 and τ_3 decrease with the water content (dm/m %). This might bring to the conclusion that water partially fills up or blocks the free volumes. Considering that water can be viewed as a poison for positronium, the τ_3 and ϕ values of the hydrated samples cannot be properly used.

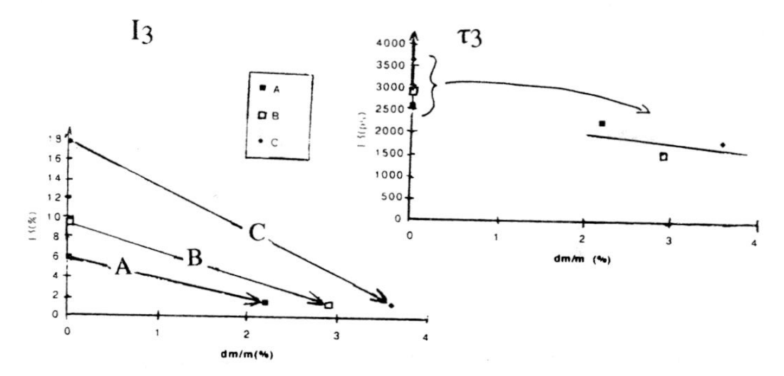

Figure 7- Variations of I_3 and τ_3 with water content

From all these results, it is possible to settle some interesting correlations between the free volume diameters and the gas permeation outputs. The following table groups together the two series of data obtained on the same samples (as cast or annealed without degradation).

Table IX-Comparison of permeation and P.A.T. data

Samples	Permeation			Free volumes
	P_e (H_2)	P_e (CH_4)	α (H_2/CH_4)	ϕ (Å)
A	25-30	0.11-0.18	165-280	5,4-6.1
A annealed 330°C	50	0.33	150	6.2
A annealed 400°C	55	0.28	195	5.9
B	66	0.39-0,43	150-165	5.1
C	70-213	2-18	12-40	6.4-8.1
C annealed 330°C	400	23	17	7.7

The permeabilities of both gases increase in proportion with the average size of the elementary free volumes, be the variations due to the methylation degree on the diamine (A to C) or to the annealing treatments. More precisely, the methane permeability jumps from values inferior to 0.5 Barrers to values superior to 2 Barrers when the ϕ value exceeds 6.3 Å. This jump involves the sudden drop in selectivity from values superior to 150 to values inferior to 40.

The figure 8 illustrates the 6.3 threshold which appears to be the diffusion threshold of the methane.

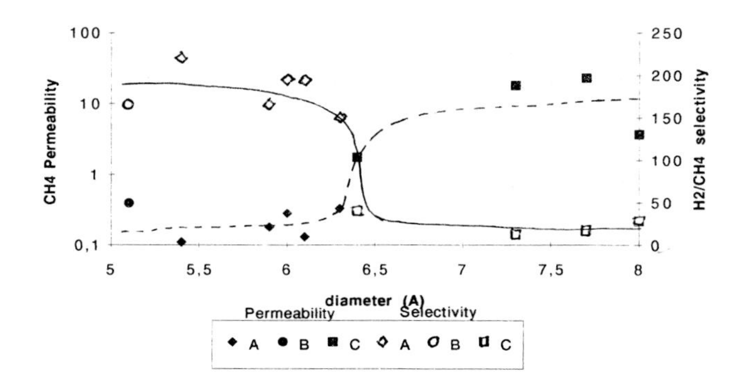

Figure 8-Methane permeabilities and H_2/CH_4 selectivities versus the mean size of free volumes for non (A) mono (B) and tri (C) methylated polyimides

CONCLUSION

The studied polyimides exhibit very different transport properties for slight structural variations. The H_2/CH_4 selectivity is much higher for the non or mono methylated materials than for the trimethylated. Such a selectivity is mainly due to the differences in the methane permeabilities and, within these permeabilities, the diffusion factor appears to be the most important one.

It was shown that the various transport properties did not easily correlate with the molecular movements as observed through NMR and internal friction analyses. On the opposite, satisfying correlations were made between the transport properties and the free volumes characteristics that were studied by PAT. The most important outcome emphasizes a methane diffusion threshold connected with an elementary hole diameter of 6.3 Å. The non and mono methylated polyimides offer a distribution of holes which is mainly positionned under the threshold. Only the non methylated one preverses its hole distribution after an above Tg heating followed by a rapid quenching. On the contrary, there is always a sufficient amount of free volumes around 6.3 Å in the trimethylated samples to account for the low H2/CH4 selectivities and the different possible distributions in size from one sample to another can explain the wider dispersion in the permeation values.

The presence of water was proved to obstruct or partially fill up the free volumes : a reduction in the concentration is evidenced by positron annihilation technique and a linear decrease in permeabilities of both gases is also emphasized through transport experiments.

REFERENCES

1 : P..MICHAUD, Y. CAMBERLIN, M. BALME
European technical symposium on polyimides,STEPI, Edited by J.M. Abadie and B Sillion, Vol 1, A-1/1 (1989).

2 : Craig C. SCHUCKERT, G.B. FOX, B.T. MERRIMAN
European technical symposium on polyimides,STEPI, Edited by J.M. Abadie and B Sillion, Vol 1, B-3/1 (1989).

3 : J.Van.HELMOND
European technical symposium on polyimides,STEPI, Edited by J.M. Abadie and B Sillion, Vol 1, C-6/1 (1989).

4 : D. WILSON
High Performance Polymers, 3, N° 2, p 73-87, (1991)

5 : A. DESCHAMPS, A. DRIANCOURT, J.C. MILEO
in "Polyimides and other high température Polymers" Edited by M.J.M. ABADIE and B. SILLION. Elsevier Scince Publishers B V Amsterdam, p 497-506, (1991)

6 : S.A. STERN, U. MI, M. YAMAMOTO and A.K. St CLAIR
Journal of Polymer Science, Part B : Polymer Physics, 27, p 1887-1909, 28, p 2291-2304 (1989).

7 : T.H. KIM, W.J. KOROS, G.R. HUSK K.C. O'BRIEN
Journal of Applied Polymer Science, 34, p 1767-1771 (1987)

8 : R.T. CHERN, W.J. KOROS, M.B. HOPFENBERG, V.T. STANNETT
In Material Science of synthetic Membranes, by D.R. Lloyd Editor, American Chemical Society Symposium Series, 269, p 25-46, (1985)

9 : G.A. KUMINS and T.K. KWEI
J. Crank and G.S. Park -,"Diffusion in Polymers", New York University Press p 107-140,(1968)

10 : S.A. STERN and H.L. FRISCH
Ann. Rev. Mater. Sci., 11, p 523-550, (1981)

11 : R.J. PACE and A. DATYNER
Polymer Engineering and Science, 20, N°1, p.51, (1980)

12 : T.H. KIM, W.J. KOROS, G.R. HUSK and K.C. O'BRIEN
Journal of Membrane Science, 37, p 45-62. (1988)

13 : T H KIM, W.J. KOROS
Journal of Membrane Science, 46, p 43-56 (1989)

14 : W.J. KOROS, G.K. FLEMING, S.M. JORDAN, T.H. KIM.and H.H. HOEHN
Prog. in Polym. Sci., 13, (1988)

15 : M. PINERI
Polymer, 16, P. 595-600.(1975)

16 : F. BIDDLESTONE, A.A. GOODWIN, J.N. HAY and G.A.C. MOULEDOUS
Polymer,V. 32, N° 17, p 3119-3125.(1991)

17 : Z. SUN, L. DONG, Y. ZHUANG, L. CAO, M. DING and Z. FENG
Polymer, vol 33, N°2, p 4728-4731 (1992)

18 : V.V. VOLKOV
Polymer journal, 23, N°5, p 457-466 (1991)

19 : K. TANAKA, M KATSUBE, K.-I OKAMOTO, H. KITA, O. SUEKA and.Y. ITO
The chemical society of Japan, 65, p 1891-1897, (1992)
20 : J. C. ABBE, G. DUPLÂTRE and J. SERNA
International Conference on Positron Annihilation, p 796-798 (1988)
21 : M. ELDRUP
Annalytical Chemistry Fr, 10, p. 681-692.(1985)

MANUFACTURING

Polyetherketones - New, High-Performance Polymers for Ultrafiltration Membranes [1]

H.M. Colquhoun, K. Roberts, A.F. Simpson, and T.M.C. Taylor
North West Water Group PLC, P.O. Box 8, The Heath, Runcorn, WA7 4QD, UK

G.C. East, J. E. McIntyre, and V. Rogers
Department of Textile Industries, The University, Leeds, LS2 9JT, UK

Summary

Polyetherketones such as PEK and PEKEKK (below) are semi-crystalline aromatic thermoplastics with outstanding thermal and hydrolytic stability, resistance to oxidative degradation, and ability to withstand contact with organic solvents. Indeed, because of their crystallinity they are insoluble in conventional organic solvents at ambient temperatures. Fabrication of ultrafiltration membranes from this type of polymer would clearly be of great potential value, but their lack of solubility means that membrane formation *via* the normal organic / aqueous phase-inversion process is impracticable.

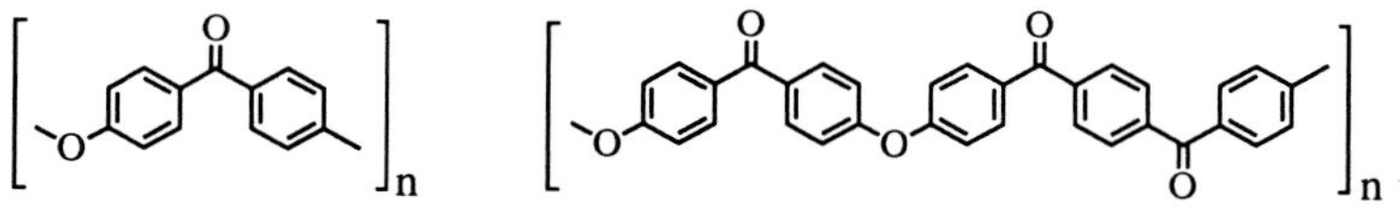

Polyetherketones (PEK and PEKEKK)

These polymers are however soluble without degradation in very strong acids such as liquid hydrogen fluoride, concentrated sulphuric acid, and trifluoromethanesulphonic acid, which dissolve the polymer by protonation of the carbonyl groups. We have now shown that sulphuric acid solutions of such polymers can be used to fabricate UF membranes in both flat-sheet and hollow-fibre configurations. The properties of these membranes and their potential for aqueous and non-aqueous molecular separations are discussed.

Introduction

Semi-crystalline aromatic polyetherketones show considerable promise as high-performance engineering polymers because of their unique combination of mechanical toughness, high modulus, thermo-oxidative and hydrolytic stability, resistance to organic solvents, and retention of physical properties at very high temperatures (up to 250°C).[2] A number of such polymers, shown below, are currently available either as fully commercial materials or as development-products. They are, in effect, crystalline analogues of the (amorphous) aromatic polyethersulphones which have proved so successful in the past as materials for ultrafiltration membranes.[3]

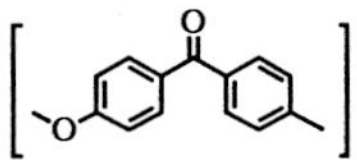

PEK (*ICI*)

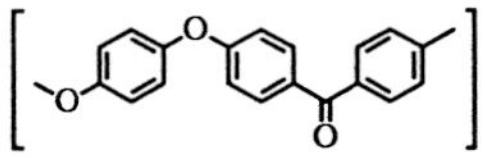

PEEK (*ICI*)

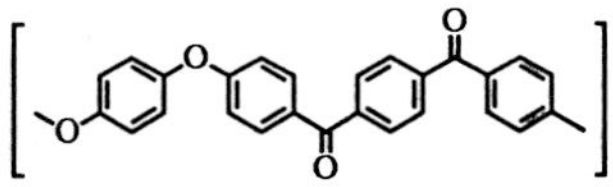

PEEKK (*Hoechst*)

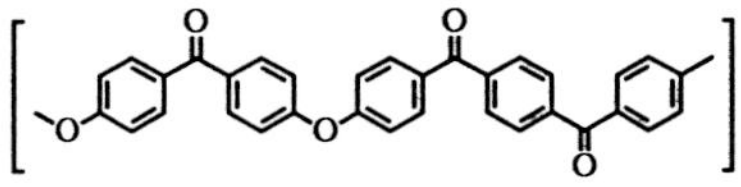

PEKEKK (*BASF*)

Figure 1. Currently available poly(aryletherketones)

The crystalline regions of polyetherketones behave essentially as reversible, physical cross-links between the polymer chains in the amorphous phase. Even though the achievable weight-fraction of crystalline material in such polymers is rarely more than 40%, the crystalline regions restrict the mobility of chains in the amorphous phase and lead, overall, to enhanced mechanical strength, creep-resistance, and ability to withstand contact with aggressive organic solvents. Crystalline polyaryletherketones normally melt only at very high temperatures (typically 350 - 400° C), and dissolve only in solvents which interact *chemically* with the carbonyl groups and so provide a powerful thermodynamic driving force for dissolution. Fabrication of high-stability ultrafiltration membranes from such polymers would clearly be of great potential value, but their lack of solubility means that membrane formation *via* the normal organic solvent / aqueous coagulant (phase-inversion) process is not feasible.[4]

Aromatic polyetherketones are however soluble without degradation in certain very strong acids such as liquid hydrogen fluoride and trifluoromethanesulphonic acid, which dissolve the polymer by protonation of the carbonyl groups. This type of polymer will also dissolve in concentrated sulphuric acid, by far the cheapest and least hazardous strong-acid solvent available, but polyetherketones such as PEEK, which contain the hydroquinone residue, are susceptible to progressive, if slow, sulphonation in sulphuric acid (Figure 2), leading eventually to highly sulphonated non-crystalline polymer.[5] Ultrafiltration membranes cast from sulphuric acid solutions containing sulphonated PEEK have been reported,[6] but such membranes, as expected, exhibit only low levels of polymer crystallinity.[1]

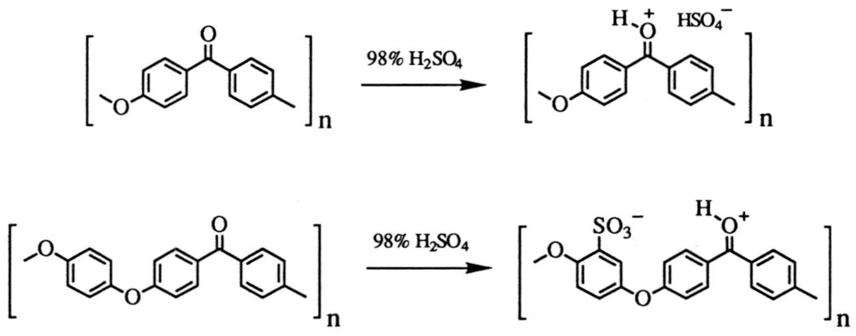

98% H_2SO_4

98% H_2SO_4

Figure 2. Effect of dissolution in concentrated sulphuric acid on PEK (fast, reversible protonation) and on PEEK (protonation *and* slow, irreversible sulphonation)

On the other hand, PEK and PEKEKK are readily soluble in 98% sulphuric acid *without* sulphonation, since every aromatic ring in these polymers is deactivated towards electrophilic substitution by the presence of a (protonated) carbonyl group (Figure 2). We therefore set out to investigate whether such sulphuric acid solutions of crystallisable polyetherketones can be used to fabricate polymeric UF membranes with extreme chemical stability and resistance to physical degradation.

Results and Discussion

A. Flat-sheet membranes

Exploratory casting work was carried out using solutions containing 5 - 20 wt% of PEK (ICI *Victrex-PEK,* grade 220G) in 98% sulphuric acid, which were hand-cast onto glass plates in an atmosphere of dry nitrogen using a stainless steel doctor-blade, set to give a film thickness of 200 μm. After gelation in water, the resulting asymmetric membranes were evaluated in terms of pure-water permeability and polymer crystallinity.

From these initial studies it was established that the optimum range of PEK concentration for membrane fabrication is approximately 7 - 14 wt%. Above 15 wt% polymer the water-flux of the membrane is very low, and below 6 wt% the mechanical integrity of the membrane is poor. As with UF membranes produced from other hydrophobic polymers, it is difficult to restore the flux of an asymmetric PEK membrane once it has been allowed to dry out.

Evaluation of polymer crystallinity in these membranes showed, interestingly, that although PEK and PEKEKK are known to crystallise readily from the melt, gelation from solution gives material of rather low crystallinity (ca. 15 - 20%). A proprietary post-treatment has however been developed which promotes further crystallisation (up to ca. 40%) and thus stabilises the membrane.[7] This treatment normally reduces the water flux by a factor of 2 to 3, and the molecular weight cut-off (MWCO) is also significantly lowered (see below), both results indicating that the polymer-densification resulting from crystallisation acts to reduce the effective pore-size of the membrane.

In Figure 3, the upper DSC trace (heating scan at 10°C min^{-1}) is from a typical 'as-made' PEK membrane. The glass transition temperature (T_g) of PEK at ca. 156° C is followed immediately by a strong 'cold-crystallisation' exotherm, and finally by an endotherm at the melting point of the polymer. The difference in magnitude between endotherm and exotherm represents the heat of fusion of crystallites in the as-made membrane, from which the degree of crystallinity of the polymer can readily be calculated.[8] In contrast, DSC analysis of a post-crystallised membrane (lower trace) shows no evidence of a cold-crystallisation exotherm , confirming that the polymer has already reached its maximum crystallinity. Membranes made from PEKEKK behave in very similar fashion.

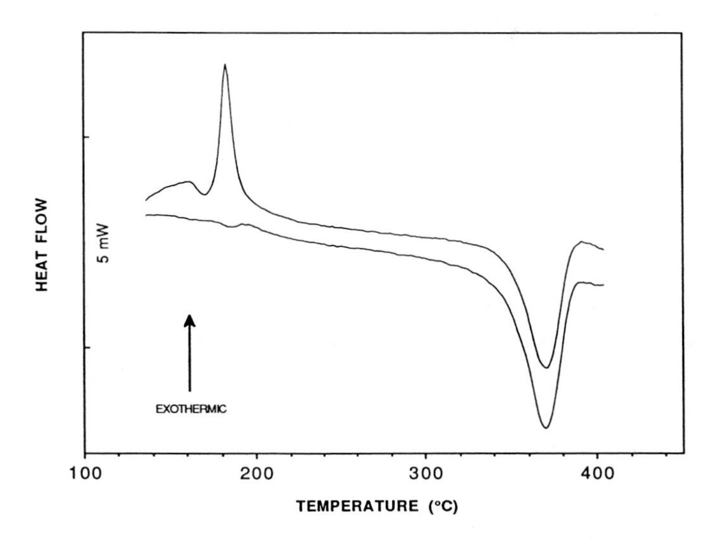

Figure 3. DSC traces for PEK membrane "as-made" (upper trace) and "crystallised" (lower trace)

A series of fibre-reinforced membranes for more detailed evaluation was prepared by casting PEK solutions in 98% sulphuric acid onto various non-woven backing-papers [polyester, polyethylene, glass fibre, and polyphenylenesulphide (PPS)]. The most satisfactory results were obtained using thermally bonded *Ryton* ™ PPS paper from Freudenberg. This paper was not appreciably attacked by concentrated sulphuric acid and was evenly wetted by the casting solution. Such PPS-reinforced membranes were characterised in terms of their morphology (by scanning electron microscopy), crystallinity (by DSC) , water-flux at 2 bar, and MWCO using the dextran challenge method.[9]

Reinforced membranes were obtained by casting 250μm films of PEK solutions in concentrated sulphuric acid onto PPS paper and gelling the membranes in water at 20°C. Scanning electron microscopy revealed good penetration into the backing-paper, a substantially macrovoided membrane morphology and a uniform, well-defined surface-skin some 0.15 μm in thickness. A typical membrane 'as-made' had a water flux of 200 $l/m^2/hr$ at 2 bar, and a MWCO of 15 kilodaltons (15k). After crystallisation, these values were reduced to 70 $l/m^2/hr$ and 5k respectively. Steam-sterilisation of the membrane at 121° C led to no deterioration of performance. The performance of the membrane was similarly unaffected by successive permeation at 2 bar pressure and 20°C of a) acetone, b) toluene, and c) perchloroethylene (an hour each), solvents which would lead to immediate failure of, for example, a polysulphone-based membrane. Not even more chemically-tolerant polymer membranes such as those made from polyvinylidene fluoride would survive this type of solvent treatment.

It was by now clear that ultrafiltration membranes with useful flux and selectivity could be successfully fabricated from crystalline aromatic polyetherketones such as PEK and PEKEKK, and that moreover the exceptional durability of this type of polymer was reflected in the properties of the membranes. We therefore next investigated the production of a polyetherketone *hollow-fibre* membrane since, in operation, this configuration minimises the number of system components in contact with aggressive feeds.

B. Hollow-fibre membranes

After extensive optimisation of process parameters, double-skinned, hollow-fibre membranes (0.4 - 0.7 mm internal diameter) were successfully spun from solutions of PEK and PEKEKK in 98% sulphuric acid, using aqueous coagulants. The dark red, viscous, spinning solution (typical rotational viscosity ca. 100 poise) was degassed under vacuum for 30 minutes and then spun continuously as hollow fibre using a corrosion-resistant, stainless steel, tube-in-hole spinneret. After passing through the coagulation bath, the hollow fibre was taken up on an electrically-driven drum before being washed acid-free and crystallised. Fibres were potted in nylon tubing and evaluated in terms of water-flux, MWCO (dextran challenge) and burst pressure.

Although membranes with adequate water-fluxes and useful molecular weight cut-offs (30 - 80k) are obtained from the above process, various proprietary additives may also be incorporated into the spinning solution and/or the coagulants, and these can markedly alter the membrane's performance. Water-fluxes are then typically 50 - 150 $l/m^2/hr$ at 2 bar, for MWCO's of 5 - 15k. Molecular weight distributions (GPC) for a 0.5% aqueous mixed-dextran feed and permeate from a crystallised PEK membrane with a nominal MWCO of 10k are shown in Figure 4. In the permeate trace the relatively sharp transition from rejection to transmission suggests a narrow distribution of effective pore-size.

Crystallisation of a hollow-fibre PEK membrane also produces a dramatic enhancement of mechanical strength, leading to a typical increase in burst pressure from 8 bar for the as-made fibre to 12 bar after crystallisation.

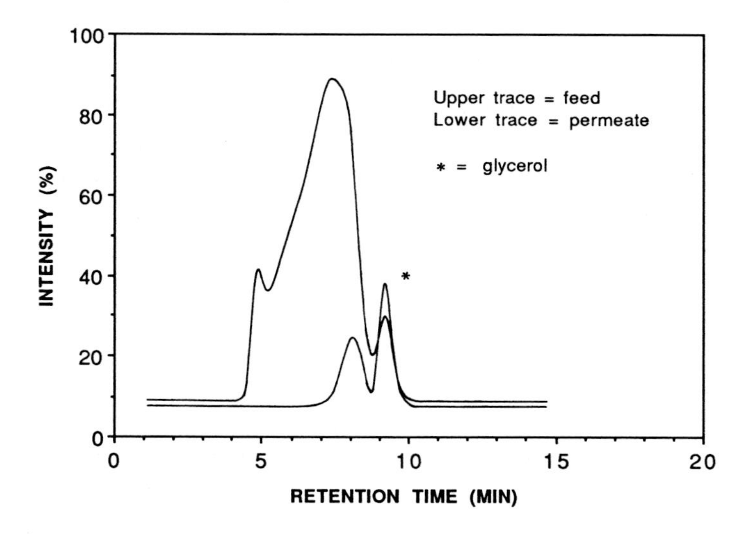

Figure 4. GPC traces for a feed solution comprising 0.5% mixed-MW dextrans (10k - 5,000k) in aqueous phosphate buffer, and for the permeate from ultrafiltration of this solution with a nominal 10k MWCO PEK membrane.

A PEK membrane (MWCO 12k) was also used to fractionate a mixture of polystyrenes (nominal MW's 4k, 12k, 24k, 32k and 50k) in toluene solution (0.5% initial polymer concentration). GPC traces for the feed and permeate are shown in Figure 5, which indicates essentially 100% retention of polystyrene with MW greater than 24k, 97% retention of the 12k fraction and 65% retention of 4k material. With pure toluene as feed a steady flux of 120 $l/m^2/hr$ (1 bar) was observed, and with toluene containing 0.5% polystyrene the flux was 100 $l/m^2/hr$, falling to 80 $l/m^2/hr$ as the polystyrene concentration rose to ca. 1%. Although the membrane is visibly swollen by solvents such as toluene, the effect appears to be fully reversible, so that as the feed is exchanged from toluene to acetone and finally to water, the fibres rapidly revert to their original dimensions. At least in the short term (hours), fibre-permeability and -selectivity are not appreciably affected by such solvent contact. Lifetime solvent-tolerance tests are currently in progress.

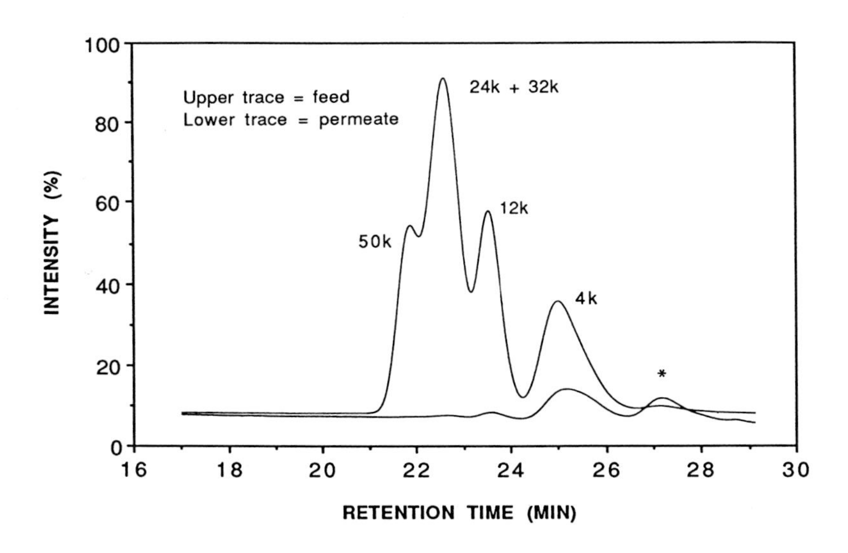

Figure 5. GPC traces for a feed solution comprising 0.5% mixed-MW polystyrenes in toluene, and for the permeate from ultrafiltration of this solution with a nominal 12k MWCO PEK membrane. The unidentified impurity (*) in the permeate may have been extracted by toluene from the peristaltic tubing used in this experiment.

A scanning electron micrograph of a pre-production hollow fibre membrane in diagonal cross-section is shown in Figure 6. The fibre is double-skinned, significantly macrovoided, with a circular lumen cross-section, a wall-thickness of 0.15 mm, and an internal diameter of 0.65 mm. The provisional specification for this membrane indicates a burst pressure of 12 bar under aqueous conditions, a nominal MWCO of 10,000, and a pure-water flux of 120 $l/m^2/hr$ at 2 bar trans-membrane pressure.

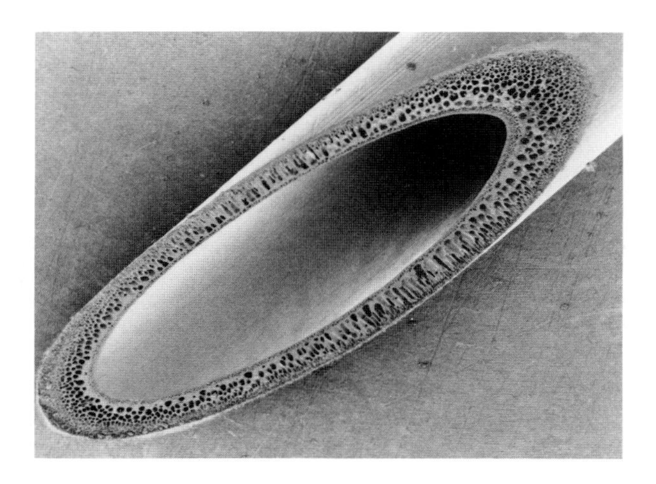

Figure 6. SEM (diagonal cross-section , x 35) of a hollow-fibre PEK UF membrane

Conclusions

Ultrafiltration membranes with good performance and exceptional resistance to physical and chemical degradation can be cast from solutions of PEK or PEKEKK in sulphuric acid. Such polymers confer intrinsic oxidative- and hydrolytic stability on the membrane, and their semi-crystalline morphology provides sufficient chemical resistance for solvent-based membrane processes to be carried out. Only ceramic UF membranes, with their attendant disadvantages of fragility and cost, would appear to exceed polyetherketone membranes in terms of chemical resistance and potential range of application.

Acknowledgements

We wish to thank Dr M.S. Le for invaluable advice on the characterisation and testing of UF membranes. Electron microscopy was carried out by Dr G.T. Finlan and Mrs P. Ball, and non-aqueous GPC analyses were provided by A.J. Handley and Mrs B.A. Gorry.

Notes and References

1. H.M. Colquhoun and T.M.C. Taylor, E.P.A. 382,356 (1990), to ICI.

2. M.J. Mullins and E.P. Woo, *J. Macromol. Sci., Rev. Macromol. Chem.*, 1987, **C27**, 313

3. M. Cheryan, *Ultrafiltration Handbook*, Technomic Publishing, Lancaster (USA), 1986, pp. 39 - 42.

4. Fabrication of UF membranes from an organic-soluble *amorphous* aromatic polyetherketone has however been reported [E. Drioli and H. Zhang, *Chim. Oggi,* 1989, **7**(11), 59].

5. J.B. Rose, U.S. 4,268,650 (1981), to ICI. See also; X. Jin, M.T. Bishop, T.S. Ellis, and F.E. Karasz, *Br. Polym. J.*, 1985, **17**, 4.

6. P. Zschocke and D. Quellmaltz, D.E. 3,321,860 (1984), to Berghof.

7. H.M. Colquhoun , patent filed (1992), to ICI.

8. H.J. Zimmermann and K. Könnecke, *Polymer*, 1991, **32**, 3162

9. G. Schock, A. Miquel, and R. Birkenberger, *J. Membr. Sci.*, 1989, **41**, 55.

Modified Polysulfones as Membrane Electrolytes

R. Nolte, K. Ledjeff, Fraunhofer-Institut für Solare Energiesysteme, Oltmannsstr. 22, 7800 Freiburg, Germany

M. Bauer, R. Mülhaupt Freiburger Materialforschungszentrum, Stefan-Meier-Str. 31, 7800 Freiburg, Germany

1. Introduction

Ion exchange membranes are usually used as separators in electrochemical processes. A second very interesting aspect is the application of these membranes as solid polymer electrolytes in modern energy conversion technologies /1/. Water electrolysers and hydrogen/oxygen fuel cells using polymeric ionic conductors offer the possibility for the efficient utilization of solar energy and hydrogen technology /2/. The principle of a system for the long time storage of energy with hydrogen technology is shown in Fig. 1.

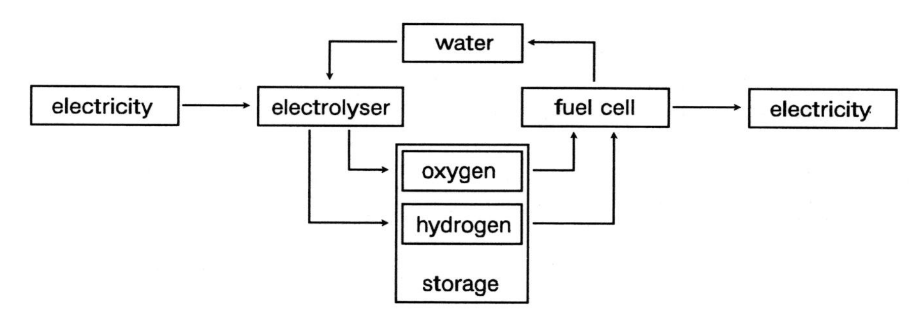

Fig. 1: Long-time energy storge with hydrogen technology using an electrolyser and a fuel cell.

Electricity, e.g. produced by solar systems, is used to split up water into hydrogen and oxygen. These gases can be stored and now offer the possibility of a long-time storage of the chemical energy. When needed the fuel cell combine hydrogen and oxygen to water again, producing directly electricity. The use of membranes in these components is demonstrated in

Fig. 2, which shows the principle of a solid polymer fuel cell. The basic reactions are the anodic oxidation of hydrogen to protons (electrons are produced) and the cathodic reduction of oxygen to water (electrons are needed). The transport of protons between the two electrodes has to be done by an appropriate electrolyte. Instead of corrosive liquid electrolytes like acids or bases the solid polymer fuel cell (SPFC) uses a solid polymer electrolyte. This is normally an ion exchange membrane, bearing negatively charged, fixid ionic groups within its structure, which are responsible for a good proton conductivity of the membrane. On both sides the ion exchange membrane is coated with a catalyst layer, where the electrochemical oxidation and reduction take place as mentioned above. The end plates and the porous current collectors, both electrically conductive, provide the electric connections and ensure the homogeneous gas distribution over the membrane surface.

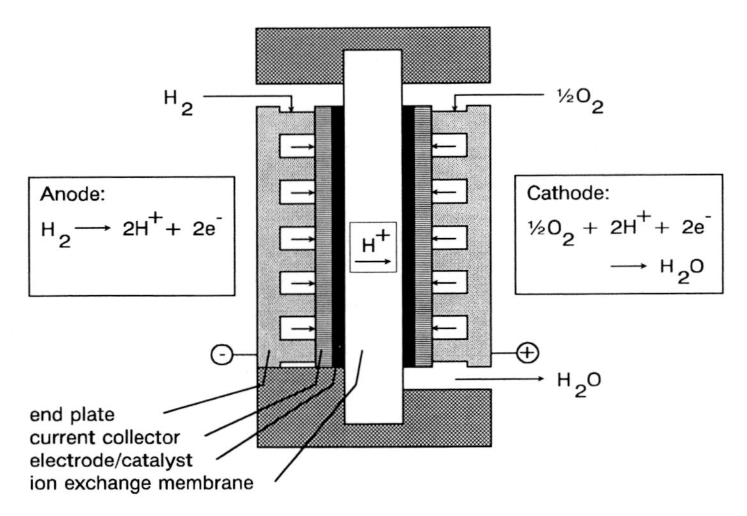

Fig. 2: H_2/O_2-fuel cell using a polymer electrolyte

The membrane properties are a main criterion for a good cell performance and the membrane materials have to fulfil a number of requirements like chemical resistance, thermal and dimensional stability and good ionic conductivity. Only perfluorinated membranes have been successfully applied /3/. The disadvantages of these materials are a high price and difficulties in the processability. So there is a demand for novel stable polymers, which combine good membrane properties with lower cost and improved processability.

For this purpose commercially available polymers, in particular poly(arylen ether sulfones) which are stable against oxygen, hydrolysis and temperature /4/, have been functionalized by sulfonation and tested as solid polymer electrolytes. Another objective of the work was to improve the thermal stability of the ion exchange groups. It is desirable to work at elevated temperatures because the electrode kinetic is much faster. The aromatic sulfonation is a reversible reaction, consequently the sulfonic acid groups are not stable at elevated temperatures. Therefore phosphonic acid groups have been attached to the poly(arylene ether sulfone) and the thermal behaviour of this material was investigated.

2. Sulfonated polymers

Victrex^{TM} PES 5200P (PES) as poly(arylene ether sulfone) has been sulfonated with SO_3 by a slurry procedure similar to procedures described by Coplan and Götz /5/. Membranes were casted from DMF-solution with three materials (ISE3, ISE7, ISE10) having different degrees of sulfonation. ISE3 has a degree of sulfonation of 34 %, ISE7 72 % and ISE10 94%. The materials resp. membranes were characterized using FTIR-spectroscopy, 1H-NMR-spectroscopy, viscosimetry, titration, thermal analysis, measurement of ionic conductivity, permselectivity, swelling, current/voltage plot and a life time test in an electrolysis cell. Table 1 summarizes the characteristic values for the sulfonated materials and the resulting membranes.

		ISE3	ISE7	ISE10
Ion exchange capacity	meq/g	1.3	2.5	3.0
Degree of sulfonation	%	34	72	94
Glass transition temperature	°C	308	316	333
Permselectivity (0.5/1m KCl)	%	74	87	90
Resistivity (0.5m KCl)	Ω·cm	1460	131	76
Swelling	%	11	56	74

Table 1: Characteristic values for the sulfonated PES materials

The values of ionic conductivity and permselectivity are within the range of commercial membranes. A life time test of 300 h was run using the membrane as polymer electrolyte in a water electrolyser at a constant current of 1 A/cm^2. Only a small increase in the cell potential was observed during the measuring period, indicating a promising long-time stability of the system. The swelling behaviour of the membrane is important for the mechanical stability. The ISE10 membrane with a degree of sulfonation of about 90 % shows an uptake of water of 74 % at room temperature and >400 % at 80°C, which means a significant deterioration of mechanical stability. To solve the swelling problem, the membrane has been crosslinked during the casting process, using 1,1'-carbonyl-diimidazol for the activation of the sulfonic acid groups and a diamine for the crosslinking reaction. Table 2 shows the results for the crosslinked and non-crosslinked membrane casted from ISE10 material.

		non-crosslinked	crosslinked
Resistivity	Ω·cm	76	160
Permselectivity	%	90	80
Swelling	%	420	180

Table 2: Characteristic values for the crosslinked and non-crosslinked membrane

The swelling could be reduced by about 50 %. The values for the ionic conductivity and the permselectivity remained in an acceptable range.

3. Phosphorylated polymers

In order to improve the thermal stability of the ion exchange groups, phosphonic acid groups were attached to the backbone of Udel[TM] P-1700 (PSU) polymers. This modification was done by metalation of the aromatic ring with n-butyl lithium, followed by the reaction either with $POCl_3$, P_4O_6 or P_4O_{10} and if needed by a final oxidation/hydrolysis step to obtain the phosphonic acid. In addition some of these modified polymers were sulfonated.

polymer modification	IEC before treatment meq/g	IEC after treatment meq/g	IEC loss %
$POCl_3$	1.95	1.92	1.3
$POCl_3/ClSO_3H$	8.86	5.51	37.8
P_4O_6	1.83	1.80	1.6
$P_4O_6/ClSO_3H$	8.83	5.81	34.2
P_4O_{10}	1.95	1.93	1.0
$P_4O_{10}/ClSO_3H$	8.54	5.32	37.3

Table 3: Thermal stability (water, 150°C, 24 h) of poly(arylene ether sulfones) containing phosphonic and sulfonic acid groups

To test the thermal stability in water, the materials were exposed to water at 150°C (pressure: 4 bars) for 24 hours and the ion exchange capacity was measured before and after the treatment. Table 3 summerizes the results for the modified poly(arylene ehter sulfones). Sulfonated materials showed a remarkable decrease in the ion exchange capacity (34-38 %). In contrast the loss in ion exchange capacity for the polymers containing only phosphonic groups was negligible (1-2 %).

4. Summary

Poly(arylene ether sulfones) were sulfonated with regard to their application as solid polymer electrolyte in electrolysers and fuel cells. Membranes of these materials had similar ionic conductivities and permselectivities compared to conventional perfluorinated products. The high value for the uptake of water has been diminuished by a crosslinking reaction using 1,1'-carbonyldiimidazole/diamine. Sulfonated poly(arylen ether sulfone) membranes are able to combine the good electrochemical properties of the perfluorinated materials with low cost and advanced processability.

As alternative to the reversible sulfonation, the phosphorylation of poly(arylene ether sulfone) was investigated. Materials with phosphonic acid groups showed a significantly higher temperature stability compared to the sulfonated polymers. Membranes of this material should have an interesting potential for applications at elevated temperatures.

5. Literature

/1/ K. Ledjeff, J. Ahn, D. Zylka, A. Heinzel, "Ion Exchange Membranes as Electrolyte for Electrochemical Energy Conversion", Ber. Bunsenges. Phys. Chem. 94, 1005 (1990)

/2/ A. Heinzel, K. Ledjeff, "Self-Sufficient Solar House: Hybrid Energy Storage System", ISES Solar World Congress, 2543 (1991), Denver

/3/ D. Watkins, K. Dircks, D. Epp, C. De La Franier, "Canadian Solid Polymer Fuel Cell Development", Proceedings of the 4th Annual Battery Conference on Applications and Advances (1989), California State Univ., Long Beach

/4/ C. A. Linkous, "Development of Solid Electrolytes for Water Electrolysis at IntermediateTemperatures", Hydrogen Energy Progress IX - Proceedings of the 9th World Hydrogen Energy Conference Vol. 1, 419 (1992), Paris

/5/ M. J. Coplan, G. Götz, "Heterogeneous Sulfonation Process for Difficultly Sulfonatable Poly(ethersulfone)", US Pat. Nr. 4,413,106 (1983)